Buckingham Bay
Cape York
PACIFIC OCEAN
aningrida
Gove
Weipa
Blue Mud Bay
LAND
GULF OF CARPENTARIA
McArthur R.
Mitchell R.
Cooktown
GREAT BARRIER REEF
CORAL SEA
Gregory R.
Karumba
Cairns
Normanton
RRITORY
Leichhardt R.
Gilbert R.
Townsville
nnant Creek
Charters Towers
Bowen
Flinders R.
Mount Isa
Warburton R.
Mackay
Broad Sound
Shoalwater Bay
Capricorn Coast
ce Springs
Diamantina R.
Rockhampton
Gladstone
Finke R.
Hervey Bay
QUEENSLAND
Cooper's Creek
Great Sandy Strait
STRALIA
Noosa
Cunnamulla
Lake Eyre
Lake Grace
Moreton Bay
BRISBANE
Culgoa R.
Gold Coast
Lake Torrens
Lake Callabonna
Lake Frome
Clarence R.
Darling R.
Macquarie
Namoi R.
Broken Hill
Armidale
EYRE
NEW SOUTH WALES
PENINSULA
Whyalla
Hunter R.
Parkes
R.
Port Stephens
St Vincent Gulf
Lachlan R.
Newcastle
Murray
R.
Mildura
Murrumbidgee
Spencer Gulf
ADELAIDE
Swan Hill
R.
CANBERRA
SYDNEY
Kerang
Kangaroo Island
The Coorong
VICTORIA
Jervis Bay
Horsham
Shepparton
Bendigo
Geelong
MELBOURNE
TASMAN SEA
Portland
Gippsland Lakes
Warrnambool
Port Phillip Bay
Corner Inlet
BASS STRAIT
King Island
Furneaux Islands
Cape Portland
St Helen's
HOBART
TASMANIA
GREAT DIVIDING RANGE

# SHOREBIRDS IN AUSTRALIA

# SHOREBIRDS IN AUSTRALIA

**Brett A. Lane**

*with colour illustrations by*
Jeff N. Davies

Nelson

FOR NINKS

Brett A. Lane,
Royal Australasian Ornithologists Union,
GPO Box 5236 BB,
Melbourne, Vic. 3001, Australia.

Nelson Publishers
Thomas Nelson Australia
480 La Trobe Street Melbourne Victoria 3000

First published 1987


National Library of Australia
Cataloguing in Publication data:

Lane, Brett A.
   Shorebirds in Australia.

   Includes index.
   ISBN 0 17 006824 2.

   1. Shore birds — Australia. I. Davies, Jeff.
   II. Royal Australasian Ornithologists Union. III. Title.

598'.33'0994

Printed in Hong Kong

The Royal Australasian Ornithologists Union
thanks the bird-watchers of Australia for their
efforts during this study. Without their com-
mitment and support in gathering new and
exciting information, this book would not
have been possible.

The Australian National Parks and Wildlife Service provided much of the funding for the Wader
Studies Programme and the Utah Foundation gave financial help during the preparation of this book.
Their support is gratefully acknowledged.

The Royal Australasian Ornithologists Union (RAOU) was founded in 1901 to advance knowledge
and conservation of the birds of the Australian region. It is the internationally recognised organisation of
ornithologists of Australia and nearby countries. Its members are people from all walks of life who have
an interest in birds, as well as professional ornithologists. Projects and activities include the annual
Congress and Campout, the annual Scientific Day, the Nest Record Scheme and research on waders,
waterbirds, birds of prey and endangered species such as the Orange-bellied Parrot. Donations to
support the Union's research activities are tax deductible in Australia.

The RAOU publishes the *Emu*, a quarterly journal of world standing, and holds a comprehensive
ornithological library. Bird observatories are maintained at Eyre in Western Australia, Rotamah Island
in Victoria and Barren Grounds in New South Wales. With *Shorebirds in Australia* completed, the
RAOU is now compiling *The Handbook of the Birds of Australia, New Zealand and the Antarctic*.

Membership of the Union is open to anyone. Enquiries should be sent to: The Secretary, GPO Box
5236BB, Melbourne, Vic. 3001, Australia.

# CONTENTS

# LIST OF ABBREVIATIONS

| | |
|---|---|
| ANU | Australian National University |
| AOU | American Ornithologists' Union |
| AWSG | Australasian Wader Studies Group |
| BTO | British Trust for Ornithology |
| CSIRO | Commonwealth Scientific and Industrial Research Organization |
| in prep. | in preparation |
| MAPS | Migratory Animal Pathological Survey |
| NEDECO | Netherlands Engineering Consultants |
| NSW | New South Wales (Australia) |
| NT | Northern Territory (Australia) |
| NWT | northwest territories (Australia) |
| pers. comm. | personal communication |
| pers. obs. | personal observation |
| OSJ | Ornithological Society of Japan |
| OSNZ | Ornithological Society of New Zealand |
| OUP | Oxford University Press |
| PNG | Papua New Guinea |
| Qld | Queensland (Australia) |
| RAOU | Royal Australasian Ornithologists Union |
| SA/S.Aust. | South Australia |
| Tas. | Tasmania (Australia) |
| US Army R&D Group | United States Army Research and Development Group |
| UWA | University of Western Australia |
| Vic./Vict. | Victoria (Australia) |
| VORG | Victorian Ornithological Research Group |
| VWSG | Victorian Wader Study Group |
| WA/W.Aust. | Western Australia |
| WBSJ | Wild Bird Society of Japan |
| WSG | Wader Study Group |

# LIST OF PLATES

*(The plates appear between pages 54 and 55. The keys to the plates appear between plates 16 and 17.)*



*Note.* Birds are shown at different stages, juvenile and adult, breeding and non-breeding.

# INTRODUCTION

From 1981 to 1985, over 700 bird-watchers visited beaches, mudflats and inland swamps throughout Australia. These people were involved in a programme of surveys and expeditions designed to learn more of the largely undescribed life of shorebirds in Australia.

Shorebirds, or waders, make up about 10 per cent of Australia's species of birds. Most breed in the Northern Hemisphere and migrate southwards each year through Asia and the Pacific Islands to Australia. Some occur only in Australia, and one breeds in New Zealand and crosses the Tasman Sea. Nearly all live gregariously on coastal and inland wetlands.

Expeditions to remote places — such as Eighty Mile Beach in northern Western Australia and the coast of the Gulf of Carpentaria in Queensland — revealed previously unknown congregations of shorebirds. The estimated world populations of some species had to be revised when very large numbers were discovered at a number of sites. Nearly a million shorebirds visit the vast northern Australian mudflats each year, some for only a short time before flying directly across the central and western deserts to southern Australia to join hundreds of thousands more.

The study, organised by the Royal Australasian Ornithologists Union (RAOU), produced a wealth of new knowledge about shorebirds in Australia, and was undertaken as a part of Australia's obligations under the Japan—Australia Migratory Birds Agreement, signed in 1974, which obliges these countries to research and conserve their shared shorebird populations.

This book reports the results of the study and brings together other information on the biology of shorebirds in Australia. The group studied and described here includes thick-knees (Burhinidae), the Painted Snipe (Rostratulidae), oystercatchers (Haematopodidae), lapwings and plovers (Charadriidae), stilts and avocets (Recurvirostridae), sandpipers, snipes, godwits, curlews and their allies (Scolopacidae), as well as the pratincoles (Glareolidae) and phalaropes (Phalaropodidae).

Chapter 2 gives an overview of shorebird distribution in Australia and examines physical factors influencing this. Chapter 3, written by Peter Dann, presents what is known about shorebird feeding ecology in Australia, an important biological factor, which determines in part where they occur. Chapter 4 looks at the migratory and movement strategies of shorebirds. Chapters 3 and 4 also describe how information was gathered and illustrate the important contributions both amateurs and professionals can make to this type of research. Chapter 5 forms the bulk of the book and consists of 45 detailed species accounts for the regularly occurring, more abundant species (30 migrants and 15 residents). Chapter 6 has shorter descriptions of the rarer shorebirds (26 migrants). Chapter 7 describes the shorebirds and their habitats in the ten regions of Australia each subject to generally similar climatic and other physical conditions. The final chapter (Chapter 8) examines how human activities have affected shorebirds and the actions required for their conservation.

# THE DISTRIBUTION OF SHOREBIRDS IN AUSTRALIA

In Australia, shorebirds are found on a diversity of wetlands, from the ephemeral salt lakes of the deserts to the wide intertidal mudflats of the coast. To some extent most wetlands provide the muddy or sandy margins and shallow water that most shorebirds require. A handful of species do not rely on wetlands, but spend much of their time on grassy or arid plains away from tree cover.

The distribution of shorebirds in Australia is determined by the availability of suitable habitat. The creation of shorebird habitat is the result of several factors, including wave energy, climate, sea-level changes, tidal range, mangroves and human activities on the coast and in inland areas, the origins of lakes and swamps, water salinity and rainfall. This chapter discusses these topics under the headings:

- Where shorebirds are found in Australia.
- Coastal shorebird habitat.
- Inland shorebird habitat.

This discussion of the general pattern of shorebird-distribution leads into an account of how shorebirds use wetlands (Chapter 3) and how they migrate and move between them (Chapter 4).

## Where shorebirds are found in Australia

In the non-breeding months (August–April), about 80 per cent of the almost 2 million migratory shorebirds that visit Australia occur in three main regions: the north-west coast between Broome and Port Hedland; the coast of north-eastern Arnhem Land and the Gulf of Carpentaria; and the south-eastern coast and lakes between Eyre Peninsula and Corner Inlet. (See Figure 2.1; Table 2.1.)

TABLE 2.1. Maximum counts at zones that held more than 10 000 shorebirds at some time between 1981 and 1985. (A) = aerial count; (G) = ground count.

| ZONE | MAXIMUM COUNT |
|---|---|
| Eighty Mile Beach, WA | 337 500 (A) |
| SE corner, Gulf of Carpentaria, Qld | 250 000 (A) |
| The Coorong, SA | 236 000 (G) |
| Roebuck Bay, WA | 170 900 (G) |
| *Lake Eyre, SA | 123 400 (A) |
| NE Arnhem Land, NT | 79 000 (A) |
| Port Hedland Saltworks, WA | 66 800 (G) |
| Port Phillip Bay, Vic. | 64 100 (G) |
| St Vincent Gulf, SA | 64 000 (G) |
| *SE coast & lakes, SA | 63 100 (G) |
| Shark Bay, WA | 50 000 (A) |
| Corner Inlet, Vic. | 46 200 (G) |
| Spencer Gulf, WA | 37 000 (G+A) |
| Cape Keraudren–Port Hedland, WA | 30 000 (A) |
| LaGrange Bay, WA | 23 800 (A) |
| Peel Inlet, WA | 23 300 (G) |
| SW corner, Gulf of Carpentaria, NT | 21 900 (A) |
| Darwin region, NT | 20 900 (G+A) |
| W coast, Eyre Peninsula, SA | 19 900 (G) |
| Moreton Bay, Qld | 18 200 (G) |
| Western District lakes, Vic. | 17 600 (G) |
| Horsham region lakes, Vic. | 16 700 (G) |
| *Far NW region, NSW | 16 700 (G) |
| Kangaroo Island, SA | 13 900 (G) |
| Hervey Bay–Great Sandy Strait, Qld | 13 600 (G) |
| Kerang–Swan Hill region, Vic. | 13 400 (G) |
| S coast, Gulf of Carpentaria, NT | 12 700 (A) |
| Swan coastal plain, WA | 12 500 (G) |
| *Esperance region, WA | 11 900 (G) |
| Westernport Bay, Vic. | 11 200 (G) |
| Pilbara coast, WA | 11 100 (G) |

(* = predominantly non-migratory species: mostly Banded Stilt and Red-necked Avocet)

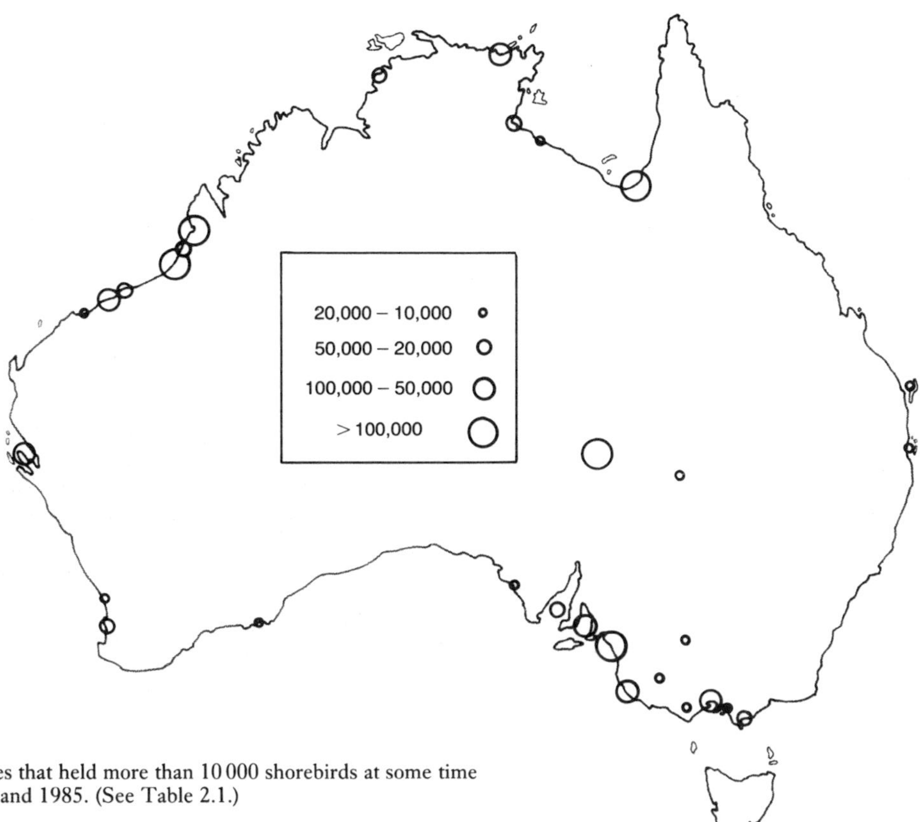



FIG. 2.1. Sites that held more than 10 000 shorebirds at some time between 1981 and 1985. (See Table 2.1.)

Most species are not spread uniformly around the continent. The Red-necked Stint, for example, occurs in largest numbers in the north-west and southern parts of Australia; the Lesser Golden Plover is found frequently only on the eastern and south-eastern coasts; and the Whimbrel is common only on the north and east coasts. Detailed information on the distribution of each species is given in Chapter 5.

## Coastal shorebird habitat

Three principal types of coast have been identified: rocky shores, beach and dune shores, and tidal flat shores (104). The first two are used by a few species — such as resident Sooty Oystercatcher and Hooded Plover and migratory Sanderling — but flat tidal shores with their extensive intertidal muddy areas, rich in invertebrate prey, hold the largest numbers of shorebirds.

Australian coastal shorebird habitat differs markedly from the well-studied habitats in temperate regions of the Northern Hemisphere. In Britain, for instance, habitats are predominantly estuarine; that is, where freshwater from rivers mixes with the sea, creating a transitional zone rich in living organisms. In Australia, by contrast, aridity has resulted in a lack of high-discharge rivers and little estuarine habitat. Only a small area of southern tidal shore is estuarine, in the immediate vicinity of small creek- and river-mouths. In the north during the wet season (December–April),

river discharges can be high, creating temporary estuarine conditions. Most coastal shorebird habitat in Australia is in a marine environment.

## WAVE ENERGY

One of the most important factors influencing the formation of intertidal mudflats is wave energy: that is, the amount of energy transmitted to the shore by ocean waves. The more energy waves have, the rougher the sea and the more turbulent is the water at the shore. Mudflats form only in quiet water.

The southern coast, from Shark Bay in the west to Fraser Island in the east, is subject to high wave energy as it is exposed to ocean swell generated in the mid-latitude storm belt of the Southern Ocean (40°-60°S). Here there are no tidal mudflats exposed to the open ocean. Instead, mudflats form in embayments and inlets sheltered from ocean swell. (See Figure 2.2.) These sheltered pockets are of two types: geological basins and barrier systems. The former are created by earth movements or erosion of softer rock, and the latter by the accumulation of sand (the only sediment not washed out to sea by high energy waves) across river-mouths, coastal plains or geological basins. In the south, the largest numbers of shorebirds occur on the largest sheltered embayments and barrier systems. (See Figure 2.1.)

Fig. 2.2. Anderson's Inlet, Victoria, showing the inlet sheltered from breaking ocean swell by a long, sandy barrier. Photo: J. Hooper

The northern half of the continent lies in tropical seas, where wave energy is low, with the occasional localised exception generated by a tropical cyclone. The north-west and north coasts are not exposed to swell generated farther south, and the prevailing winds are offshore. The central and northern Queensland coast is sheltered by the Great Barrier Reef from waves generated by south-easterly trade winds. Consequently, finer sediment is not washed out to sea, and the northern Australian coast has much more tidal mudflat than the southern coast, despite being exposed to the open ocean. (See Figure 2.3.)

## CLIMATE

The formation of mudflats on the coast depends on the accumulation of sediment eroded from nearby catchments and delivered to the sea by rivers. Temperature controls the rate at which rock is weathered to form transportable sediment: the higher the temperature the faster the weathering. Rainfall determines the rate at which sediment is washed from the land and transported to the sea by rivers.

In cooler southern Australia, sediment yields are lower than in the north, but high rainfall ensures a steady build-up of mudflats in sheltered bays and inlets. In tropical northern Australia, rates of weathering are much higher and, during most wet seasons, rivers carry much sediment to the sea. Often, wide depositional plains form at the coast, and mudflats constitute their seaward edge. The south-eastern shores of the Gulf of Carpentaria are a striking example of this. (See Figure 2.4.)

The extensive depositional plains of coastal northern Australia are a very important habitat for large numbers of some species of shorebirds. During the wet season (December–April), they flood and become unavailable to most species. At this time these birds move inland to the grassy plains of northern Australia.

## SEA-LEVEL CHANGES

In the last million years there have been a number of major sea-level fluctuations, caused by periods of colder and warmer climate. During 'ice ages', water was locked up in polar and alpine ice-caps, but during warmer periods the ice-caps melted and sea-levels rose. The most recent fluctuation in sea-level occurred

FIG. 2.3. Intertidal mudflats exposed to the open ocean along Eighty Mile Beach, north-western Australia. Photo: C. Minton

FIG. 2.4. Aerial view of the south-western shore of the Gulf of Carpentaria showing part of the extensive depositional plain and the intertidal mudflats at its seaward edge. Photo: B. Lane

between at least 10 000 and 6000 years ago. (See Figure 2.5.) Earlier evidence from Australia is lacking, but about 10 000 years ago the sea-level was 25 metres below its current position. By 6000 years ago it had risen to its present level (453). During periods of lower sea-level, erosion of what is now the continental shelf occurred, in the same fashion as the present land surface is being eroded. With the rise in sea-level, many river valleys and low-lying coastal plains were flooded. These flooded areas were the first stage in the development of today's coastal shorebird habitat.

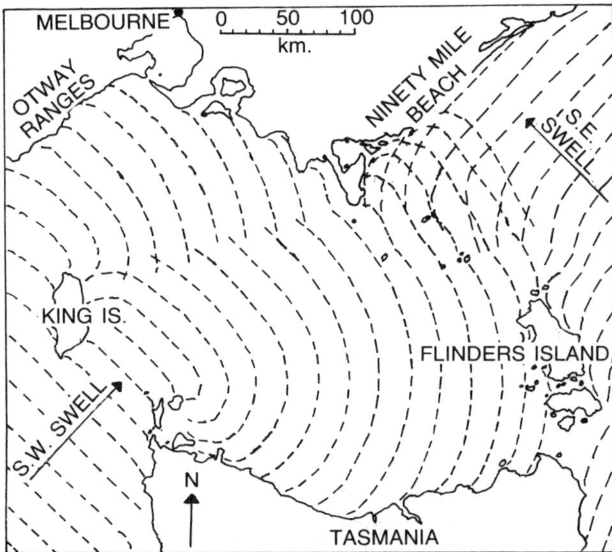


FIG. 2.6.    The predominant ocean swell pattern in Bass Strait. Note how many parts of the Victorian coast are aligned parallel with the swell. (Redrawn from J.N. Jennings & J.A. Mabbutt, 1967. *Landform studies from Australia and New Guinea*. Canberra: Australian National University Press. With permission from the publisher.)

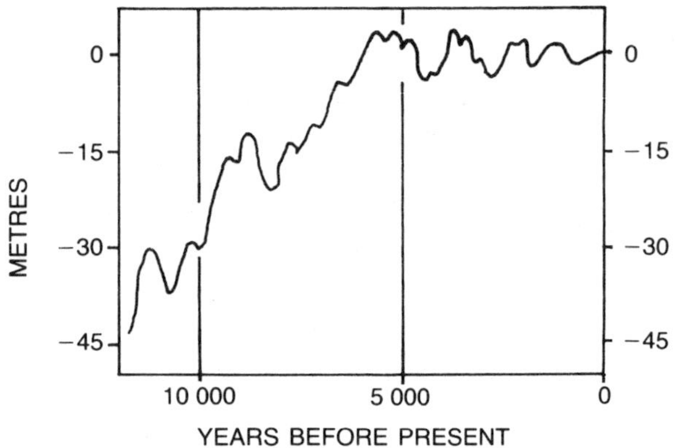


FIG. 2.5.    The most recent change in sea-level. (Redrawn from E.C.F. Bird, 1976. *Coasts*. Canberra: Australian National University Press, 2nd ed. With permission from the publisher.)

In southern Australia, the most recent rise in sea-level led to a considerable amount of sand being carried shoreward by high energy waves. This was deposited as beaches and dunes. Barriers of sand aligned parallel to the prevailing ocean swell accumulated across the mouths of drowned river valleys and basins. They protected pockets of quiet water, where sediment from creeks and rivers accumulated as mudflats. The correspondence between ocean swell patterns and shorebird habitat on the Victorian coast is shown in Figure 2.6.

their shores from ocean swell, and mudflats are formed.

Tidal ranges are generally higher in northern Australia. (See Figure 2.7.) Where tidal range is highest, mudflats are most extensive, provided the shore is gently sloping. The three areas of the north coast that are gently sloping and have the highest tidal range hold the largest concentrations of shorebirds. Near the Kimberleys, the tidal range is the highest in Australia, but mudflats are less extensive as the land drops steeply into the sea. Few shorebirds are recorded here.

## TIDAL RANGE

Another fundamental influence on the development of tidal-flat shores is tidal range. High tidal rise and fall produces strong currents. In far eastern Victoria, drowned river valleys are seasonally blocked by sandy barriers: low river discharge in summer and weak tidal currents allow sand to accumulate. Here, tidal range averages 1.2 metres. Farther west, tidal range averages 2.5 metres, and there are permanent openings in barrier systems because the strength of tidal currents stops sand from accumulating. Tidal range is higher still in the South Australian gulfs, and there are no sandy barriers. Here, the shape of the gulfs protects

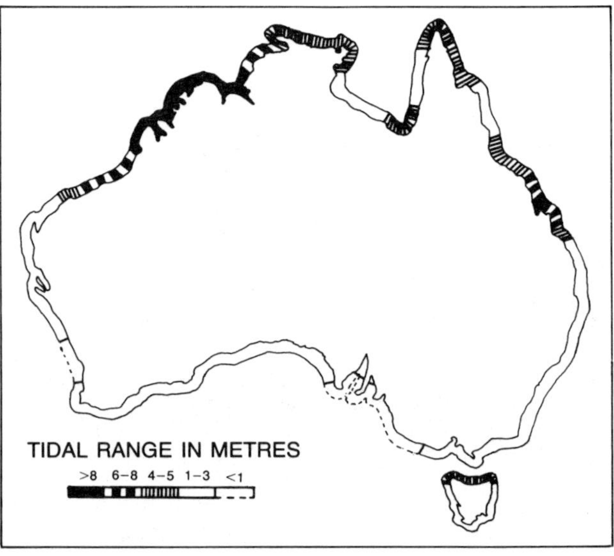


FIG. 2.7.    Variation in tidal range along the Australian coast.

FIG. 2.8.   Mangroves along the coast near Proserpine in eastern Queensland. These forests of salt-tolerant trees occupy a substantial proportion of the intertidal area on many parts of the Australian coast and prevent some shorebirds from feeding there. Photo:  J. Hooper

## MANGROVES

On the tropical north coast, mangrove forests are a major feature of tidal shores. Their distribution has been documented in detail, and the factors affecting the development of mangrove forests have been examined (147, 341). They reach their maximum extent in north-eastern Queensland where rainfall is highest and least seasonal and cover a substantial proportion of the intertidal area, rendering it unavailable for many shorebird species. (See Figure 2.8.) Some, however, such as the Whimbrel, Common and Terek Sandpipers, and Grey-tailed Tattler do feed in mangrove forests at low tide. Numbers of shorebirds are higher on the Gulf coast and in south-eastern Queensland. The drier, more seasonal climate of the Gulf coast and the cooler climate of south-eastern Queensland restrict the development of mangroves to narrower fringing forests.

## HUMAN ACTIVITIES

In addition to the natural coastal environment, man-made habitats are utilised by shorebirds. Many thousands of shorebirds in the three main regions of coastal habitat in Australia occur on artificial wetlands such as sewage treatment works and evaporative saltworks. In Port Phillip Bay, for instance, about 75 per cent of the shorebirds occur on three saltworks and the large sewage treatment works near Werribee. No historical information is available on the status of shorebirds in such areas, so the precise changes in shorebird numbers cannot be determined. Suffice it to say that such areas can be very important habitat.

# Inland shorebird habitat

Away from the coast, most shorebirds occur on lakes and swamps. The shape of these water bodies, the salinity of the water and seasonally varying rainfall and runoff greatly affect the biological characteristics of lakes and swamps (26) and their suitability for shorebirds.

## THE ORIGIN OF INLAND WETLANDS

Lakes and swamps, with the gently sloping margins most preferred by shorebirds, form in a variety of ways (26). Three types of lakes hold good numbers of shorebirds: those formed by earth movements; those formed by the action of rivers; and those resulting from wind erosion.

Lakes formed by movements of the earth's crust, which alter drainage patterns, are usually fairly large and can have gently sloping margins. Examples include Lake George in New South Wales and the large inland salt lakes of South Australia (lakes Torrens, Eyre, Grace, Callabonna and Frome). They often form the terminus of closed drainage systems (in the case of Lake Eyre on a very large scale) where salt leached from the surrounding catchment accumulates. When Lake Eyre filled in 1984, it held over 130 000 shorebirds, mostly Red-necked Avocets and Banded Stilts.

Lakes form by different sedimentation patterns along the lower reaches of rivers; where a river flows out onto a flat plain it meanders and deposits sediment. River channels can change the shape of the plain, and shallow basins formed in this way can fill with water. The lakes round Kerang, in northern Victoria, and those along the lower reaches of the Darling River, in western New South Wales, are examples of this type of lake, and large numbers of shorebirds are known to occur in the former.

Lakes form in more arid parts of Australia in basins created by wind erosion. Examples include the many lakes in the goldfields region of Western Australia.

The large South Australian salt lakes partly owe their existence to this process as well as to earth movements.

## WATER SALINITY

Many of Australia's inland wetlands are saline, with often high concentrations of salt derived from underlying sedimentary rock laid down under the sea millions of years ago.

Different plants and invertebrates have varying salt tolerances, so the salinity of a wetland affects its biological characteristics (26, 81) and its suitability for shorebirds. Reeds and rushes cover a large proportion of the shoreline of freshwater lakes. (See Figure 2.9.) Higher salinity restricts the growth of tall aquatic plants and saline lakes have characteristically open mudflats on which shorebirds can forage. (See Figure 2.10.)

## RAINFALL

Seasonal patterns of rainfall and runoff cause fluctuations in the water-level of wetlands. On the Swan coastal plain in Western Australia, winter rains fill lakes and swamps that are all but emptied by high

FIG. 2.9.   The edge of a freshwater swamp (Cherry Lake, near Melbourne), showing the rank growth of reeds and rushes along its shore. This can prevent shorebirds from feeding around the edges.   Photo: J. Hooper

Fig. 2.10.   The characteristically open shore of a salt lake: Point Cook, near Melbourne. Photo: J. Hooper

summer evaporation. These wetlands are unsuitable when full, because the mudflats are covered by water. By summer, water-levels drop to expose rich feeding areas, and shorebirds use them until they dry out, usually in autumn. The species that use these wetlands change as the water recedes (93). In the lakes on the lower reaches of rivers flowing inland from the eastern highlands, water-levels are affected more by the rainfall in the mountains and, increasingly, the discharge from man-made reservoirs than by local rainfall.

have altered the nature of wetlands used by shorebirds.

Recent information (38) suggests that some species of shorebirds that are not dependent on wetlands (e.g. Banded Lapwing and Australian Pratincole) have moved into parts of the semi-arid region that have become more open due to overgrazing by domestic stock.

## Human Activities

In common with coastal areas, the impact of human activities has affected the distribution of shorebirds in inland Australia. The construction of innumerable dams throughout agricultural areas has expanded the area of habitat for some shorebirds, e.g. Black-fronted Plover (38). Sewage treatment works have also provided habitat in regions where previously it did not exist. Flood mitigation and water conservation works on major river-systems (such as the Murray) have probably had some impact on shorebird populations and their distribution, but more investigation is needed to determine if the changes caused by these activities

# THE FEEDING BEHAVIOUR AND ECOLOGY OF SHOREBIRDS

by Peter Dann*

Shorebirds occur in most of the non-forested habitats of Australia. They are commonly seen in greatest concentrations in coastal wetlands, but are also found in a wide range of habitats away from the coast: from the dry gibber plains and salt lakes of arid areas to tropical woodland, grassland and southern farmland. Many factors determine when and where shorebirds are found. Within the physical constraints outlined in the previous chapter, the local distribution of shorebirds is influenced by food availability and feeding adaptations. Feeding requirements underlie much of what is observed, even to the extent of governing the particular sections of paddock, swamp or beach used.

Apart from their widespread distribution, shorebirds are also noteworthy because some species migrate as far as the tundra regions of the Northern Hemisphere to breed before returning to Australia each spring. How they do this is considered in Chapter 4. Large reserves of fat are necessary for these migrations, and this fat must be accumulated prior to and during migration. In addition, daily energy costs and those of breeding and moulting must be met.

In recent years there have been many scientific studies of shorebird feeding in the Northern Hemisphere. Much of this work has been designed to test the predictions of optimal foraging theory. This theory proposes that an animal should forage in an energy-efficient manner, which maximises its survival and therefore its genetic contribution to subsequent generations (fitness). Optimal foraging can be considered in four ways (374): choice of prey, selection of feeding patch, allocation of time and optimal patterns and speed of movement. Efficient feeding strategies enable shorebirds to store enough energy to survive and reproduce. Therefore, a thorough knowledge of their feeding ecology throughout the year is central to an understanding of why they migrate and occur where they do. An excellent treatment of this subject in European shorebirds is the account by R. J. O'Connor in Prater (369).

In Australia there have been very few studies of shorebird feeding, and it is not possible to discuss optimal foraging in an Australian context. This chapter presents what is known about feeding ecology and behaviour of shorebirds in Australia.

Research on feeding is necessary to explain shorebird life-histories, and until this information is available, it will not be possible to protect and manage Australian shorebird populations effectively.

## Methods of studying the feeding of shorebirds

Discovering what shorebirds eat generally involves much more than simply watching them feeding. Only when observing large shorebirds (such as the Eastern Curlew) can food items be identified as they are caught and swallowed. Unless the smaller shorebirds are very close, it is not obvious when they have obtained an item of food, let alone what kind of organism it might have been.

It is relatively simple to discern what some shorebirds have been eating. Sooty Oystercatchers forage mainly on rocky shores during the breeding season and feed their young on various species of limpets, whelks and chitons. It is easy to collect the hard shells of these prey items for they are discarded near by after the young oystercatchers have eaten the soft parts. (See Figure 3.1.)

* This chapter is by Peter Dann, Department of Zoology, University of Melbourne, Parkville 3052, Victoria. Present Address: Penguin Reserve Committee of Management, Box 403, Cowes 3922, Victoria.

FIG. 3.1.   Remains of hard-shelled prey brought to young Sooty Oystercatchers on King Island, Tasmania. Photo: P. Dann

Another species that discards pieces of its prey on the sandbanks and mudflats when foraging is the Eastern Curlew. The Curlew feeds at low tide and leaves conspicuous marks in the sand where it has probed for food. It probes into long burrows of the ghost shrimps, *Callianassa australiensis*, and is successful in capturing a shrimp about once in every twenty probes. When a Curlew pulls a ghost shrimp out of its sandy burrow, sometimes washing the sand off it in a nearby pool, it manipulates the shrimp for swallowing and usually breaks off its large claw. By following the tracks of a Curlew across a sandbank, the number, size and even sex of the ghost shrimps taken by the Curlew can be determined from the characteristics of the claws left on the sand. (See Figure 3.2.)

Most shorebirds do not leave such useful indications of what they eat, and other methods must be used to study their diets. Many insects, crustacea, molluscs and some polychaete worms have indigestible hard parts, which may be excreted in faeces or regurgitated in the form of pellets. The remnant pieces of legs, carapaces or shells present in faeces and pellets can be compared to whole organisms collected by the investigator in the same place. These remains provide much useful qualitative information on the diets of shorebirds, but those prey lacking hard parts (e.g. some polychaete worms) will not be present in faeces or pellets, because they are completely digested.

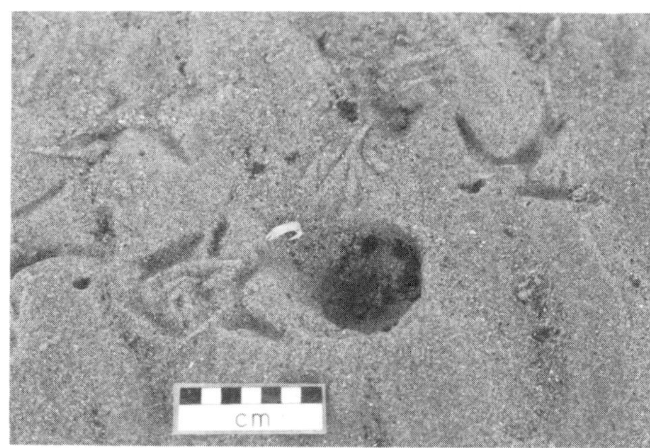

FIG. 3.2.   A claw left after an Eastern Curlew has removed a ghost shrimp from its burrow. Note the curlew footprints and the hole where the curlew has probed. Photo: P. Dann

The Sooty Oystercatcher, Eastern Curlew and Whimbrel in southern Victoria regurgitate pellets at high-tide roosts. It is likely, but not yet recorded, that other species do likewise. Pellets of Sooty Oystercatcher at Wilson's Promontory contained pieces of barnacles and mussels, and those of Whimbrel at Rhyll Inlet on Phillip Island in Victoria contained crab claws. (See Figure 3.3.) Faeces have been used to determine the diet of the Ruddy Turnstone and Hooded Plover in

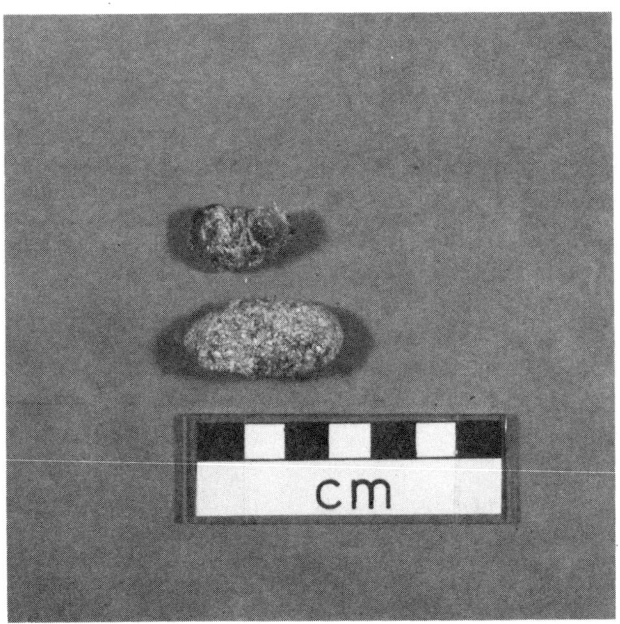

FIG. 3.3. Examples of pellets of Whimbrel (upper) and Eastern Curlew (lower) collected from a high-tide roost at Phillip Island, Victoria. Photo: P. Dann

southern Victoria (392, 394). The Turnstone had eaten small molluscs and a few barnacles, and the Hooded Plover had fed upon sandhoppers and molluscs.

Examining the stomach contents of shot or trapped shorebirds is another way of studying their diet. The effects of collection on the local shorebird population must be balanced against the ultimate aim of knowing how better to preserve them. The numbers and sizes of the food species can be measured in terms of relative abundance in each stomach sample, relative contribution to the total volume of each sample or the proportion of samples containing each prey type. Emetics and stomach-flushing techniques have been used with some success to obtain gut contents of other groups of birds and may be useful for studies of shorebirds. Stomach contents do not exactly represent the diets of birds, because different prey are digested at different rates.

It is extremely difficult, if not impossible, in many instances, to identify prey remains from faeces, pellets or gut contents, if the possibilities are not narrowed down by a good understanding of the potential food organisms available in the immediate area. There may be many hundreds of species of some groups of invertebrates (e.g. amphipods, polychaetes, ants, beetles) or seeds in the region, but a complete reference collection from the study area should reduce the identification of potential prey to a choice between just a few species.

In many circumstances, careful observation of the feeding behaviour of shorebirds, combined with a knowledge of what potential food is present and how this food is distributed and behaves (if animal), enables inferences to be drawn about their diet. One or more of the techniques mentioned above can be used to confirm the validity of such inferences.

Behaviours most useful to record in order to discover the food of a shorebird include: the depth in the soil or water at which the bird is feeding; whether the food may be detected by sight (run after or clearly picked up) or by touch (many repetitive motions during feeding with a few pauses to swallow food); and whether the food is taken in particular areas, such as near seaweed or at the edge of water.

How the behaviour of a shorebird combined with a knowledge of the potential food in the area can be used to determine a shorebird's diet is illustrated by Whimbrel in Rhyll Inlet on Phillip Island. The Whimbrel feeds at low tide on bare and lightly vegetated mud by running in quick bursts and capturing well-dispersed prey from the surface of the mud, away from the edge of the water. A walk across these flats reveals that only one animal occurs on the mud surface: the sentinel crab (*Macrophthalmus latifrons*) (pers. obs.). An examination of Whimbrel pellets left at high-tide showed that they contained mostly claws of sentinel crabs, confirming the suspicions of the observer.

## The food of shorebirds

The diet of each shorebird, where known, is described in detail in the species accounts (Chapter 5), and so a few general comments will suffice here. Shorebirds take a wide variety of foods, but animals are by far the most important. Invertebrates form the majority of shorebird food. Sometimes vertebrates (such as small snakes and frogs) are eaten by Bush Thick-knees (472); Banded Stilts and Red-necked Avocets have been recorded eating small fish (351). Seeds are the main parts of plants consumed, and berries are taken by some species of Palaearctic migrants on their breeding grounds. Various examinations of stomach contents of both species of lapwing have revealed grass and other plant fibre (61, 256, 472) but it appears unlikely that this material was a source of food, but rather was taken incidentally while the birds were feeding on seeds or ground-frequenting invertebrates. The Inland Dotterel has been recorded eating the succulent tips of some shrubs in north-western New South Wales, but these were probably taken for their water content, because the birds were not observed to drink (291).

Most shorebirds appear opportunistic in their feeding habits, and feed on most suitable items as they are encountered (375). Consequently there can be large differences in the diets of a species at different times of year (373) or at different sites. The wide range of prey taken by Sooty Oystercatchers feeding in intertidal areas in southern Victoria is demonstrated in Table 3.1.

The size of prey eaten by a shorebird generally increases with the size of the shorebird (166). Exceptions are the Red-necked Avocet and Banded Stilt, specialist feeders, which consume much smaller prey than would be predicted from their body size.

TABLE 3.1. The range of prey taken by Sooty Oystercatchers in intertidal areas in southern Victoria (69)

| | | |
|---|---|---|
| Mollusca: | *Cellana tramoserica* | Limpets |
| | *Patella peroni* | |
| | *Patelloida alticostata* | |
| | *Austrocochlea constricta* | Snails |
| | *Turbo undulatus* | |
| | *Nerita atramentosa* | |
| | *Thais orbita* | |
| | *Ischnochiton australis* | Chitons |
| | *Plaxiphora albida* | |
| | *Xenostrobus pulex* | Mussels |
| | *Austromytilus rostratus* | |
| | *Onchidella patelloides* | Sea elephant |
| | *Mactra pura* | Pipi |
| Crustacea: | *Carcinus maenas* | Crabs |
| | *Paragrapsus quadridentatus* | |
| | *Pilumnus rufopunctatus* | |
| | *Chamaesipho columna* | Barnacles |
| | *Catomerus polymerus* | |
| | *Austromegabalanus nignescens* | |
| | *Orchestria* sp. | Sandhopper |
| Annelida: | Polychaete spp. | Worms |
| Urochordata: | *Pyura* sp. | Sea squirt |

# Feeding behaviour

## WHEN SHOREBIRDS FEED

When some shorebirds feed is ultimately determined by light or tides, while others feed regardless of either of these factors. In Australia, shorebirds may be divided into two groups on the basis of when they feed: those that feed during both the day and night (most shorebirds, including Palaearctic migrants); and those that generally feed at night (Thick-knees, snipes and probably the Inland Dotterel). There does not appear to be an Australian shorebird that feeds solely during daylight.

Shorebirds that frequent intertidal habitats in southern Australia feed at low tide during the day or night. The relative amounts of day and night feeding may vary, depending upon their energy requirements for moult, migration and breeding and their feeding success during the previous low tide. Feeding at night has been shown in some species to be less successful than feeding during the day (Greenhalgh in 175), presumably because the species concerned were more proficient at detecting their prey visually than by touch. All sandpipers have extremely sensitive bills, rich in nerve endings, which enable them to find food concealed in mud, sand or water (45). Their bills are constructed in such a way that they are able to grab and extract items from well below the surface. The Eastern Curlew locates the burrows of its main prey, the ghost shrimp, by sight. Thus, while it is possible for it to feed on ghost shrimps on moonlit nights, it cannot find shrimp burrows on dark nights, when it abandons

the sandbanks and goes to the submerged eelgrass meadows near the low-tide level. Here, it stitches its open bill in and out of the eelgrass and finds crabs by touch (101). One of the disadvantages of feeding here is that the time available for feeding is much less, because the eelgrass is soon covered by the rising tide and the water becomes too deep for it to continue feeding.

Similarly, the Double-banded Plover depends upon sight to detect, pursue and capture its prey on mudflats, and on dark nights it switches to tactile methods of prey location. It is difficult to observe this plover at night, even with an infra-red telescope or image intensifier, but by examining its footprints and feeding marks, some idea can be gained of its behaviour. Figure 3.4A shows the tracks made by a plover feeding during daylight by chasing small crabs across the surface of the mud. The crabs are picked off the surface, and few bill marks are made. However, on dark nights (Figure 3.4B) the footprints are closer together, suggesting that the bird is walking rather than running and there are small patches of probing holes dotted along the tracks. The bird pauses and probes into the mud many times, searching for food by touch rather than sight.

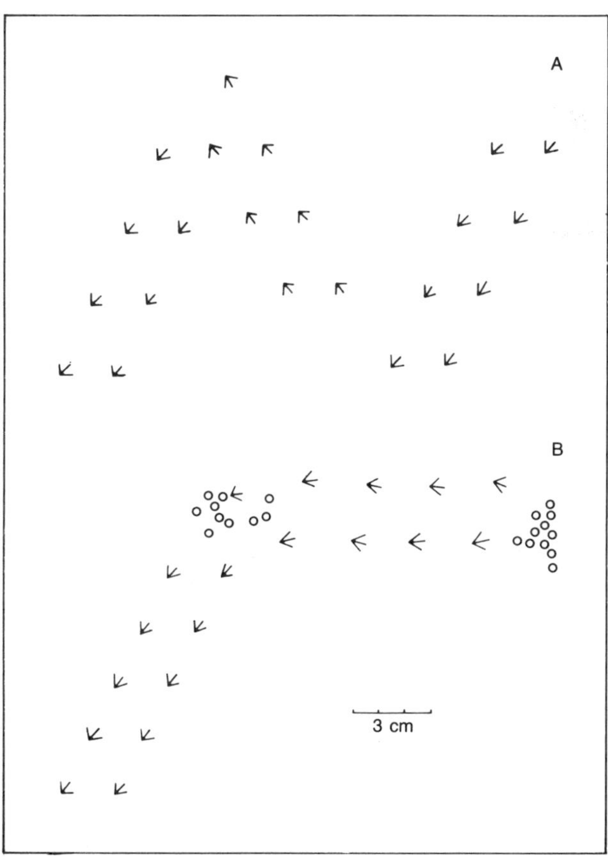


FIG. 3.4. Tracks made by a Double-banded Plover feeding during the day (A) and at night (B) on the same mudflat. Note that the plover is not running at night (footprints closer together) and is probing into the mud.

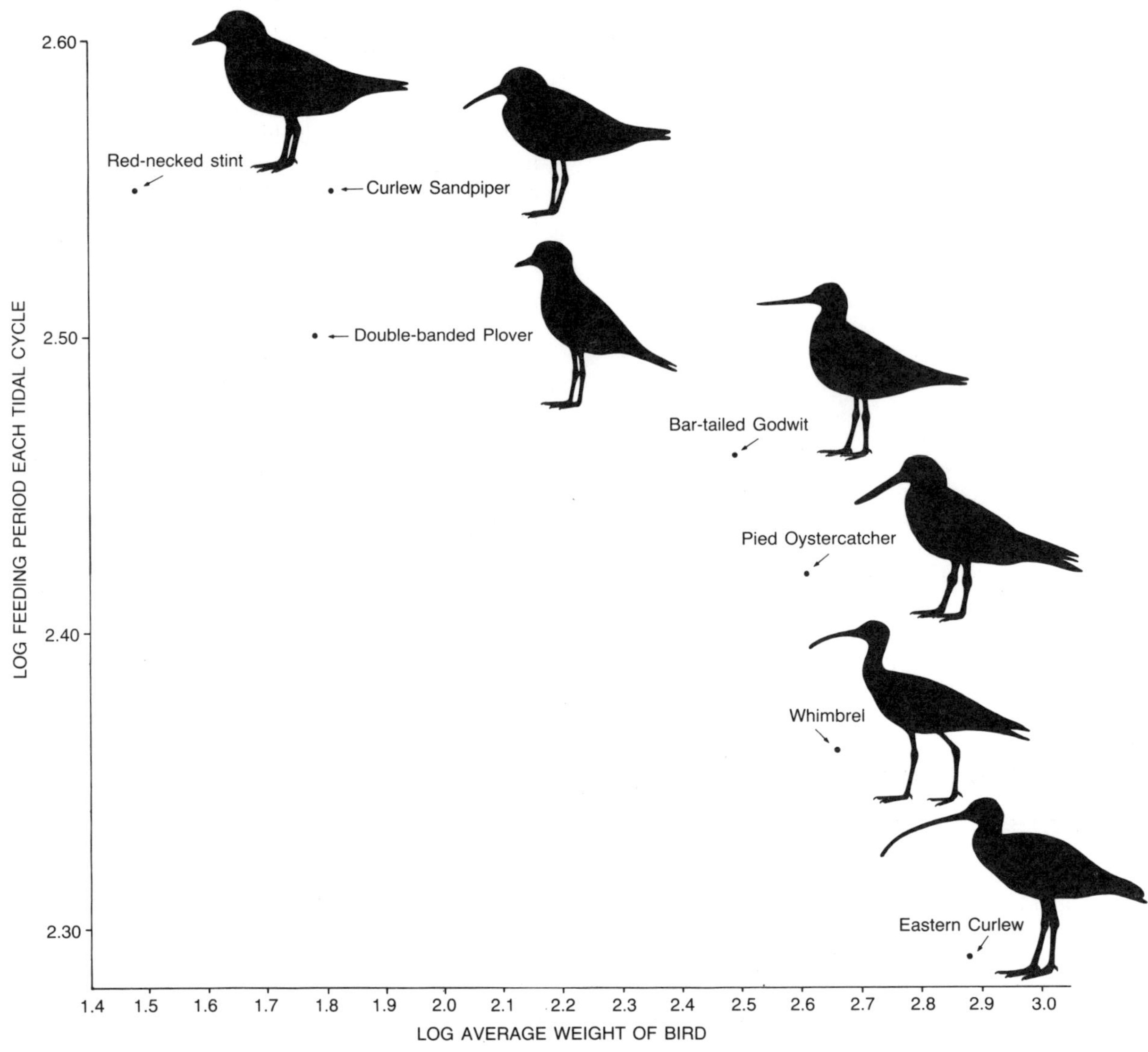


FIG. 3.5.  The relationship between the average weight (in grams) of the shorebird and the time (in minutes) spent feeding each tidal cycle. (101)

The amount of time spent feeding by shorebirds depends on their energy requirements and the availability of food. Smaller shorebirds feed for longer each tide cycle than do larger shorebirds (Figure 3.5). This trend has also been found in seabirds (356). The reasons for this are not known, but it has been shown theoretically that larger birds should be more efficient at feeding and storing fat (46). Smaller shorebirds often feed for as long as the mudflats are exposed. They may continue feeding in non-tidal areas over high tide, particularly if winds have been strong, probably to compensate for poor feeding success and higher energy costs while feeding. On mangrove-fringed coasts, numbers of small shorebirds may be limited by the few areas of mangrove-free mudflat for feeding close to high tide; larger species do not suffer this limitation. This was suggested for Westernport Bay in Victoria (264), and may also apply to other parts of southern Australia and much of northern Australia.

The feeding periods of terrestrial shorebirds may be more variable than those of intertidal species. The Masked Lapwing in southern Victoria feeds for 2.5 hours per day in April and for 10 hours per day (plus an unknown amount of night feeding) in August. In April, the lapwings are in large flocks feeding on the very abundant field cricket (*Teleogryllus commodus*) and, as breeding and moult (which demand more energy) have finished, their need to feed is not as great. In August, they are in territorial pairs in preparation for breeding and are feeding on a less abundant prey, earthworms, the availability of which is greatly influenced by rainfall.

## WHERE SHOREBIRDS FEED

Those shorebirds that forage in intertidal areas roost in flocks at high tide, moving out to feed as the tide ebbs.

At Rhyll Inlet on Phillip Island in Victoria, the Red-necked Stint and Curlew Sandpiper feed in cohesive mixed-species flocks and usually follow the same route each tidal cycle (Figure 3.6A). By contrast, the Eastern Curlew from the same roosting flock feeds singly in separate territories or in small loose flocks. The territorial birds probably defend the same area each tide and remain there until the tide rises (Figure 3.6B), while the others do not hold territories and feed on mudflats not held by territorial birds. The potential advantages of feeding in flocks include possible exchange of information about food sources (482) and reduced risks of predation: safety in numbers (30).

Some shorebirds may be less successful in obtaining food when hunting too close together (164), because prey is disturbed by adjacent birds and moves deeper into the mud where the shorebirds cannot reach it. Double-banded Plovers fly from their roosts in flocks, then spread out rapidly over exposed mudflats to defend small feeding territories. There is some social cohesion between territorial birds, and they flock when disturbed or changing feeding areas. The purpose of these small territories may be that other plovers feeding close by would disturb prey. The plovers feed on sentinel crabs, which, when disturbed, escape being eaten by retreating down burrows. Curlews, being larger, probably have fewer predators than small shorebirds, such as the Double-banded Plover, and feed on organisms that are sensitive to movement, hence the contrast in dispersal pattern. The areas defended by territorial curlews are better feeding grounds, and territorial behaviour may provide a good feeding area with minimal interference from other curlews and the opportunity to monopolise a profitable feeding area over a period of weeks or months. This has been suggested as an explanation for similar behaviour in Pied Wagtails (*Motacilla alba*) (105).

After shorebirds have dispersed from a high-tide roost, their distribution within a feeding habitat can be examined in detail. Five zones can be recognised in a tidal habitat (375): Zone A, which consists of areas of sand and mud above the water's edge not retaining a surface film of water; Zone B, which consists of areas of sand and mud above the water's edge retaining a surface film of water; Zone C, which consists of the water's edge and an area 0.5 metre on either side (modified 101); Zone D, which is the area between 0.5 and 1.5 metres below the water's edge (modified 101); and Zone E, which is the area beyond Zone D.

At Werribee, in Victoria, the three abundant species of small sandpipers fed in different zones (Figure 3.7).

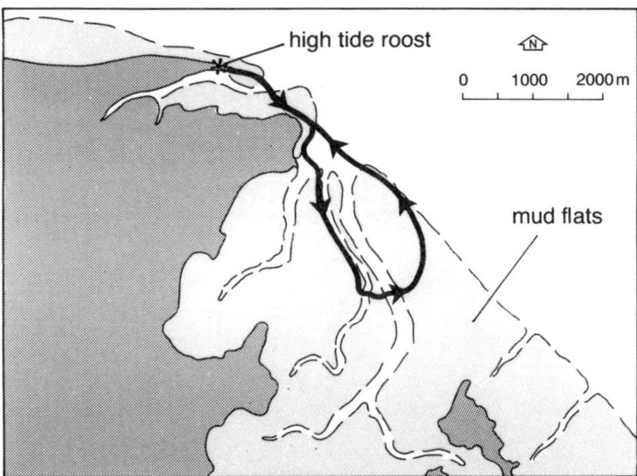


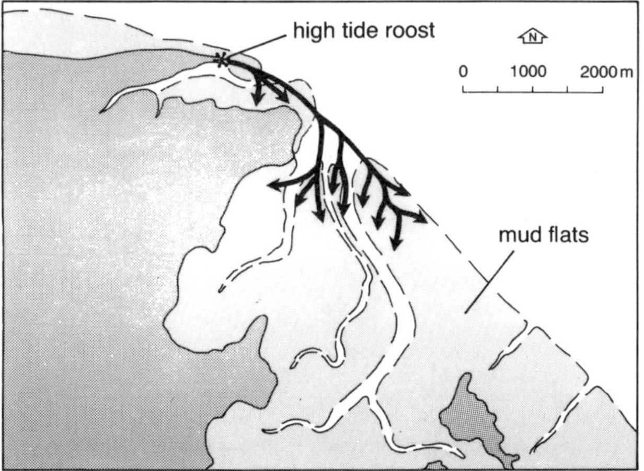


FIG. 3.6. The dispersal of mixed flocks of Red-necked Stints and Curlew Sandpipers (A) and territorial Eastern Curlews (B) from a high-tide roost on Phillip Island.

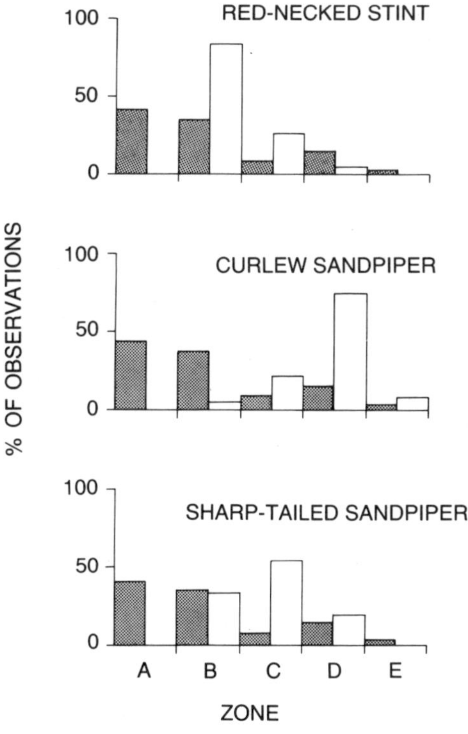


FIG. 3.7. The feeding zones used by three species of small sandpipers at Werribee, Victoria. The dark bars represent the expected distribution if each species was feeding at random and the white bars represent the observed distribution. (103)

This figure shows data collected over a period of an hour either side of low tide. The three species were not distributed in a random manner over the mudflats, rather they showed significant preferences for particular zones. Most Red-necked Stints fed in Zone B (wet mud); the majority of Curlew Sandpipers fed in Zone D (shallow water); and almost half the Sharp-tailed Sandpipers were in Zone C (the water's edge). The last showed a more even feeding distribution than the other two species.

Similar feeding distributions have been found for the Red-necked Stint and Curlew Sandpiper in Tasmania (461). In non-tidal habitats both species fed more in the water than they did in tidal areas. These preferences for different zones are probably due to different feeding adaptations (morphological and behavioural) in sandpipers in relation to the type, size and behaviour of their prey. This also reduces interference between species. This may be important when the mudflats are first exposed by the tide, and all birds are feeding in a very small area.

Small-scale differences in feeding distribution occur in mixed-species flocks of inland shorebirds. On Lake Eyre North, South Australia, the Red-necked Avocet and Banded Stilt feed together, but the stilts are significantly farther from the shore. This is probably because the stilts can swim in any depth of water and still feed on floating prey, whereas the avocets take prey from the bed of the lake (251).

## How Shorebirds Feed

Shorebirds are morphologically diverse and occur in many different habitats. As a result they obtain food in a variety of ways, and some of these are illustrated in Figure 3.8. These methods vary from repetitive sweeps and probes to active pursuit and capture. Typically, shorebirds feed while walking or wading, although some may do so while swimming (e.g. Banded Stilt, phalaropes) or flying (e.g. pratincoles). The Australian and Oriental Pratincoles are unusual in being the only species of shorebirds in Australia that take some of their food (primarily insects) on the wing. Many shorebirds have long legs, so they can walk in deep water when hunting. The Comb-crested Jacana has exceptionally long toes, permitting it to walk on waterweed, where it obtains its diet of insects and seeds.

A species may have a repertoire of feeding methods, depending on the circumstances. Some differences in day and night feeding methods have already been mentioned. These differences are due to changes in the manner that prey are detected, but different prey may also require different methods of handling. Sooty Oystercatchers hammer limpets (*Cellana tramoserica*) off rocks and eat them on the spot, whereas after hammering a dog-whelk (*Thais orbita*) in a similar way, it lodges it in a suitable crack and drills it further to dislodge the soft parts of the animal (69). The

complicated sequence of feeding actions during these attacks on dog-whelks is shown in Figure 3.9.

The depth at which a shorebird obtains food from soil, mud or water depends on two factors: the length of its beak and how far it can insert its beak. Organisms frequenting soil, mud or water have particular vertical as well as horizontal distributions and may also move vertically depending on tide or weather. The availability of prey to different species therefore depends on the length of the bill. The bill lengths of some intertidal shorebirds and typical depths of their main prey are illustrated in Figure 3.10. Not surprisingly, the Eastern Curlew feeds on deeply burrowing ghost shrimps, whereas shorter-billed species feed on small crustaceans, worms or molluscs obtained close to or from the surface.

In some Australian shorebirds, the female has a longer bill than the male: Curlew Sandpiper (460),

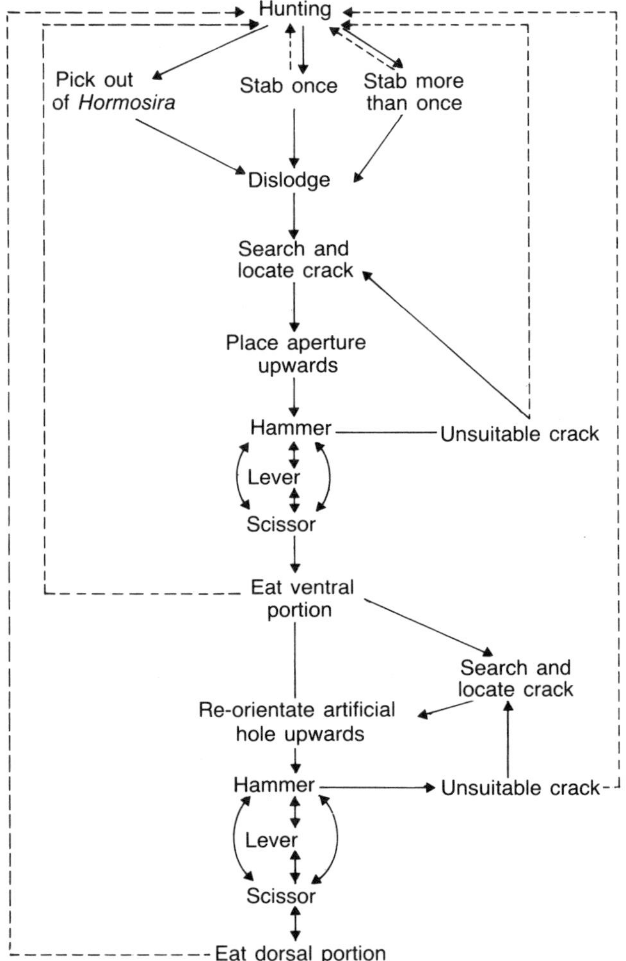


Fig. 3.9. A summary of the main sequence of actions during attacks by Sooty Oystercatchers on dogwhelks. (69)

Fig. 3.8. The diversity of feeding methods used by shorebirds. ►

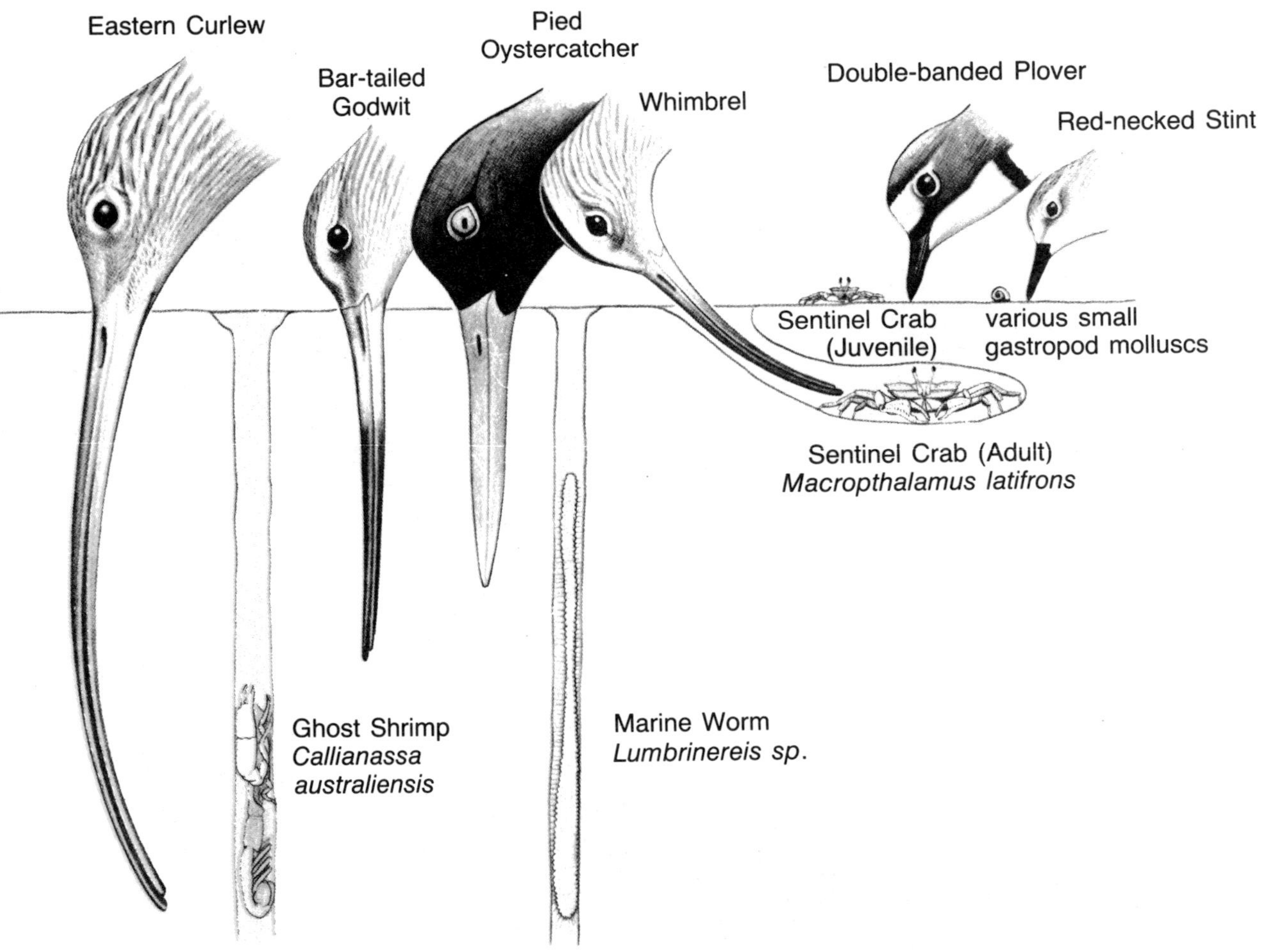


FIG. 3.10.    The bill lengths of some intertidal shorebirds in relation to the typical depths of their main prey.

Bar-tailed Godwit (384), Eastern Curlew (101), Sooty Oystercatcher (pers. obs.). There are several reasons why this sexual dimorphism might exist. It may have evolved to reduce competition between the sexes for food by allowing them to obtain food differently (398). A study showed that there was almost as great a difference in diet between the sexes of Pectoral Sandpiper and Dunlin on the breeding grounds as there was between the two species (203). At the end of the breeding season, just before southward migration, the diets of male and female Dunlins diverge even further, perhaps because of dwindling food supplies and the increasing energetic demands of moult and pre-migratory fattening.

The sexual dimorphism in bill length in the Eastern Curlew is associated with significant differences in diet. Many of the longer-billed females feed alone and defend territories on sandbanks and mudflats where much of their prey inhabits deep burrows. On the other hand, the majority of males feed in loose flocks in areas of mudflat pools and tidal eelgrass (Zosteraceae), where their prey lives near the surface.

In the Sooty Oystercatcher at Wilson's Promontory in Victoria, the longer-billed female feeds on small mussels and barnacles, while the shorter-billed male feeds on large mussels (Table 3.2) (69). This difference may be due to males having a more robust, less sensitive beak than females and thus using it more for hammering than for prising and stabbing. The European Oystercatchers (*Haematopus ostralegus*) that feed by hammering open cockles lose their bill sensitivity,

TABLE 3.2.  The differences in diet between the sexes of Sooty Oystercatchers at Wilson's Promontory, Victoria (69).

| PREY | MALE % | FEMALE % |
|---|---|---|
| Dogwhelk | 6 | — |
| *Xenostrobus* (small mussel) | 7 | 28 |
| *Austromytilus* (large mussel) | 75 | 6 |
| Barnacles | 12 | 66 |
| Number of items eaten | 144 | 71 |

whereas those that feed by probing for marine worms retain it (339).

An extension of feeding on different foods in similar areas is to feed in geographically distant places. In the Curlew Sandpiper, the difference in bill length may be a product of the sexes being concentrated in different localities in the non-breeding season (23) and hence experiencing different feeding conditions.

Conversely, sexual differences in the size and proportions of the bill may be related to dimorphism in body size for courtship display and therefore may be a secondary effect of competition for mates (398).

The other factor affecting the depth at which the shorebird feeds is how far it inserts its bill into the substrate. Three methods of feeding can be easily recognised: pecking — taking a food item from the surface of the substrate; jabbing — inserting the bill into the substrate for less than half its length; and probing — inserting the bill into the substrate for more than half its length (201). Most shorebirds may use all three methods, but pecking is probably most frequent.

The Red-necked Stint inserts its bill to different depths, depending on whether it is in or out of the water (461). When on dry mud (Zone A), most peck, but they jab and eventually probe as they feed in progressively wetter substrates (Figure 3.11). The reasons for this shift are not clear, but may relate to the ease of penetrating wetter substrates as well as the vertical distribution and behaviour of prey in each zone.

Estimates of the average depth at which shorebirds feed include: Red-necked Stint, 3.4 mm; Sharp-tailed Sandpiper, 9.0 mm; and Curlew Sandpiper, 14.0 mm (103).

The rate of feeding is another aspect that varies greatly. Many apparent feeding actions are largely searching movements or unsuccessful capture attempts, so the actual number of food items eaten is much less than the number of apparent feeding actions observed.

Plovers, with their characteristic pause-run-peck sequences, have a lower rate of attempts at prey capture than sandpipers, which rapidly probe and peck. The difference is apparently due to the methods used in detecting prey, with plovers depending more on sight and the sandpipers probably using touch. It would be expected that sandpipers would have more developed sensory areas in their bills than plovers, but no difference has yet been confirmed.

## Interactions Between Species

The morphological and behavioural differences in shorebirds result in species feeding in the same area, but consuming different prey or different sizes of the same prey. At Werribee, Victoria, all three species of small sandpipers eat similar food, particularly polychaete worms (*Ceratonereis erythraeensis*) (97). The percentage of overlap in diet between Red-necked

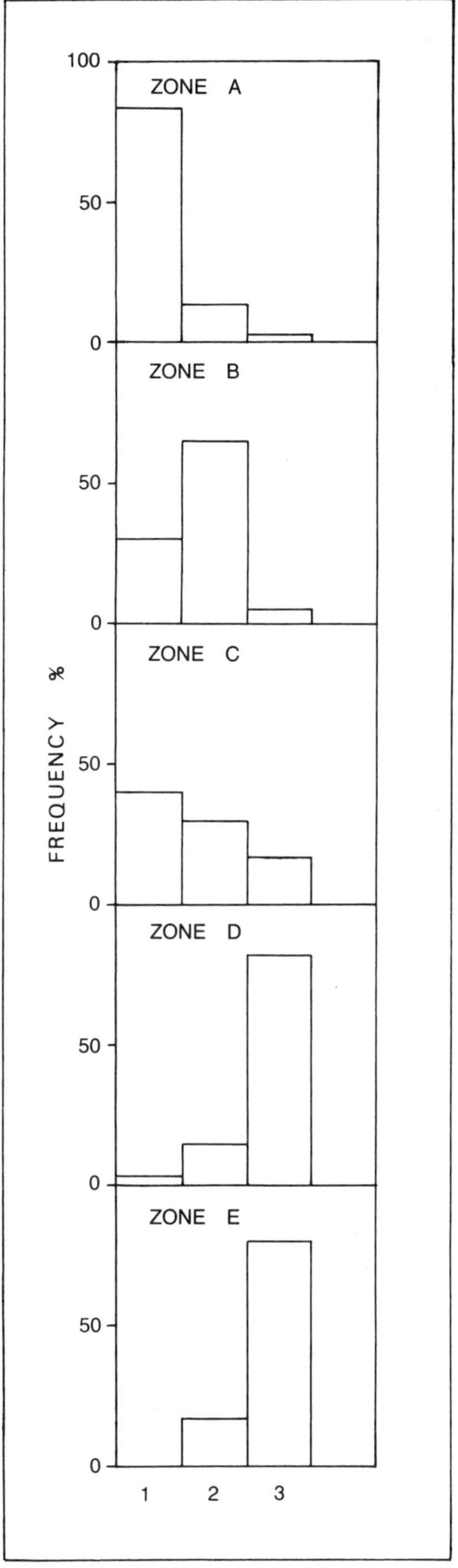



FIG. 3.11.   The feeding methods of Curlew Sandpipers in relation to feeding zone. 1 pecking; 2 jabbing; and 3 probing (461). See text for explanation of zones.

Stint and Curlew Sandpiper (57 per cent) is greater than between stints and Sharp-tailed Sandpiper (53 per cent) and between Curlew Sandpiper and Sharp-tailed Sandpiper (42 per cent). A large proportion of these overlaps is the polychaete worm *C. erythraeensis*, which is digested quickly, and only its jaws are found in shorebird stomachs. The size of these jaws is an index of the size of worms eaten (103). The relative abundance of jaws of different lengths found in the stomachs of these birds is shown in Figure 3.12. Stints eat significantly smaller worms than Sharp-tailed and Curlew Sandpipers, which take worms of similar sizes. No differences are found in the sizes of the other prey eaten by the three species (103). When diet overlaps, competition is reduced by such choice of prey size. Interference between different species of shorebird in the form of aggression is rare (375).

Interference between shorebirds and other species occurs at times. At Phillip Island, Victoria, Silver Gulls robbed food from Bar-tailed Godwits by chasing them and snatching it from them. This reduced the feeding rate of the godwits by 36 per cent and may have detrimentally affected pre-migratory fattening (95).

## Conclusion

Studies of the feeding ecology and behaviour of shorebirds assist greatly in understanding their distribution and movements. Until more research on feeding has been carried out in Australia, active management of shorebird populations will not be possible. With the present reduction and modification of feeding areas, it is necessary to understand the processes determining where shorebirds occur and hence predict the likely effects of such changes to their habitats.

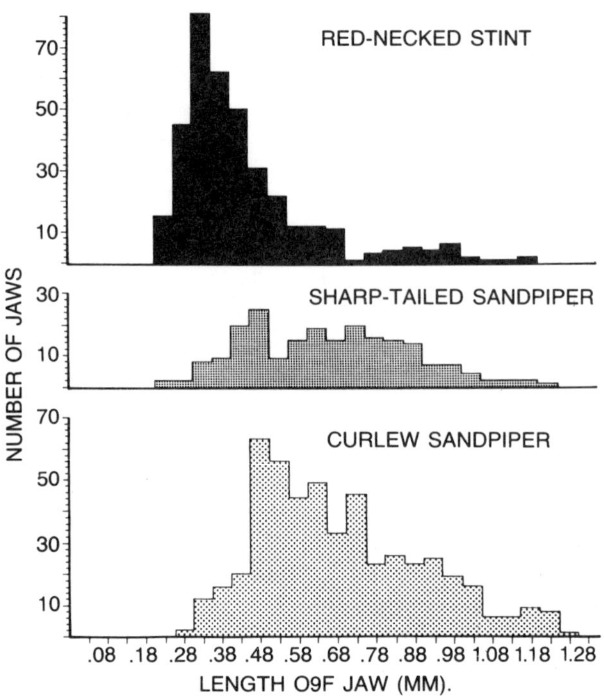


FIG. 3.12.   The sizes of the polychaete worm *C. erythraeensis* taken by three species of small sandpiper at Werribee, Victoria. Those eaten by Red-necked Stints were significantly smaller than those taken by Curlew Sandpipers and Sharp-tailed Sandpipers (103)

# SHOREBIRD MIGRATION AND MOVEMENTS

The sudden departure and arrival of birds, often in large flocks, with the passing of the seasons has long fascinated people. Many explanations have been offered, perhaps the most novel being that for the Barn Swallow (*Hirundo rustica*), which was thought to spend the winter months hibernating in mud at the bottom of ponds. In the eighteenth century the German ornithologist J. L. Frisch disproved this theory by capturing swallows and tying threads dyed with a water-soluble dye to their legs. In spring they returned with their threads still coloured, and Frisch concluded that they did not dive to the bottom of ponds in the winter (303). This is the first known example of individual birds being marked to test a theory about bird behaviour. Since then, bird-banding (i.e. marking individual birds with numbered metal leg-bands) has answered many questions about bird behaviour, particularly about their movements.

Australia has over seven hundred species of birds, but only a small proportion are strict migrants; that is, most of their population moves annually between traditional breeding and non-breeding areas. Sacred and Buff-breasted Paradise kingfishers (*Halcyon sancta, Tanysiptera sylvia*), the Torresian Imperial-Pigeon (*Ducula splorrhoa*) and the Clamorous Reed-Warbler (*Acrocephalus stentoreus*) are north-south migrants, many crossing to New Guinea and nearby islands for the southern winter. Many more species may yet be found to be migratory.

The remarkable trans-equatorial flights of the shorebirds are among the most spectacular migrations of any birds. Individual shorebirds return to the same stretch of beach each year, having flown from the same patch of marsh in their breeding grounds in the Siberian wilderness.

The RAOU study has found out much about shorebird migration and movements in Australia, and the results of this investigation are presented in this chapter by examining why they must migrate, how they do it, and how these facts have been discovered. More detailed information on the migration and movements of each species can be found in Chapter 5. This chapter discusses shorebird migration and movements in four section, as follows.

*Shorebird breeding* describes the breeding grounds of Australia's migratory shorebirds and shows how they have adapted to the often difficult conditions of high-latitude areas of the Northern Hemisphere.

*Shorebird migration* describes how shorebirds migrate, by examining how the information needed to piece together the story of their long flights is obtained.

*Migration within Australia* looks at what is known of their migration within Australia.

*Movements of shorebirds that breed in Australia* examines how resident shorebirds respond to factors that affect the availability of their habitat.

## SHOREBIRD BREEDING

Migration across great distances to a strict schedule is one adaptation in shorebirds to the extreme seasonality of conditions on their high-latitude breeding grounds in the Northern Hemisphere. There, during the northern summer, very rich food resources are available for limited periods. This has led to the evolution of a breeding schedule, whereby courtship, laying, incubating, hatching and fledging occur in a relatively short period and are timed to ensure maximum reproductive success. Courtship usually starts in early June, soon after arrival, laying in mid-June, hatching from the beginning of July and departure from mid-July onwards.

## Where Australian shorebirds breed

Most shorebirds that migrate to Australia come from breeding grounds in the eastern USSR, Mongolia and northern China, an area covering 30° of latitude and a number of distinct vegetation zones and climatic regions. They breed in areas ranging from the Gobi Desert of Mongolia north to the tundras and shingle beaches of the Arctic Ocean. The range of vegetation types and their distribution in Asia is shown in Figure 4.1. The breeding distribution of the more common species is described in detail in Chapter 5; the rare species in Chapter 6.

The deserts and steppes of southern Siberia, Mongolia and northern China consist of open stony country with a varying coverage of grasses and shrubs. This is the preferred breeding habitat of some of the plovers. Despite its usual association with coastal wetlands during the non-breeding months in Australia, the Large Sand Plover breeds on this stony country away from

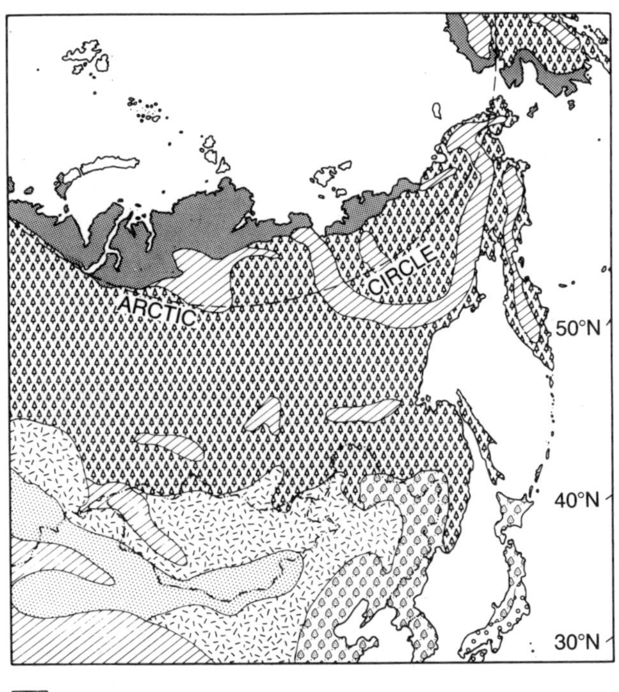


Tundra

Mountain Vegetation

Coniferous Forest

Deciduous Forest

Grassland

Desert

Sub tropical

FIG. 4.1. The vegetation of north-eastern Asia.

water, only staying around lakes when it first arrives and during courtship (237). Similarly, the Oriental Plover is rarely seen near wetlands on its breeding grounds in the Gobi Desert (237).

The *Tringa* sandpipers, such as the Marsh Sandpiper and Greenshank, breed in better watered areas of grassy steppe, especially near small bogs and marshes (85). They also breed in the vast northern coniferous forest or taiga zone, which stretches north from the steppes to the arctic tundra. The coniferous forest belt is not uniform. Fire and storm damage and differences in drainage, altitude and rainfall produce a mosaic of forest types in different stages of development. The Little Curlew, for example, breeds in recently burnt areas that are quite open, prompting the indigenous Yakutsk people to call them 'fire-birds' (239). The forest zone also supports breeding populations of species such as the Wood Sandpiper, Long-toed Stint and Whimbrel.

North from the edge of the coniferous forest the trees are replaced by shrubs, and the shrubs eventually disappear to be replaced by mosses, lichens, herbs and sedges, stony rises and boggy country. Stretching to the shores of the Arctic Ocean, this is the true arctic habitat: tundra. It is frozen for nine months of the year, and a layer of permafrost impedes drainage during the summer thaw, so innumerable bogs and marshes form, particularly on the coastal plains and river deltas, such as those of the Kolyma, Yana and Indigirka rivers. (See map on back endpaper.)

The small sandpipers, familiar to observers in southern Australia, breed in the tundra (325). Tundra is a diverse mixture of vegetation types, and each species breeds in a characteristic place. The Sharp-tailed Sandpiper places its nest among dense sedges in the wettest parts of the tundra. The Red-necked Stint nests in stony, shrubby parts of the tundra in foothills away from coastal plains, flying some distance to wet areas to feed. The Great Knot breeds in drier alpine tundra, flying across country to feed in small mossy bogs.

## The ecology of shorebirds on the breeding grounds

Apart from contrasts in the type of country used in the breeding and non-breeding months, the behaviour of shorebirds is also markedly different. On the non-breeding grounds, they live in tightly packed flocks, often large, which stay together for months. When breeding, however, they are spread over millions of square kilometres of country, each pair defending its own patch of ground. In Australia, the abundance of waders is measured in terms of numbers of birds packed into a high-tide roost-site or on a single lake. In Siberia, it is measured in numbers of birds per square kilometre; a density of 50 or so per square kilometre is considered quite high (325).

Breeding begins when the snow has thawed or been blown away by spring winds (167). Summer weather varies greatly from year to year and adult shorebirds of some species arrive on the breeding grounds with substantial fat reserves (a source of energy) as insurance against possible unfavourable conditions, e.g. Turnstone (270). On arrival, many shorebirds feed on fruits of plants, produced in the previous summer, exposed by the receding snows. They switch to insect larvae, which hatch once the thaw is well under way.

In the tundra, shorebirds are the principal predators of insects, which emerge in huge numbers during the very short summer. Non-biting midges (Chironomidae) are one of the most important prey items, and the adults emerge during the first half of July, when the eggs of arctic-nesting shorebirds are hatching. Close correspondence has been found between the hatching of Turnstone eggs and the emergence of adult midges, the main food source of the chicks (328). (See Figure 4.2.) After the thaw, adult shorebirds feed on insect larvae by probing their bills into the tundra. When the young hatch, they have incompletely hardened bills and, unlike the adults, cannot probe; they must rely on active prey that lives on the surface of the ground. This limitation has probably led to the evolution in shorebirds of hatching at the same time as the emergence of adult insects (293).

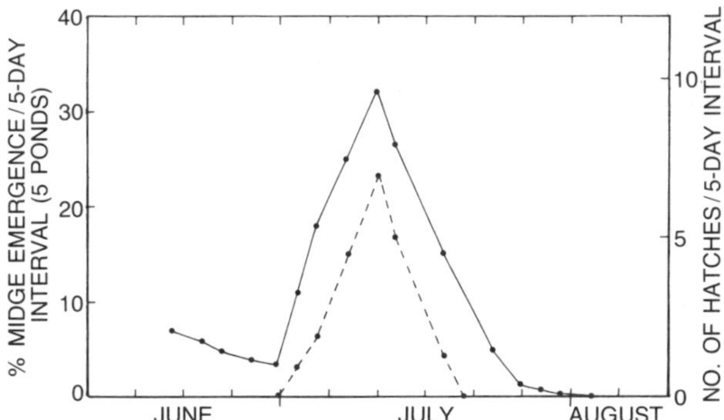

Fig. 4.2. Hatching times of Ruddy Turnstones in relation to emergence of adult Chironomid midges, arctic Alaska (———— adult midges, − − − Turnstone hatching) (Redrawn from D.N. Nettleship, 1973. *Ibis* 115: 202−17. With permission from the publisher and author.)

Differences in weather on the breeding grounds lead to differences in the breeding success of shorebirds from year to year. Many young may be produced in years with favourable weather. Only a few pairs will breed successfully if the weather is bad. Breeding may be precluded in a late thaw. If conditions deteriorate after breeding begins, it will be unsuccessful.

Annual fluctuations in the breeding success of shorebirds in the Northern Hemisphere can be recorded in Australia by determining the proportion of immature birds in non-breeding flocks. Winter numbers are also a good index of breeding success, because individuals of most species do not return to the breeding grounds until they are eighteen months old. Figure 4.3 illustrates the correspondence between summer and winter counts and proportion of juveniles in flocks of Curlew Sandpipers in coastal Victoria between 1981 and 1985.

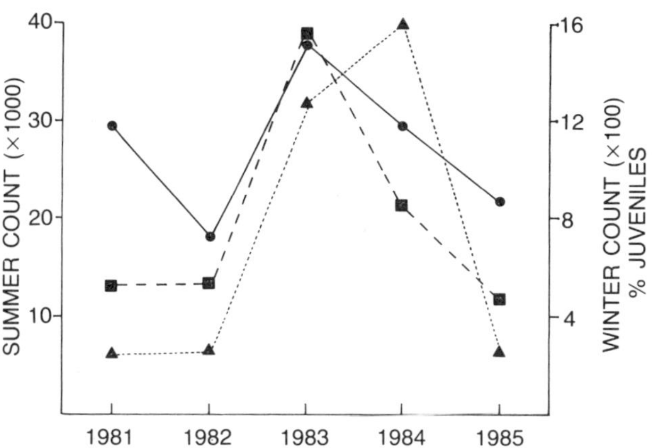

Fig. 4.3. Changes in summer and winter counts of Curlew Sandpipers and in the proportion of young birds in flocks in coastal Victoria from 1981 to 1985. (———— summer count; . . . . . winter count; − − − percentage of young birds).

In the northern summer of 1983, shorebird breeding in the tundra of the Taimyr Peninsula in central Siberia was not very successful (Tomkovich, cited in 360). This may have been due to predation by Arctic Foxes (*Alopex lagopus*). Lemmings (*Dicrostonyx groenlandicus*) were particularly abundant on the Taimyr Peninsula in the northern summer of 1982, leading to an increase in the population of their main predator, the Arctic Fox. In 1983 there were many fewer Lemmings, and the foxes may well have switched to feeding on breeding shorebirds. Curlew Sandpipers breed there, and this may account for the drop in the proportion of immature birds in flocks between 1983 and 1984 as shown in Figure 4.3.

## An Australasian migrant

The contrasts between the breeding and non-breeding areas of shorebirds are striking. Some examples have been given above. A particularly interesting example occurs in the Southern Hemisphere in the Double-banded Plover. This species is unusual in that it is the only shorebird in Australia to migrate east-west, breeding in New Zealand and migrating across the Tasman Sea.

In New Zealand, many breed in the high country, some distance from the sea, on gravel bars in the wide rivers that flow from the glaciers in the alps of the

FIG. 4.4   Breeding area of the Double-banded Plover: Cass River, near Lake Tekapo, South Island, New Zealand. Photo: B. Lane

South Island (Figure 4.4). In winter, these areas are very cold; snow lies on the ground for weeks and most ponds and swamps freeze. In the southern winter, some flocks migrate to south-eastern Australia, where they inhabit coastal mudflats, beaches, saltmarshes, and occasionally pasture and inland salt lakes. Others remain in New Zealand for the winter. After the thaw, Double-banded Plovers begin arriving in the New Zealand high country. They reach their maximum numbers in August, breed from late August to January, and most depart in February and March (363).

# SHOREBIRD MIGRATION

Shorebirds leave the breeding grounds and fly southwards, stopping on the way at areas that provide sufficient food for them to prepare for the next leg of the migration; these areas are links in a chain of sites between the breeding and non-breeding grounds. Millions of shorebirds leave the breeding grounds in northern Eurasia and North America in July and fly to coastal and inland wetlands in all continents. It has been estimated that about 6.5 million shorebirds spend the non-breeding months on the coast of Europe (369). A further 2 or 3 million pass through western Europe to western and southern Africa. Preliminary estimates suggest that about 5 million shorebirds migrate from eastern Siberian and Alaskan breeding grounds to Asia, the Pacific Islands, Australia and New Zealand (346). Of these, about 2 million reach Australia, and the remaining birds stay in eastern Asia or go to New Zealand. The number of shorebirds in other migration systems, such as that between North and South America, has not been estimated.

Different species and even separate breeding populations within species may use different migration routes and non-breeding areas. The breeding areas of Asian/Australasian shorebirds are spread over 30° of latitude and 60° of longitude in northern Asia. In Australia, the non-breeding area covers 30° of latitude and 40° of longitude. The possible combinations of breeding areas, migration routes and non-breeding sites for the 30 or so species of shorebirds that occur annually in Australia are enormous. The task of unravelling this tangle of routes through Asia and the Pacific into Australia is only just beginning. The remainder of this section illustrates how this task can be undertaken.

## Techniques for studying shorebird migration

A variety of techniques are used to determine how shorebirds migrate between their breeding and non-breeding grounds. Counting shorebirds, banding and colour-dyeing them and using radar to watch them have all been applied with great success. This section describes the techniques and gives examples from Australia as well as other parts of the world.

## COUNTS

Information from counts shows what species are present, how many of each there are and when they occur in an area.

Counts during the non-breeding months after migration (December–February), such as the national counts conducted in Australia as part of the RAOU study, are an essential first step in clarifying migration routes. In Australia, some species that spend the non-breeding months on the Pacific coast do not occur in such large numbers on the north-west coast, e.g. Lesser Golden and Mongolian Plovers. Some species common in the south and on the north-west coast are far less abundant on the Pacific coast, e.g. Red-necked Stint, Curlew Sandpiper and Sanderling. These results suggest that there are at least two major migration routes into Australia; one via north-western Australia to southern Australia and the other via north-eastern Australia. Some species use only one route, others may use both. A third route may exist via Arnhem Land and the Gulf of Carpentaria, but this possibility requires more investigation.

Another way to piece together the migration jigsaw is to count shorebirds regularly and frequently at one place to determine changes in numbers due to arrival and departure. Such information from a number of places enables the timing of movement and location of migration routes to be determined. In western Europe and Africa, work like this has shown that Red Knots that breed in Siberia leave their non-breeding sites in South Africa in late April, pass through western France in mid-May, Holland, West Germany and Denmark in late May and arrive on the breeding grounds from early June onwards (112). This is illustrated in Figure 4.5. Similar counts in Asia (306, 477) and Australia have shown that the southward migration of most species of shorebirds starts in mid-July and finishes by December. Northward migration from Australia starts as early as mid-February in some

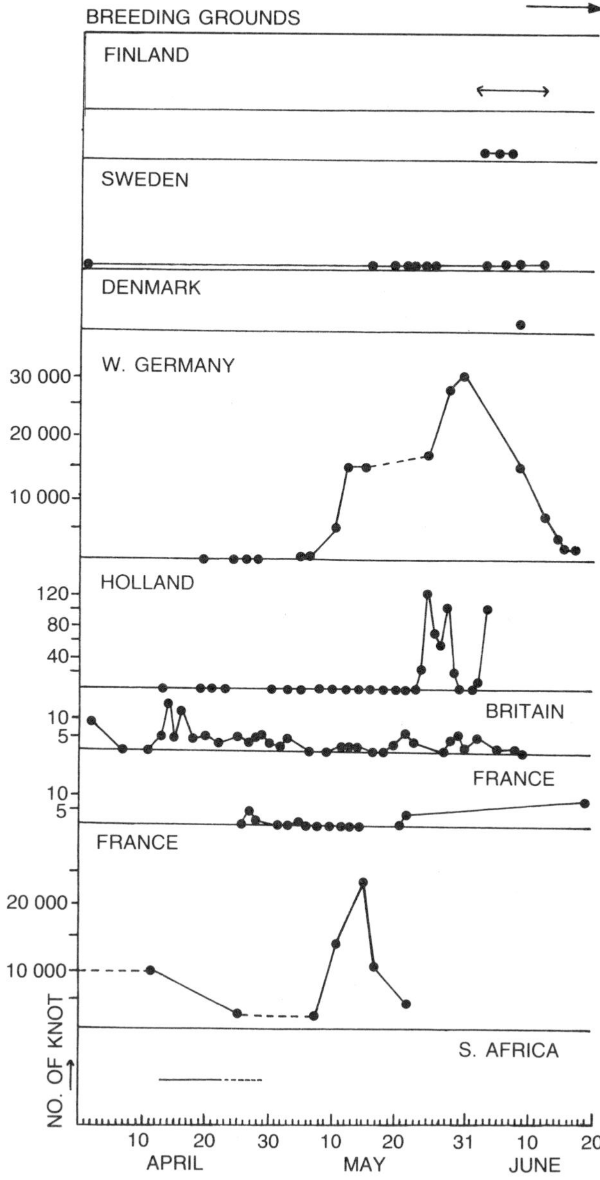


FIG. 4.5. Changes in numbers of Red Knots recorded at a number of African and European sites during their northward migration. (Redrawn from W.J.A. Dick, 1979, *WSG Bulletin* 27: 8–13. With permission from the author and the publisher.)

species, but more usually in mid-March. They pass through Asia between March and May and arrive on the breeding grounds in early June. The pattern of change in numbers at the southern end of a migration route, in this case Werribee-Avalon area in Victoria, is shown in Figure 4.6.

Regular counts of Sharp-tailed Sandpipers at coastal and inland sites in eastern Australia (see Figure 4.7.) show that they only pass through the inland during southward migration and do not spend the entire non-breeding period there. The January peak in numbers at some coastal sites probably represents the final southward movement of the population.

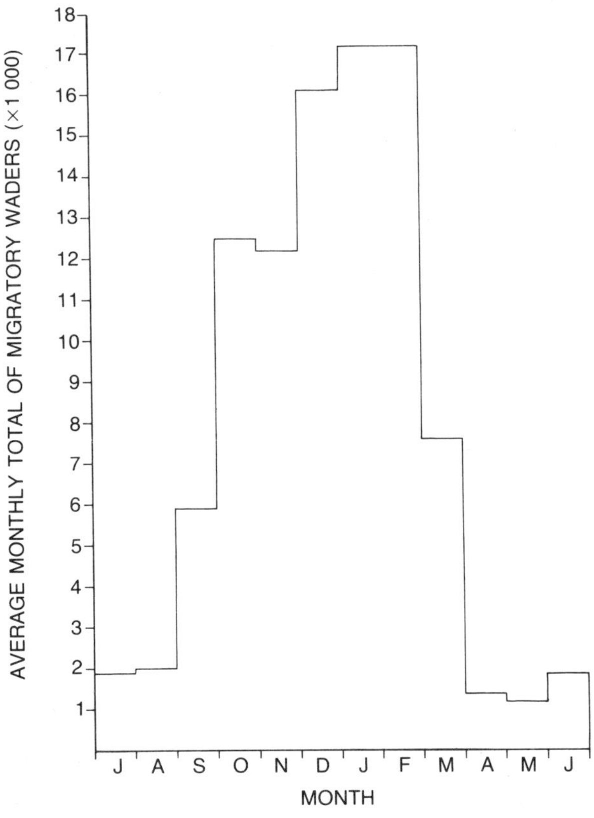

FIG. 4.6. The changes in numbers of Northern Hemisphere breeding migratory shorebirds at a site in southern Australia: Werribee–Avalon area, Victoria.

# BANDING

Banding is a widely used technique for studying bird biology. Apart from providing direct evidence of movements, studying shorebirds during banding over the course of a migration can reveal whether or not different populations, age-groups or sexes are moving into and out of an area. These differences cannot be detected by regular counts alone, as a turnover of birds occurs all the time. Banding leads to understanding the causes of changes in numbers detected during a series of counts. This section describes the way that banding can be used to discover facts about migration routes, size differences in shorebirds, their energy requirements for flight and their moult. Figure 4.8 shows the procedures involved in catching, banding and handling shorebirds.

## History

In Australia, the first bird-banding was done in 1912 on Phillip Island in Victoria by the Bird Observers' Club (55). Short-tailed Shearwaters, or Mutton Birds (*Puffinis tenuirostris*), were banded at their breeding colonies. Continued banding and the resulting international recoveries over subsequent decades resulted in their migration to the northern Pacific Ocean being discovered. In 1953 the Commonwealth Scientific and Industrial Research Organization (CSIRO) started the Australian Bird Banding Scheme. Since 1953 over 2 million birds have been banded in Australia and its territories. It was not until the early 1970s that banding

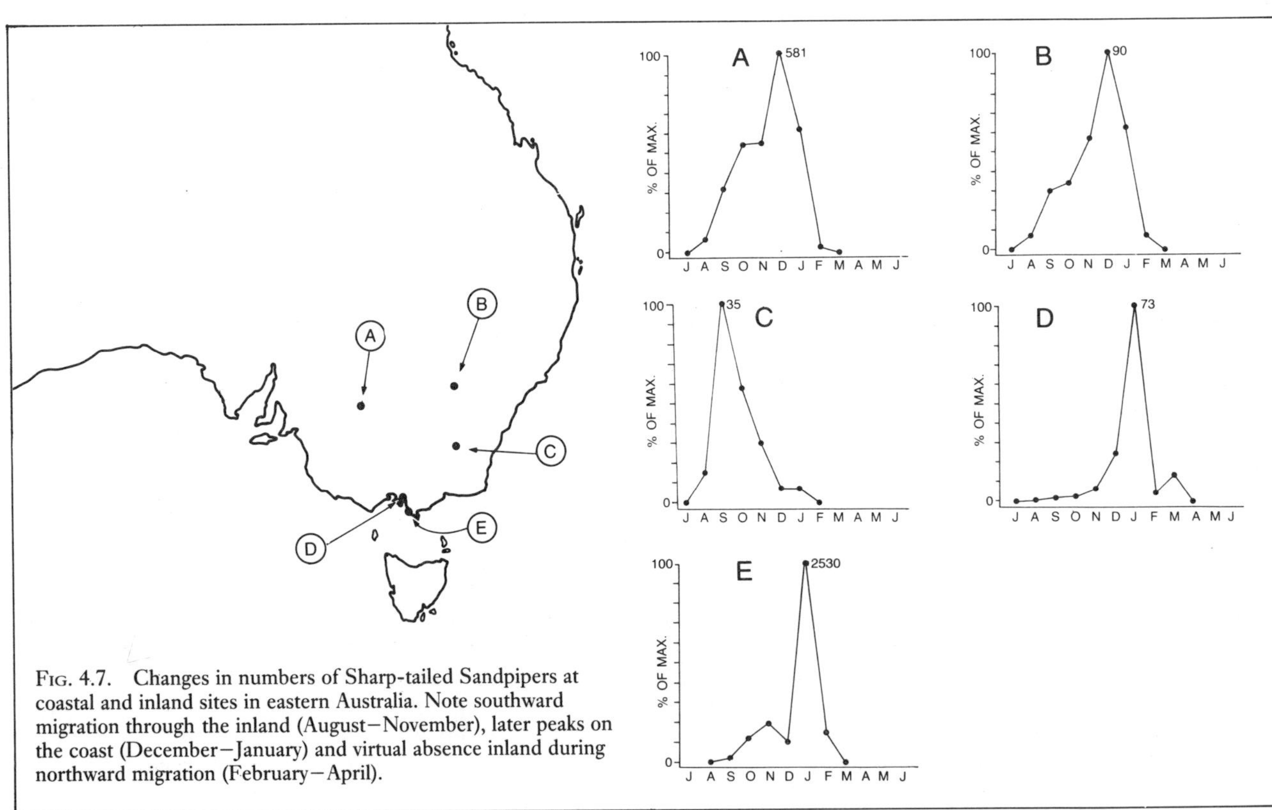

FIG. 4.7. Changes in numbers of Sharp-tailed Sandpipers at coastal and inland sites in eastern Australia. Note southward migration through the inland (August–November), later peaks on the coast (December–January) and virtual absence inland during northward migration (February–April).

FIG. 4.8.  Banding of Lesser Golden Plover. (A) A cannon-net being fired over roosting birds. (B) Bird being gently extracted from under a cannon-net. (C) A band is placed on its leg, using specially made pliers. (D) Total head length being measured. (E) Primary moult being studied. (F) Bird being weighed. Photos: J. Hooper

programmes aimed specifically at shorebirds started at the Swan estuary in Western Australia, the Hunter estuary in New South Wales, and on Kangaroo Island in South Australia (372). Since then, over 60 000 shorebirds have been banded. The scheme is now run by the Australian National Parks and Wildlife Service.

## Migration Routes

Recoveries of banded shorebirds provide the most direct evidence of the location of migration routes and the timing of migration. For instance, Dunlins, a species abundant in Western Europe, which were banded in the British Isles in the non-breeding months (November–February), have been recovered mostly in Russia and the countries round the Baltic Sea during southward migration (July–October). Those banded in the British Isles during southward migration have been recovered between France and Mauritania in western Africa in the non-breeding months and in the Baltic countries on their way south between July and October in subsequent years (177). This shows that Dunlins banded in Britain during southward migration stop there for a short time on their way to more southerly non-breeding grounds. Most pass through the Baltic countries on their way south.

In Australia, Curlew Sandpipers have been banded in large numbers. The small number of recoveries in Asia during northward migration shows that they probably pass through Hong Kong and southern China between mid-April and mid-May and through northern China and south-eastern Siberia in late May (15). They arrive on their northern Siberian breeding grounds in early June (111).

Another way of seeking direct evidence for shorebird movements is by colour-marking them. There are two kinds of colour-marking: colour-banding, by which unique combinations of coloured plastic bands and the numbered metal band are placed on the legs of a bird, and colour-dyeing, whereby a bright dye is painted onto the plumage of a bird so it can be easily seen among flocks along its migration route. (See Figure 4.9.) Some fading occurs, but usually the dye is conspicuous until the bird next moults its body plumage. Unique combinations of colour-bands or dyes for areas, dates, ages, sexes or all four can generate detailed information on migration without the need for time-consuming banding operations along the migration route. International co-ordination of colour-marking projects to eliminate overlap, and a network of observers along the migration route are essential to maximise the value of this technique. A combination of colour-banding and colour-dyeing techniques are being used to clarify the movements of Double-banded Plovers within and between New Zealand and south-eastern Australia (317).

## Size Differences

In any animal population there are differences in size between individuals. In many shorebirds, apart from

Fig. 4.9. A Double-banded Plover, colour-banded to facilitate resighting during migration. Photo: M. Barter

normal variation, there are systematic differences in size. For example, measurements of wing, bill, tarsus or total head (morphometrics) can be used to separate size-groups within a population. These size-groups may represent different sexes or be indicative of different breeding populations.

Sex size dimorphism is common in shorebirds. Sometimes the female shorebird is larger, e.g. Bar-tailed Godwit, Eastern Curlew (383, 384); sometimes the male, e.g. Sharp-tailed Sandpiper (250). In Victoria, Sharp-tailed Sandpipers arrive and depart before the females. The distribution of wing lengths from samples of Sharp-tailed Sandpipers taken at the beginning, middle and end of their stay in southern Australia shows how the sex-ratio changes in time. (See Figure 4.10.)

There is a slight difference in size between Lesser Golden Plovers that breed in Alaska and those that breed in eastern Siberia (370). Measurements of individuals in south-eastern Australia suggest that they come from Alaska (476).

## Energy Requirements

The flights shorebirds make between staging areas along the migration route require the expenditure of

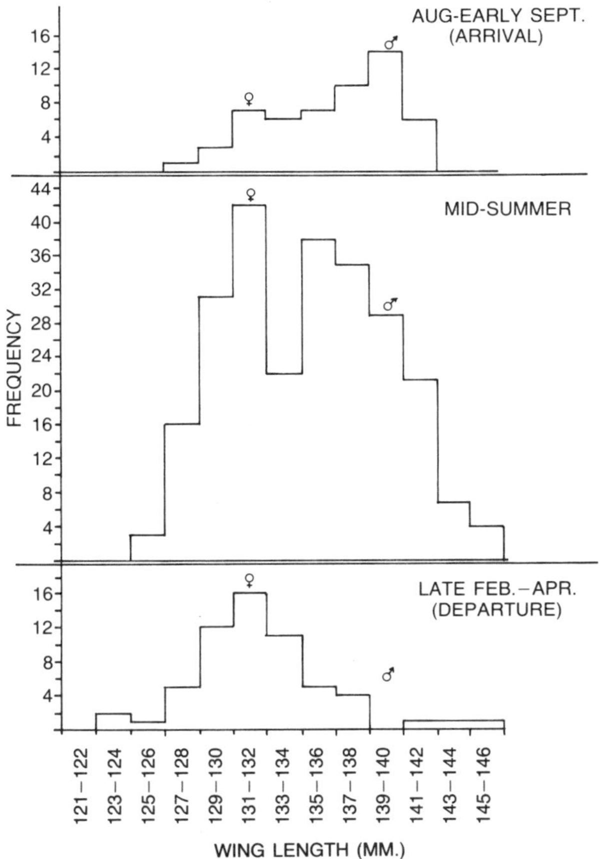



FIG. 4.10.   Frequency distribution of wing lengths in catches of Sharp-tailed Sandpipers at Werribee, Victoria, when they first arrive (August–September), in midsummer (December–February) and when they depart (late February–April). (Data courtesy of Victorian Wader Study Group)

energy that birds generate by the metabolism of fat. Fat, having more available energy per unit weight than carbohydrate or protein, is the most efficient way for animals to store energy, and its presence in shorebirds can be monitored by weighing them during banding. Before migration, shorebirds accumulate fat beneath their skin and around their internal organs. The amount of fat carried is one of the factors that determine a bird's potential flight range. In southern Australia, weight-increases due to fat accumulation are found in all adult migratory shorebirds between mid-February and late April, just before departure for the breeding grounds. The size of these increases is shown in Table 4.1. All this fat, which can make up more than one-third of the body-weight of some individuals, is used in the few days of often non-stop flight to the first staging area on the way to the breeding grounds. Many shorebirds can accumulate sufficient fat in 10–14 days at a staging area to enable them to undertake the next stage of their migration. In Britain, for example, Curlew Sandpipers, which weigh about 65 grams, stop during migration and increase in weight by about 3.9 grams per day (443).

**Flight Ranges**

Various methods have been used to determine the flight range of shorebirds, based on how much fat they carry before migrating. The most widely used formula until the late 1970s (296) was based on the following principle:

$$\text{Flight Range} = \frac{\text{Weight of fat carried} \times \text{Speed} \times \text{Energy content of fat.}}{\text{Energy used during migratory flight.}}$$

TABLE 4.1.   Pre-migratory weight increases of selected species of shorebirds in southern Victoria. (Data courtesy of Victorian Wader Study Group)

| SPECIES | MEAN MID-SEASON WEIGHT (GM) | MEAN DEPARTURE WEIGHT (GM) | ADDITION % |
|---|---|---|---|
| Grey Plover | 230 | 380 | 65 |
| Lesser Golden Plover | 127 | 177 | 39 |
| Mongolian Plover | 70 | 106 | 51 |
| Double-banded Plover | 57 | 80 | 40 |
| Ruddy Turnstone | 100 | 172 | 72 |
| Bar-tailed Godwit | | | |
| (male) | 290 | 416 | 43 |
| (female) | 350 | 520 | 48 |
| Red Knot | 120 | 175 | 46 |
| Great Knot | 155 | 267 | 72 |
| Sharp-tailed Sandpiper | | | |
| (male) | 72 | 94 | 30 |
| (female) | 53 | 69 | 30 |
| Red-necked Stint | | | |
| (mean) | 29.5 | 37.7 | 28 |
| (max.) | | 42 | 42 |
| Curlew Sandpiper | 53 | 88 | 66 |

The crucial assumption is the amount of energy used in flight, and this is difficult to determine. Many other factors (such as the size of the bird and its wing loading) must be taken into account. This formula has since been modified to allow for the loss of weight during migration as fat is used up, thus extending the theoretical flight range (451).

The flight range must be related to wind, the most important influence on a bird's path over the ground. Somehow migrating birds must compensate for changes in wind strength and direction during their migration; this is especially important for shorebirds and other long-distance migrants that fly through more than one of the earth's major wind belts. It was not until radar could be used to observe migrating birds directly that this problem could be examined in detail. This topic is discussed later.

**Moult**

Moult, the replacement of worn feathers, is an essential function in birds. Data on active moult, collected during banding, generally indicate whether birds are migrating through an area or staying for longer. Adult shorebirds replace their flight feathers (wing and tail, the most important ones for migratory flight) each year. Flight feathers require more energy to replace than other feathers, being bigger, and are moulted at the time of year when other energy demands, such as breeding and migration, are minimal. They generally do not moult and migrate at the same time. In Australia, for example, most migratory shorebirds that have been studied start the moult of their flight feathers soon after they have arrived, about October, and complete it in February (476). They accumulate fat for their northward migration after completing moult.

On the Malay Peninsula in 1983, adult Curlew Sandpipers were banded in August and September during southward migration (348). Birds moulting their primary feathers made up 66 per cent of adults examined. The weights of non-moulting adults were higher than those of the moulting birds. Indeed the weights of some indicated that they were carrying sufficient fat to fly non-stop the 3600 kilometres to the north-west coast of Australia, where they are known to congregate before moving to southern Australia.

## Radar Studies

Shorebird migration routes can be further identified by using radar to monitor the direction in which shorebirds fly. Flocks show on radar screens as small, slow moving, bright spots. Time-lapse photography, movie or video can be used to resolve the direction and flight speed of flocks. (See Figure 4.11A.)

In Australia, the Bureau of Meteorology radar at Broome in northern Western Australia was used in March and April 1985 to watch flocks of departing shorebirds (248). The results showed that most flocks departed in a sector between north-west and north-north-west of nearby Roebuck Bay, flying out over the Indian Ocean. Data on pre-migratory fattening, collected during banding studies on Roebuck Bay, combined with this knowledge of flight speed and direction, enabled the destinations of shorebirds to be predicted. These ranged from southern Borneo for the Rednecked Stint, Sumatra, Malaysia and the Gulf of Thailand for Curlew and Terek Sandpipers, and the southern coast of China for larger species, such as the Turnstone and Grey-tailed Tattler. This analysis assumed that birds continued to fly at the same over-the-ground heading and at the same speed until their fat reserves were exhausted.

Radar studies over the north-western Atlantic Ocean have shown that such assumptions are not necessarily justified. Wind can be 'used' by birds to assist their migration. Thousands of shorebirds that breed in the Canadian arctic gather on the eastern coast of North America before flying across the Atlantic Ocean and the Caribbean Sea to non-breeding grounds in South America. This flight has been studied using radars on mainland North America, on ships in the North Atlantic Ocean, at Bermuda, and on the islands of the Caribbean Sea (492). Birds departing from the eastern coast of North America head out to sea in a south-easterly direction, as if heading towards western Africa. As their migration progresses, their direction changes. Over the Caribbean they start moving south-westwards towards the north coast of South America. This change in direction occurs because they encounter north-easterly trade winds. Calculations show that they may not change their heading at all; only their direction over the ground changes.

**Departure**

Shorebirds, and indeed other migratory birds, generally do not migrate when wind or other weather conditions are unsuitable. Rather, they wait until conditions are right before departing. The migrants that left the eastern coast of North America waited until a cold front had crossed the coast (492). In this way they maximised their chances of meeting north-easterly winds over the Caribbean. Migrants that departed at other times were more likely to be carried off course by unsuitable weather conditions over the North Atlantic. At Broome in Western Australia the heaviest northward departure of shorebirds occurred on evenings when strong south-easterly tail-winds were blowing. If a head-wind was blowing, little departure was observed. (See Figure 4.11B.)

To reach their destination, birds probably need to 'know' only two things: when to depart (weather conditions) and the direction in which to fly. A bird making an error in either of these would not reach its destination, would not return to breed, and could not pass this behaviour on to future generations.

It is difficult to understand how birds determine their direction of flight. Years of research have

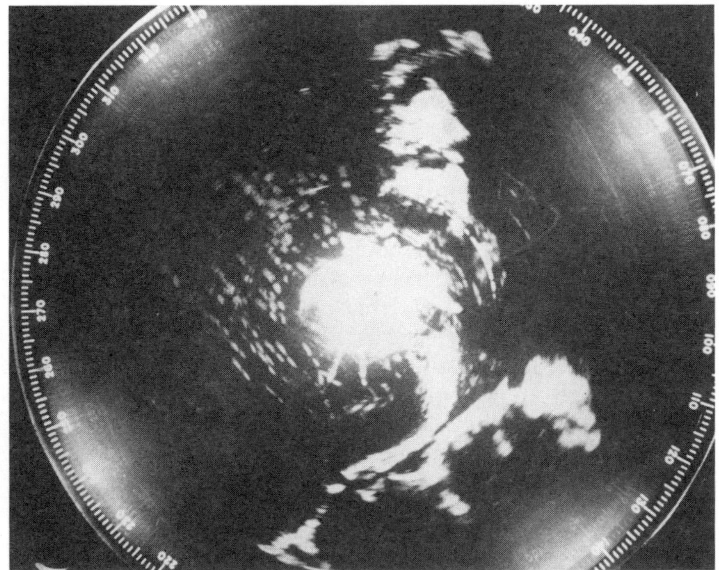

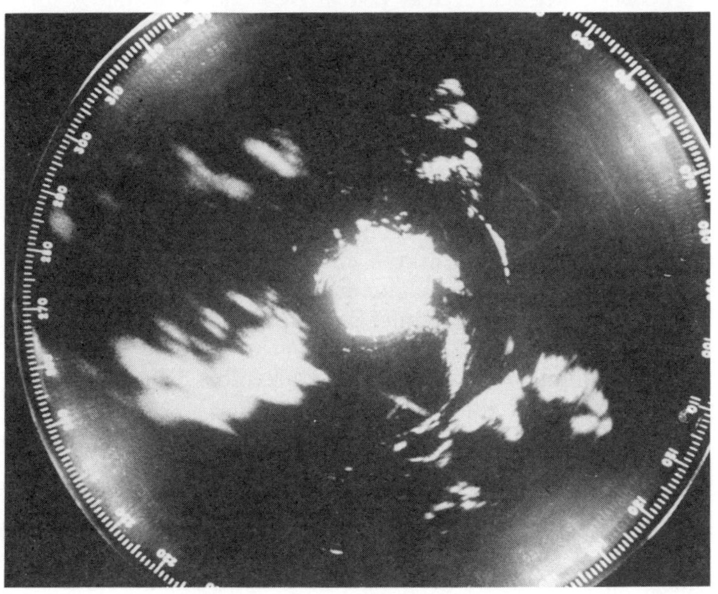

FIG. 4.11. (A) Time-lapse photograph of the screen of the radar at Broome in north-western Australia on the evening of 16 April 1985. Wader flocks appear as short then long streaks due to the camera shutter being left open for one minute, closed for one then open for two minutes. Skies were clear and a south-easterly wind was blowing. (B) Similar photograph taken on the evening of 17 April 1985 showing the lack of migrating flocks. The white smudges are rainshowers; the wind had changed to a north-westerly. Photo: A. Jessop

produced a large literature on the subject of migratory bird orientation, but a full understanding of it still eludes scientists. Griffin (169) gives a readable review up to the 1960s. Since then, the sensory capabilities of birds have been studied in more detail to investigate what cues birds can use to orientate themselves. One possibility that has arisen is that migrants find their way by sensing the earth's magnetic field (389).

# MIGRATION WITHIN AUSTRALIA

Migratory shorebirds first arrive in northern Australia in large numbers in late August and early September. Many also arrive in southern Australia at this time. The southward migration within Australia is slower; some species do not reach their maximum numbers in southern Australia until January, and depart northwards only a few weeks later. Some start their northward migration as early as mid-February, but most depart in March and early April. By the end of April most have left Australia, but some species linger in the north until early May. More detailed information on the migration of the common species can be found in Chapter 5.

Each species of shorebird may have its own unique migration route, or indeed a number of them, and research is needed to describe more fully the migration of shorebirds in Australia. This section describes the general features of shorebird migration in Australia as far as is known; that is, where they migrate, their flights across the central and western deserts, and their use of inland wetlands.

The first review of shorebird migration in Australia was made by Thomas in 1970 (458). Working with much less evidence than is now available, he concluded that north-western Australia was a possible major arrival point for southward migrating shorebirds. During the RAOU study the north-western coast between Broome and Port Hedland was comprehensively surveyed and found to be the most important arrival point for shorebirds migrating into Australia from the breeding grounds.

## Southward migration

Counts on the north-west coast suggest that some species stay for the non-breeding months, e.g. Bar-tailed Godwit and Large Sand Plover. Others move on to southern Australia, e.g. Sharp-tailed Sandpiper and Red-necked Stint.

Regular counts also show that many shorebirds migrate southwards down the east coast, e.g. Grey-tailed Tattler. How the Gulf of Carpentaria is used is not known in detail, but tens of thousands of Red Knots certainly pass through there. As there are many fewer Red Knots in south-eastern Australia during the non-breeding months, the Gulf birds could only be migrating through to New Zealand, where they are abundant (387, 388). Flocks of migrating shorebirds of a number of species can be found on much of the eastern and southern coasts during southward migration. These flocks are probably birds that have just flown from northern Australia. The great distance of coastline on which they occur suggests that shorebirds migrate southwards across the continent on a broad front. Most have moved on to larger coastal wetlands in southern Australia by the end of the migration.

Many shorebirds normally associated with the coast occur on inland wetlands during southward migration, showing that they cross the continent. Since Thomas's review, much new information has been obtained from *The Atlas of Australian Birds* project (1977–81) and from this study. Table 4.2 summarises all inland occurrences of shorebirds observed during the *Atlas* project. Of the 11 species that occur in large numbers on the southern coast, 5 were frequently reported inland. The remaining 6 species, occasionally reported inland, probably do not fly across Australia in large

TABLE 4.2. Inland occurrence of migratory shorebirds (excluding Latham's Snipe and Australian Pratincole) during the *Atlas of Australian Birds* project (1977–81). (Probable trans-Australian migrants marked thus: *)

| SPECIES | NO. INLAND RECORDS |
|---|---|
| **PREDOMINANTLY NORTHERN** | |
| Mongolian Plover | 1 |
| Large Sand Plover | 1 |
| Oriental Plover[1] | 43 |
| Whimbrel | 5 |
| Little Curlew[1] | 18 |
| Grey-tailed Tattler | 4 |
| Black-tailed Godwit[1] | 23 |
| Red Knot[2] | 4 |
| Great Knot | 1 |
| Broad-billed Sandpiper | 1 |
| Oriental Pratincole[1] | 20 |
| **PREDOMINANTLY SOUTHERN** | |
| * Lesser Golden Plover[2] | 24 |
| Eastern Curlew[2] | 2 |
| * Sharp-tailed Sandpiper[2] | 437 |
| * Red-necked Stint[2] | 212 |
| * Curlew Sandpiper[2] | 132 |
| **BOTH NORTHERN & SOUTHERN** | |
| Grey Plover[2] | 6 |
| * Wood Sandpiper | 90 |
| * Common Sandpiper | 174 |
| * Greenshank[2] | 403 |
| * Marsh Sandpiper | 157 |
| Bar-tailed Godwit[2] | 11 |
| Sanderling[2] | 1 |
| **EXCLUSIVELY COASTAL** | |
| Ruddy Turnstone[2] | — |
| Wandering Tattler | — |
| Terek Sandpiper | — |

[1] = common in inland Northern Australia
[2] = common on the southern coast

numbers. A further 3 species reach southern Australia in small numbers, but do not normally occur on the coast, and 4 species that are abundant in northern Australia move to the wetlands and plains of inland northern Australia.

Further evidence for the trans-Australia flights of some migratory shorebirds comes from banding. Estimates of the ground speed of shorebirds, obtained from the radar studies at Broome, range from 60 to 90 kilometres per hour, depending on the wind (248). At this speed a Red-necked Stint would take 34–50 hours to cross Australia; that is, about 2 days of non-stop flying over inhospitable desert. This is comparable with the well-known migration of birds, including shorebirds, across the Sahara Desert in Africa (318, 319, 320). Colour-dyeing of Red-necked Stints and Curlew Sandpipers on the north-west coast during August and September 1982 resulted in resightings in south-eastern Australia as soon as 4 and 5 weeks later. The short time that elapsed between dyeing in the north-west, when they had just arrived and were not ready for onward flight, and resightings in the south-east over 3000 kilometres away suggests that they did the journey in one hop.

The numerous migratory shorebirds that are seen inland during southward migration are probably 'fall-out' from flocks undertaking this trans-Australia journey. These birds use inland wetlands, the availability of which follows an annual pattern. In northern Australia, shorebirds arrive towards the end of the dry season, when inland wetlands are minimal. In southern Australia, wetlands fill in winter and spring before the arrival of the shorebirds.

The 'fallout' from shorebird flocks that migrate across Australia could be due to individuals depleting their fat reserves. This is more likely to occur towards the end of their migration than at the beginning, when they take off with good reserves. Adverse weather conditions *en route* may also force shorebirds to land in inland Australia. Some species, such as the Sharp-tailed Sandpiper, may habitually use inland wetlands as part of their migration strategy.

## Northward migration

Information on the northward migration of shorebirds out of Australia is sketchy. There were far fewer inland records of shorebirds from the *Atlas* project during northward migration. Furthermore, in southern Australia, regular counts showed that shorebirds occurred on inland wetlands mostly during southward rather than northward migration. In south-eastern Australia some species fatten sufficiently before northward departure to overfly the north coast of the continent and do not increase in numbers at northern Australian sites at this time. It seems likely that many species overfly or do not leave by the north coast. On the north-west coast some species (e.g. Turnstone, Red Knot, Red-necked Stint and Curlew Sandpiper) do stop for a time to replenish fat reserves before onward migration to south-east Asia (247).

By the time shorebirds depart northwards, many wetlands in southern Australia have dried out, but in northern Australia they arrive from farther south at the end of the wet season, when wetlands are more suitable. For example, thousands of Sharp-tailed Sandpipers use inland wetlands in south-eastern Australia during southward migration, moving to the coast once these dry out later in the year. They then occur in large numbers on the extensive, recently filled wetlands of the Gulf Country in north Queensland during northward migration (150).

On the Swan coastal plain in south-western Australia, shorebirds use inland wetlands before northward departure (February and March), for at that time they have dried out sufficiently to expose mudflats for feeding. The difference between the south-east and the south-west in this respect probably results from the more consistent annual changes in water-levels of inland wetlands in the south-west compared to the south-east.

# MOVEMENTS OF SHOREBIRDS THAT BREED IN AUSTRALIA

Birds move in response to variations in the quality of their environment and their requirement to obtain the resources necessary for survival and successful reproduction. Australia's endemic shorebirds are no different from the migratory species in this respect. The migrants return to their breeding grounds each year to exploit a comparatively predictable, seasonally abundant food source. Australia's resident shorebirds move within the continent to find abundant food supplies and conditions suitable for breeding. They move on when feeding or breeding conditions become unsuitable. Rainfall over much of inland Australia is comparatively unpredictable, and wetlands are not available to shorebirds as reliably as in the Northern Hemisphere. Movement of Australia's resident shorebirds, and indeed of most Australian waterbirds, is therefore somewhat irregular. The coastal residents are not affected much by these changes. This section describes the way that the coastal and inland shorebirds that breed in Australia move within the continent. The effects on their movements of the not uncommon droughts and floods in inland Australia are outlined.

Other than the ubiquitous Masked Lapwing, Australian resident shorebirds can be classified into four groups: those that occur exclusively on the coast, those that are found on both coastal and inland wetlands, those that live almost exclusively on inland wetlands, and those that inhabit inland open country away from wetlands. These groups are shown in Table 4.3.

TABLE 4.3. Australian-breeding shorebirds.

| OCCURRENCE | SPECIES |
| --- | --- |
| EXCLUSIVELY COASTAL | Beach Thick-knee |
| | Pied Oystercatcher |
| | Sooty Oystercatcher |
| | Hooded Plover (in eastern Australia) |
| ALMOST EXCLUSIVELY INLAND WETLANDS | Painted Snipe |
| | Red-kneed Dotterel |
| | Black-fronted Plover |
| | Banded Stilt |
| | Red-necked Avocet |
| BOTH COASTAL AND INLAND WETLANDS | Hooded Plover (in south-western Australia) |
| | Red-capped Plover |
| | Black-winged Stilt |
| ALMOST EXCLUSIVELY INLAND AWAY FROM WETLANDS | Banded Lapwing |
| | Australian Pratincole |
| | Bush Thick-knee |
| UBIQUITOUS | Masked Lapwing |

## Coastal shorebirds

Coastal wetlands are available to shorebirds all year round. Seasonal changes in productivity and the availability of food do, however, occur on the coast. Weather conditions change and, at least in parts of southern Australia, the tidal range and therefore exposure of feeding habitat vary seasonally. At Westernport Bay in Victoria, for example, the average period for which intertidal mudflats are exposed during daylight hours (when shorebirds can feed most efficiently) is lowest in winter (Parry, in 264).

Exclusively coastal species breed only in spring, summer and autumn in southern Australia and earlier in the year in the north. At other times, regular counts show that they move together and flock. These movements are local in scale. On ocean beaches in south-eastern Australia, Hooded Plovers flock in winter, occasionally moving to salt lakes immediately behind beaches, and spread out in spring and summer to breed. The oystercatchers and the Red-capped Plover, which live on the coast, also breed and flock seasonally.

## Inland shorebirds

Away from the coast, the story becomes intriguing. In the inland, Red-capped Plovers breed at any time of year, and flocks numbering hundreds can occur at any time (246). On inland habitats they have been observed breeding in loose colonies (196), whereas they breed generally as territorial pairs on the coast (385). Counts

FIG. 4.12.   A flock of Banded Stilts. They often gather in large flocks, especially when on the move. Photo: J. Hooper

show that Red-capped Plovers, Black-winged Stilts and Hooded Plovers in south-western Australia move to the coast in summer, presumably because inland wetlands dry out.

Despite such a large proportion of the continent being arid, Australia has 5 species of shorebirds that occur almost exclusively away from the coast. Of these, 4 are abundant and widespread: Red-kneed Dotterel, Black-fronted Plover, Banded Stilt and Red-necked Avocet. A substantial number of another species, the Black-winged Stilt, also occur on inland lakes and swamps. All except the Banded Stilt breed in the regularly available wetlands around the edge of the continent. They also breed on ephemeral wetlands in the far inland, which are filled occasionally by rainfall and the resulting runoff. When this happens, large-scale movements of most of the populations of these species can occur. The movements of each species are documented in detail in the relevant species accounts in Chapter 5.

Counts conducted during this study before the drought of 1982−83 and data from *The Atlas of Victorian Birds* (27) suggest that, in eastern Australia, the Red-kneed Dotterel, Black-winged Stilt and Red-necked Avocet occur in largest numbers near the coast in spring, summer and autumn. This is also the time when they breed (385). This movement towards the

coast for the breeding season probably maximises their chances of finding wetlands suitable for breeding, as those farther inland are more likely to be dry at this time. After breeding, they disperse inland to use lakes and swamps filled by winter rains.

Banded Stilts show less regular movements in many parts of their range. They are most abundant in coastal areas of South Australia and south-western Australia in summer, but move irregularly, often in flocks of tens of thousands into coastal and inland wetlands in western Victoria (3, 27). (See Figure 4.12.) There are few observations of them breeding, as they do so irregularly on the large, infrequently filled salt lakes in the more remote arid regions of southern Australia. The records that exist show that they can breed in almost any month of the year.

# The effect of drought and flood on shorebird movements

During 1982 and early 1983 most of eastern Australia experienced a severe drought. This was followed in mid and late 1983 by flooding in the Darling and Cooper drainage basins. The floods receded by late 1984. In early 1984 Lake Eyre filled for the third time this century. By the end of that year it was drying out, and in spring, summer and autumn of 1984–85, a less extensive, shorter drought affected inland eastern Australia. Between 1981 and 1985 the inland species showed changes in abundance near the coast in eastern Australia, which can be explained by movement into and out of the Darling, Cooper and Lake Eyre drainage basins. Unfortunately, most of these remote areas were not covered consistently from year to year during this study. Lake Eyre was visited when it held water.

As the drought intensified in 1982 the numbers of Red-kneed Dotterels, Black-winged and Banded Stilts, and Red-necked Avocets increased in near-coastal areas of south-eastern Australia (see Chapter 5). Very few young birds were seen in flocks, suggesting that little breeding took place that year. Soon after the drought broke in mid-1983 and the inland flooded, Red-kneed Dotterels, Black-winged Stilts and Red-necked Avocets virtually deserted near-coastal areas and moved onto the new far inland wetlands. Once these floods receded, Red-kneed Dotterels returned in large numbers to near-coastal areas, but this time there were many young birds in flocks, showing that they had bred successfully in the far inland. Red-necked Avocets, however, did not reappear near the coast until Lake Eyre started drying out rapidly in late 1984. They too had bred successfully.

The difference in the times when these two species returned from the inland may be due to different habitat preferences. The Red-kneed Dotterel lacks the nasal salt-secretion gland that most shorebirds have (292) and it generally prefers low salinity lakes and swamps. It was unable to exploit the saline Lake Eyre, and once the flood receded in the Darling and Cooper drainage basins, it moved towards the coast. The Red-necked Avocet, which can tolerate saline lakes, moved to Lake Eyre. There were 95 000 there in September 1984, when the salinity of the water was nearly four times that of the sea. The picture is less clear for the Black-winged Stilt, but the Banded Stilt did not desert near-coastal areas, despite inland flooding, until 1983, well after the drought broke. *The Atlas of Australian Birds* (38) shows that the Banded Stilt does not often occur in the Darling drainage basin. It appeared to wait until Lake Eyre flooded before leaving near-coastal areas.

These observations show that the *normal cycle* of coastward movement for the summer and return to the inland in winter can be interrupted by either a *drought cycle*, in which shorebirds concentrate on permanent wetlands near the coast (drought refuges), postpone breeding and desert the far inland, or by an *inland flood cycle*, in which they leave the coast to breed on recently filled wetlands far inland, often in large numbers and, when these recede, move to coastal areas. Furthermore, the timing of the cycles varies between species, depending on the location and timing of droughts and floods.

One migratory species, the Red-necked Stint, is also known to have responded to changes in inland wetland availability. Resightings of birds colour-dyed on the Victorian coast (476) and higher winter than summer numbers in many inland areas during this study indicate that some Red-necked Stints move inland for the winter. These are young birds that have not returned to the Northern Hemisphere to breed. The July 1984 count of Red-necked Stints in Victoria and Tasmania was only 30 per cent of previous July counts, despite better breeding productivity in 1983 (476). At the same time there were 12 000 small shorebirds, the majority of which were stints, on Lake Eyre. Clearly some migratory shorebirds while in Australia are capable of movement in response to inland flooding, even that as unpredictable as the filling of Lake Eyre. They were not responding to annually unsuitable winter conditions in Victoria and Tasmania (e.g. reduced daylight exposure of intertidal mudflats), since, if this were the case, numbers there would be low every winter. Rather, they were apparently drawn away by suitable conditions at Lake Eyre. The exploitation of an irregular event by a migratory shorebird which has evolved a regular annual cycle is most unexpected.

Questions about the movements of shorebirds in Australia remain:
*From which breeding populations do Australia's shorebirds come?*
*What are their migration routes through Asia?*
*How do shorebirds near the coast 'know' that the inland is flooded and that they might find suitable habitat there?*
Investigation of these problems would be worthwhile research projects.

# SPECIES ACCOUNTS

This chapter describes 45 species of shorebird that commonly occur in Australia. There are detailed accounts of the 30 more abundant migratory species and the 15 Australian-breeding species. (Shorter accounts are given in Chapter 6 of the 26 species that occur either irregularly or annually in small numbers (rarities).) The accounts give the following information:

Species name.
Other names.
Description.
Breeding range (migratory species only).
Asian range (migratory species only).
Status and distribution in Australia.
Movements in Australia.
Breeding (resident species only).
Feeding.
Conservation.

Descriptions are designed to aid in field identification and should be used with the colour plates. Except for two migratory and six resident species, each account contains a map showing, to the extent supported by the data, the numbers of birds that occur in different zones of Australia. (For the eight exceptional species there were insufficient data for analysis to be done.) For migratory species, an additional map shows breeding range, distribution in Asia and non-breeding range. A brief description is given below of how the data on which the accounts are based were collected and analysed followed by some helpful hints for identifying shorebirds. The species accounts follow this section.

## Gathering the facts

The aim of the RAOU study was to determine the distribution and status of migratory and resident shorebirds in Australia, particularly to identify zones of major population concentrations. Elsewhere in the world (e.g. Britain) similar objectives have been achieved by simultaneous, country-wide counts involving many amateur bird-watchers (369). The size of Australia, the limited number of observers and their uneven distribution over the continent made a simultaneous survey of all coastal and inland wetlands impossible. In consequence a variety of approaches was adopted depending on the nature of the country being surveyed. Four techniques were used:

National counts.
Expeditions.
Aerial surveys.
Regular counts.

All count results were recorded and submitted to RAOU on standard data sheets. (See Figure 5.1.)

## NATIONAL COUNTS

National counts were planned to determine the number of shorebirds occurring simultaneously in as many coastal and inland wetlands as possible. Bird-watchers visited shorebird habitats and counted the numbers of each species present. The counts were held twice a year over two consecutive weekends in summer (February) and winter (June–July). Dates were chosen so that observers could visit coastal wetlands at high tide during the day to count shorebirds concentrated at roosts. Trying to count them accurately at low tide, when they may be spread over vast mudflats, is impossible.

Many parts of Australia could not be visited during national counts, but they provided nearly comprehensive information on the status of shorebirds on the

Fig. 5.1.   Standard data sheet on which all count results were recorded. ▶

**AUSTRALASIAN WADER STUDY GROUP**

Name: ...........................................................

Address: ........................................................

..........................................................

k1  k2  State  Obs No  Site code  Record No

Site Name: ...............................  ...............................

Please sketch map on back of form

Year  Mnth  Day  Start  Finish  Tide  Habitat

Lat  Long  Acc  Site  Wn Dir  Wn Str

Count disrupted?  Total coverage?  Total species  Total birds

| Species | Total | Notes |
|---|---|---|
| Bush Thick-knee | 174 | |
| Beach Thick-knee | 175 | |
| Painted Snipe | 170 | |
| Pied Oystercatcher | 130 | |
| Sooty Oystercatcher | 131 | |
| Masked Lapwing | 133 | |
| Banded Lapwing | 135 | |
| Grey Plover | 136 | |
| Lesser Golden Plover | 137 | |
| Red-kneed Dotterel | 132 | |
| Hooded Plover | 138 | |
| Mongolian Plover | 139 | |
| Double-banded Plover | 140 | |
| Large Sand Plover | 141 | |
| Oriental Plover | 142 | |
| Red-capped Plover | 143 | |
| Black-fronted Plover | 144 | |
| Black-winged Stilt | 146 | |
| Banded Stilt | 147 | |
| Red-necked Avocet | 148 | |
| Ruddy Turnstone | 129 | |
| Eastern Curlew | 149 | |
| Whimbrel | 150 | |
| Little Curlew | 151 | |
| Wood Sandpiper | 154 | |
| Grey-tailed Tattler | 155 | |
| Wandering Tattler | 156 | |
| Tattler sp. | 847 | |
| Common Sandpiper | 157 | |
| Greenshank | 158 | |
| Marsh Sandpiper | 159 | |
| Terek Sandpiper | 160 | |
| Latham's Snipe | 168 | |
| Snipe sp. | 846 | |
| Black-tailed Godwit | 152 | |
| Bar-tailed Godwit | 153 | |
| Red Knot | 164 | |
| Great Knot | 165 | |
| Sharp-tailed Sandpiper | 163 | |
| Pectoral Sandpiper | 978 | |
| Red-necked Stint | 162 | |
| Long-toed Stint | 965 | |
| Curlew Sandpiper | 161 | |
| Sanderling | 166 | |
| Ruff or Reeve | 934 | |
| Broad-billed Sandpiper | 167 | |
| Oriental Pratincole | 172 | |
| Australian Pratincole | 173 | |

Other Waders

southern and eastern coasts and nearby inland wetlands. Furthermore, it was not practical to obtain consistent coverage of wetlands in each year of the study. The exceptions include parts of Tasmania, Victoria, New South Wales, Queensland and Western Australia. Where there were few observers, a different area was counted each year. Over the five-year period, most potential shorebird habitats were counted comprehensively at least once.

Australia was divided into zones, shown in Figure 5.2A, which represent regions with more-or-less uniform wetland types. Coastal zones often consisted of a major embayment or estuary (e.g. Westernport Bay in Victoria; Hunter estuary in New South Wales) or a stretch of coast with similar features (e.g. Mackay area of Queensland; northern Arnhem Land in the Northern Territory). Inland zones covered regions with similar wetland types (e.g. the salt lakes of western Victoria; the lakes and overflow country of far north-western New South Wales). All results are presented by these zones.

## EXPEDITIONS

Expeditions were organised to remote regions of northern Australia to count shorebirds and, in some cases, to band them. These were held in conjunction with aerial surveys to determine the composition of flocks located and counted from the air. Expeditions went to the Broome—Port Hedland area in north-western Australia in

> August—September 1981,
> April, August—September and November 1982,
> October—November 1983,
> October 1984,
> March—April 1985;

to the south-eastern corner, Gulf of Carpentaria, northern Queensland, in

> September—October 1983;

to Lake Eyre, in inland South Australia, in

> June, July, August, September and October 1984.

## AERIAL SURVEYS

Aerial surveys, using light aircraft flying at low speed and low altitude, enable the numbers of shorebirds to be counted on long, inaccessible stretches of coastline. However, species identification is difficult. Sometimes larger, more boldly marked species (such as oystercatchers, Eastern Curlew, Whimbrel, stilts and avocets, Beach Thick-knee and Black-tailed Godwit) could be identified.

Accurate counting of shorebirds from the air is difficult. Studies have shown that inexperienced obser-

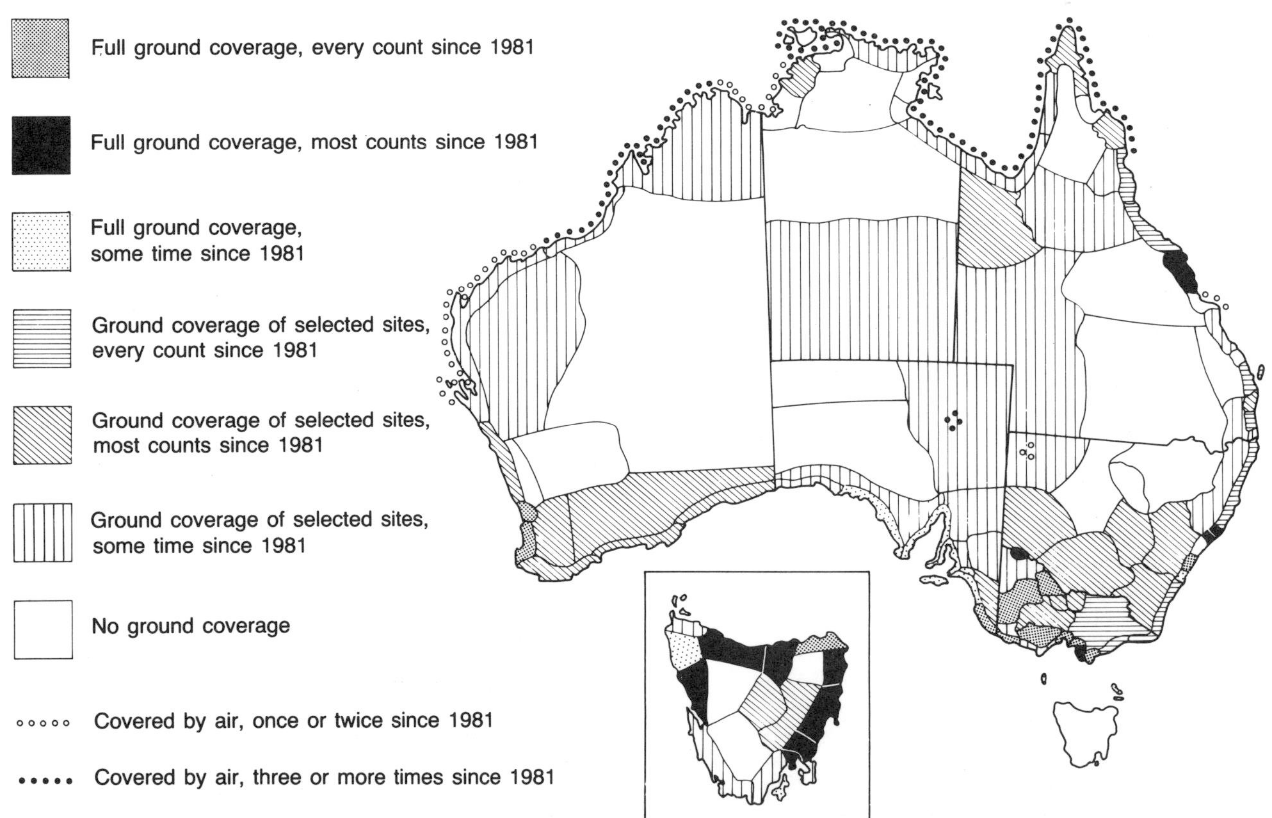


FIG. 5.2A.   The zones used in the analysis of counts and the extent and intensity of coverage between 1981 and 1985.

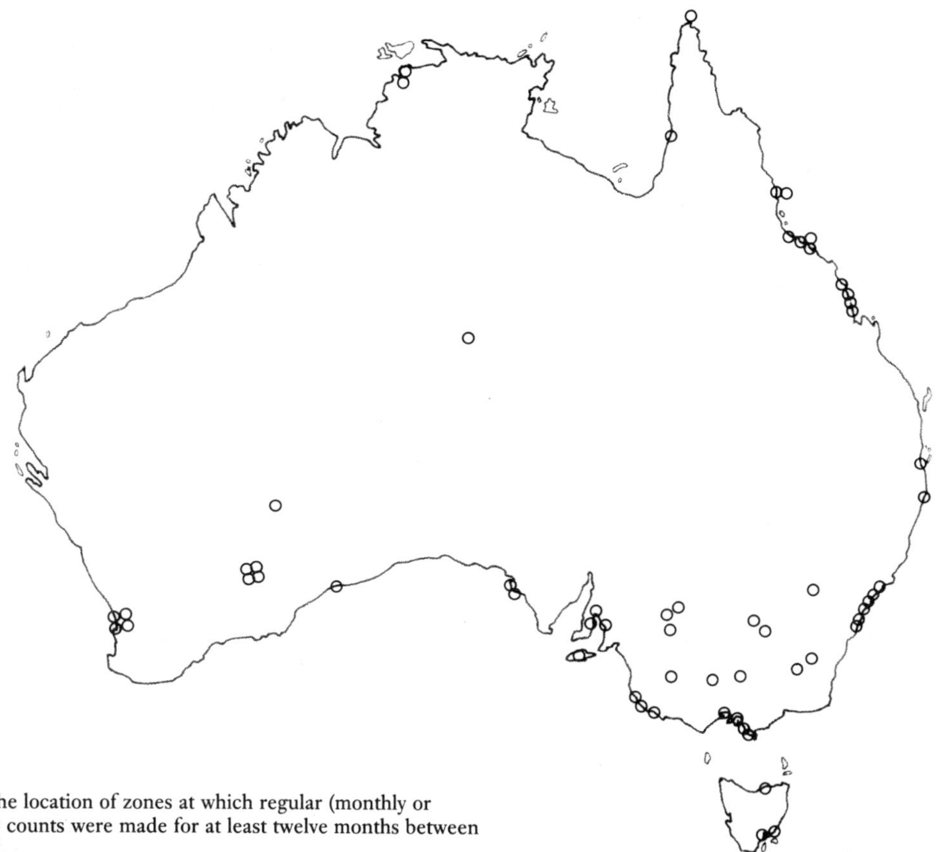

Fig. 5.2B.   The location of zones at which regular (monthly or more frequent) counts were made for at least twelve months between 1980 and 1985.

vers underestimate shorebird numbers more than experienced observers, and that the underestimate is greater for larger numbers of birds (368, 151). These results were confirmed by simultaneous aerial and ground counts on the Broome–Port Hedland expeditions. Accordingly, where aerial counts are used, the analysis is based only on those made by experienced observers.

## Regular Counts

Regular counts were conducted in many zones for periods of a year or more. (See Figure 5.2B.) Changes in numbers indicate movement into and out of an area and enable the overall timing and location of migration to be determined (see Chapter 4). The results of these counts are presented in the 'Movements in Australia' section of each species account.

## The analysis

The coverage achieved by national counts, expeditions and aerial surveys is shown in Figure 5.2A. Results from these are combined as outlined below.

The population of a species in each zone is estimated from national-count results over five years, by taking the average of the counts for areas covered consistently from year to year, where it is clear that no major flock was missed during any count. Where there are insufficient counts to provide a reliable average, the maximum number counted is used. For resident species, both winter and summer count results are used, whereas for the migrants only summer counts are used. Winter counts only are used for Double-banded Plovers: in spring and summer they are in New Zealand. To assess the relative importance of sites throughout Australia, these figures are combined with the best ground counts of species in northern Australia, mostly between September and November, before the wet season. Aerial counts of species, if available (e.g. Black-tailed Godwit), are used for areas where no ground coverage was obtained. The zones are ranked.

The counts in the north, although made up to five months before the national counts in the south, are evidence that a certain number of birds use these areas. Combining these figures gives a reasonable picture of the relative national importance of an area for a species and of the species' distribution.

It is not possible to estimate the population size in Australia of most migratory species, because they move between northern and southern Australia in the five months between northern and southern counts. Consequently, it is not possible to apply the internationally accepted criterion of a wetland being of international significance if it holds more than 1 per cent of the population of a species in a migratory 'flyway' or biogeographical region (407). This criterion has been useful in identifying wetlands for conservation under international treaties.

## Identifying Australian shorebirds

'I've never bothered trying to identify them, they're just too hard.' 'Little brown blobs way out on mudflats don't really grab me.' 'Maybe one day I'll make the effort to learn how to identify them.'

These comments, and similar ones, are often heard from bird-watchers, even some of the more experienced ones. Yet shorebirds are among the most conspicuous groups of wetland birds. They constitute 10 per cent of all species of birds found in Australia. Avocets, stilts and oystercatchers are easy to distinguish, so the negative comments usually refer to the migratory species in non-breeding plumage, which at first glance do look like 'little brown blobs'. But identifying shorebirds is no harder than learning to distinguish the ducks, the birds of prey or the thornbills.

### WHAT TO LOOK FOR

**Shape** and **size** are the most helpful guides. Two major groups of shorebirds can be distinguished by shape. Plovers (Charadriidae) have larger heads and shorter necks in proportion to their bodies than sandpipers (Scolopacidae). Searching for food mainly by sight, they have larger eyes. Sandpipers are more slender, have longer necks and smaller heads. Their eyes are smaller because they search for food by touch.

Within these groups there are size differences. Table 5.1 shows the more abundant Australian plovers and sandpipers in different size classes. The largest shorebird is the Eastern Curlew, which stands over a half-metre tall and the smallest is the 30 gram Red-necked Stint, which is only 15 centimetres long. However, the Eastern Curlew is not a larger version of the Red-necked Stint; it has a longer bill, neck and legs. The Curlew Sandpiper and Sharp-tailed Sandpiper, two species of small sandpiper abundant in southern Australia, are similar at first glance. The former is more slender, with slightly longer legs and a longer neck and bill.

The two main groups can also be distinguished by their **bill shape**, another feature that assists in identifying shorebirds. A plover has a short bill, no longer than its head, which is usually swollen towards the tip. A sandpiper has a tapering bill, often longer than its head. There is more variation in bill length and shape among sandpipers than among plovers; between the up-turned bill of the Terek Sandpiper, the straight bills of the Sharp-tailed Sandpiper and Greenshank and the down-curved bills of the Curlew Sandpiper and Whimbrel.

**Leg colour** is visible at close range in good light conditions. The bright orange legs of the Terek Sand-

TABLE 5.1. Size-groups of Australian plovers and sandpipers, ordered from largest to smallest.

| SIZE | PLOVERS | SANDPIPERS |
|---|---|---|
| LARGE | (>230 mm) | (>300 mm) |
| | Masked Lapwing | Eastern Curlew |
| | Grey Plover | Whimbrel |
| | Banded Lapwing | Bar-tailed Godwit |
| | Lesser Golden Plover | Black-tailed Godwit |
| | | Asian Dowitcher |
| | | Little Curlew |
| | | Greenshank |
| MEDIUM | (170−230 mm) | (230−300 mm) |
| | Oriental Plover | Great Knot |
| | Large Sand Plover | Grey-tailed Tattler |
| | Australian Dotterel | Wandering Tattler |
| | Hooded Plover | Latham's Snipe |
| | Mongolian Plover | Red Knot |
| | Double-banded Plover | Ruddy Turnstone |
| | Red-kneed Dotterel | Marsh Sandpiper |
| SMALL | (<170 mm) | (<230 mm) |
| | Red-capped Plover | Terek Sandpiper |
| | Black-fronted Plover | Wood Sandpiper |
| | | Sharp-tailed Sandpiper |
| | | Curlew Sandpiper |
| | | Common Sandpiper |
| | | Sanderling |
| | | Broad-billed Sandpiper |
| | | Red-necked Stint |

piper, the red legs of the Ruddy Turnstone or the yellow legs of the Grey-tailed Tattler are give-aways.

**Plumage** is different in most species of shorebirds. They may superficially resemble 'little brown blobs', being often brown or grey, but they are streaked, barred or speckled brown or grey in different ways and on different parts of their bodies. The species accounts give detailed descriptions of adult non-breeding, adult breeding and juvenile plumages (the last is usually visible until about December). The parts of a shorebird are shown in Figure 5.3 and this terminology, recommended by *British Birds*, vol. 74, pp. 239−42, is used throughout. Knowledge of the parts of a bird greatly assists in recording observations of plumage for later reference if identification cannot be made immediately.

From August to October and after February, adult shorebirds moult out of and into breeding plumage, which can often look completely different to their dull non-breeding dress. At these times shorebirds show a mixture of the two plumages.

Plumage as it appears **in flight** can also be distinctive. Wing-bars or a lack of them combined with rump and tail patterning assist in identification. The latter is particularly observable on birds flying away. The extensive white back and rump of the Greenshank or the

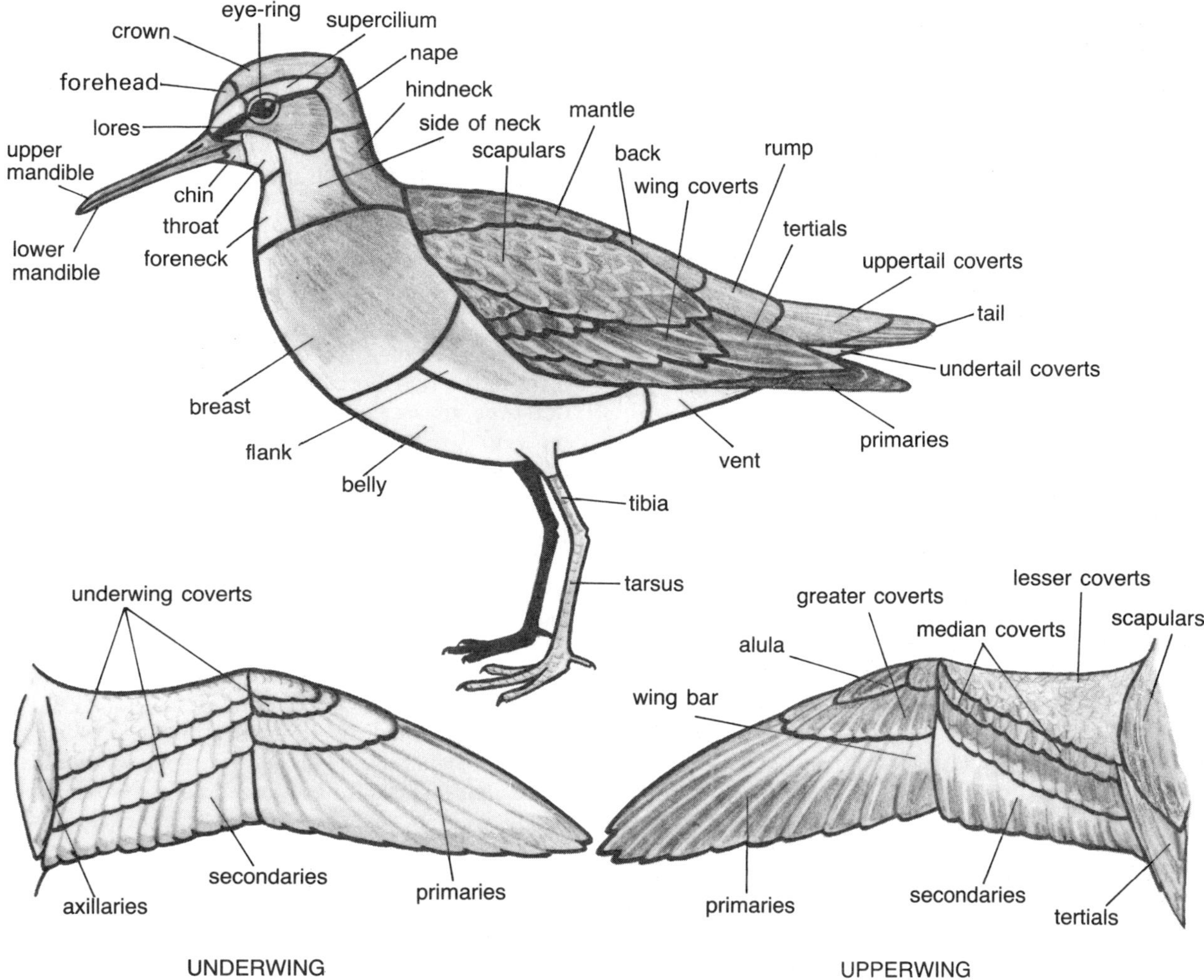


FIG. 5.3.   A generalised shorebird and its wing, showing terminology used in the field description of each species.

Curlew Sandpiper's white rump are good guides. The noticeable white wing-bars, white rump and black armpits of the Grey Plover distinguish it from the similar Lesser Golden Plover, which has uniform brown wings and upperparts. The Ruddy Turnstone has a striking wing-bar and rump pattern.

**Habitat preferences** can also give clues to a bird's identity: many shorebirds are exclusively coastal, others inhabit only inland wetlands.

All species have a distinctive **voice**, but in small sandpipers and plovers it can take some time to learn the differences. The loud 'tew tew tew' of the Greenshank or the mournful 'tee-u-wee' of the Grey Plover are especially distinctive.

Finally, there is a feature called called '**jizz**', which is difficult to define. It refers to the 'character' or 'look in the eye' of a bird, together with its behaviour. It is somewhat subjective, but with a little experience the observer will find it a useful way to distinguish birds.

Shorebirds move about a lot, and most species on the Australian list could occur anywhere. However, different areas of Australia have a distinctive complement of abundant species among which to search for the less common ones. The information in Chapter 7 indicates which species are most common in each part of Australia.

## How to interpret the maps

The maps of migratory species distribution in Asia and Australia show where they breed and where they occur outside the breeding season. The area south of the bold line across these maps (or north of it in the case of the Australian Pratincole) represents the destination of the species. They occur outside this region only during migration or when breeding.

The maps of species distribution in Australia show the twenty zones that hold the largest numbers of individuals as open circles of four sizes corresponding to four divisions of rank: the largest circles represent the top five zones and the smallest those zones ranked sixteen to twenty. Zones ranked greater than twenty are shown as small dots. For species whose populations are concentrated in less than twenty zones, the same divisions of rank and circle size are used. The table that appears near each map gives the information on which the map is based.

# BUSH THICK-KNEE

## *Burhinus magnirostris*
(Plate 18)

**Other names** Bush or Southern Stone-curlew; Stone, Southern Stone or Eastern Stone Plover; Bush Curlew; Weerloo; Willaroo.

Three races are recognised (65): *magnirostris* of southern Australia *rufescens* of north-western Australia and the Northern Territory, and *ramsayi* of North Queensland.

## DESCRIPTION

There is some geographical variation in plumage: birds in the south are paler, those in the north and inland are more rufous.

Large, thick-set shorebird with robust bill; inhabits woodlands and coastal saline plains in most parts of Australia; mostly nocturnal.

**Adult plumage** Crown, nape and hindneck, grey, finely streaked black; supercilium, white; *black line down sides of neck*; mantle and scapulars, dark grey with heavy black streaks; lesser coverts, grey; median coverts, pale grey and tapered with central black streaks; greater coverts, grey. Chin, throat and breast, buffish-grey to white, heavily streaked black; belly, flanks and vent, buffish-grey to white, sparsely streaked black. Iris, yellow; bill, thick, medium length and black; legs, long and thin, pale olive-yellow. Sexes similar.

**In flight** Back and rump, dark grey with heavy black streaks; tail, grey with black bars; pale grey median coverts contrast with rest of wing, which is grey; bases of outer primaries, white, showing as *large white wing spot*.

**Voice** A mournful 'keer-loo', often repeated and uttered at night.

**Juvenile plumage** Similarly marked, but slightly paler than adult plumage.

**Field notes** It is very cryptic and well camouflaged, often skulking away from disturbance or flying after taking a few running steps to become airborne.

**Similar species** Beach Thick-knee. The Bush Thick-knee is distinguished by having white patches in wings, heavily streaked plumage, a slightly smaller size and less robust bill. The Beach Thick-knee is exclusively coastal.

## STATUS AND DISTRIBUTION IN AUSTRALIA

It is found throughout the *Eucalyptus* woodlands of Australia. In Western Australia it also lives in *Acacia* woodlands, but it avoids the most arid parts of the continent, except around the central ranges, where there is permanent water (38,390). Occasionally it occurs on saline coastal plains in northern Australia, where these are fringed by woodland. The Bush Thick-knee has not been recorded in Tasmania since 1895.

The Bush Thick-knee mostly inhabits dry woodland, often away from water. It is widespread but sparse, and no estimate of its abundance can be made as counts only covered wetlands.

## MOVEMENTS IN AUSTRALIA

The Bush Thick-knee appears not to move far. Flocks form in the non-breeding months, and these can wander over an area of a hundred square kilometres or more (390): at the start of the breeding season they disperse locally.

## BREEDING

The Bush Thick-knee breeds in late winter, spring and summer (July–February) throughout its range, slightly earlier in the north (385, 390).

Eggs are laid on bare ground in woodlands, copses and open grassland. In coastal areas it has been recorded breeding in saltmarsh and on beaches above high-water mark. The clutch is 2 eggs, sometimes 1 or 3. Incubation lasts for between 22 and 24 days, and the young take approximately 50 days to reach adult size (149, 385). There have been no detailed studies of its breeding biology.

Factors that cause breeding failure include disturbance from human activities (including agricultural and earth-moving operations) and predation by introduced and native predators (385).

## FEEDING

The Bush Thick-knee obtains its food from the ground below woodland trees, in wheatfields and on saltpans. There have been no detailed studies of its feeding behaviour. Its diet consists of seeds, insects including dragonflies, cockroaches, mantises, earwigs, grasshoppers, crickets, bugs, beetles, caterpillars, wasps and ants, as well as spiders, centipedes and small frogs and snakes (472).

## CONSERVATION

Numbers have declined in near-coastal south-eastern and south-western Australia (38) and in eastern South Australia (39). Urban development, intensive cultivation, burning and overgrazing are thought to have caused its decline (38). It still occurs in small numbers in parts of the wheat-farming belt of Australia, where roadside and riverine remnant woodland may be important to its continued survival (195).

The retention of sufficient woodland is important if numbers are to be maintained. In already disturbed areas, such as wheat-farming regions, remnant roadside and riverine vegetation should be preserved.

# BEACH THICK-KNEE

## *Burhinus neglectus*
(Plate 18)

**Other names** Beach Stone-curlew; Reef Thick-knee.

The race in Australia is *neglectus*, which occurs as far north as the Solomons and the Malay Peninsula (65).

## DESCRIPTION

A large, heavily built, brown, grey and white shorebird with thick, medium-length bill.

**Adult plumage** Crown and nape, black; supercilium, broad and white; ear-coverts and lores, black; sides of neck, hindneck, mantle and scapulars, brown, streaked darker; *when folded, leading edges of wings show dark stripe, edged white above and below*. Chin and throat, white; breast, brown, streaked darker; belly and flanks, white. Iris, yellow; *bill, large and thick, black with yellow base*; legs, pale green-yellow. Sexes similar.

**In flight** Back and rump, brown, streaked darker; inner primaries and bases to secondaries, white, appearing as *broad white area in wing*.

**Voice** It is not as vocal as the Bush Thick-knee, but has a similar call, 'keer-loo'. When disturbed it utters a quiet 'kleee-klink'.

**Juvenile plumage** Similar to adult plumage.

**Field notes** Exclusively coastal; individually or in pairs on undisturbed beaches and tidal shores across much of northern Australia, including the islands of the Great Barrier Reef.

**Similar species** Bush Thick-knee. The Beach Thick-knee is larger and more heavily built, with more uniformly marked plumage, has a bulbous bill, dark bar on folded wing and, in flight, large white area in wing.

## STATUS AND DISTRIBUTION IN AUSTRALIA

*The Atlas of Australian Birds* shows that it occurs round the north coast of Australia, from near Onslow in Western Australia to about Red Rock on the north coast of New South Wales; reporting rates (a rough measure of relative abundance) are higher near Dampier in Western Australia, near Darwin and on Groote Eylandt in the Northern Territory, and on Cape York, round Cairns and from Proserpine to Fraser Island in eastern Queensland (38). It is more common on islands than on the mainland in the Kimberley region of Western Australia (446).

During national counts, coverage was patchy in these areas, but consistent from year to year. Only small numbers were observed: the maximum was 16 in the Mackay area in February 1983. These results, combined with a generally low reporting rate during the *Atlas* project (1977–81), indicate that they are spread along the coast in lower densities than other shorebirds, and that they do not flock.

## MOVEMENTS IN AUSTRALIA

Nothing is known of its movements, but information from the *Atlas* suggests that the Beach Thick-knee is probably sedentary (38).

## BREEDING

The Beach Thick-knee occasionally excavates a shallow scrape, but usually lays its single egg on bare ground. Nests are sited on sandy beaches located on a shore or island on the open coast, among mangroves, or in an inlet. There have been no detailed studies of its breeding biology.

A recorded cause of nesting failure is flooding of nests by extra-high tides (385).

## FEEDING

There have been no detailed studies of the feeding behaviour or diet of the Beach Thick-knee. It usually forages over intertidal mudflats and sand spits, rather than wading in shallow water, and has been seen eating crabs (380).

## CONSERVATION

The Beach Thick-knee occurs in very low densities compared with other shorebirds. Laying only 1 egg, its ability to recover from any population decline may not be great. Their numbers on Brampton Island, Queensland, declined after the introduction of a cat, but they increased again after the cat died (380).

The Beach Thick-knee is vulnerable, and there is an urgent need for more research on its biology and on factors controlling its population level.

# PAINTED SNIPE

## *Rostratula benghalensis*
(Plate 10)

**Other names**   Greater Painted Snipe.

Two races of Painted Snipe are recognised: *benghalensis* and *australis*, the latter occurring only in Australia.

## DESCRIPTION

A distinctive, strikingly marked, medium-sized snipe of freshwater wetlands. Flies fairly slowly; wings rounded; sexes different; the female is slightly larger than the male.

**Adult male plumage**   Crown and nape, mottled grey with *white central stripe*; feathers around eye and supercilium, buff; hindneck, sides of neck, mottled grey; mantle and scapulars, blue-grey, mottled white; edge of mantle shows buff 'V'; wing-coverts, grey-green with large buff spots edged dark grey. Chin, throat and upper breast, mottled grey; white line extends around sides of lower neck, bordered black; sides of breast, mottled dark grey, sometimes forming bar; lower breast, belly, flanks and vent, white. Iris, brown; bill, long, pinkish-brown with paler base; legs and feet, pale olive-green.

**Adult female plumage**   Crown and nape, rich maroon *with white central stripe*; feathers around eye and supercilium, white; hindneck, sides of neck, mantle, rich maroon; scapulars, greenish-brown, mottled black and white; wing-coverts, glossy green, barred black. Chin, white; throat and upper breast, rich maroon; sides of breast, mottled black, sometimes forming patchy breast-bar; lower breast, belly, flanks and vent, white. Soft parts as in male.

**Voice**   A loud 'chek' when flushed.

**In flight**   Flight laboured, usually stays low. Back, rump and tail, mottled grey with buff spots edged black. Wings darker in female; spotted, with darker bases to primaries in both sexes.

**Juvenile plumage**   Like adult male plumage, but wing-coverts duller and tipped buff.

**Field notes**   Very cryptic and well camouflaged; found in wetlands with rank vegetation or samphire; occasionally seen in flocks after breeding.

**Similar species**   Latham's Snipe. The Painted Snipe is distinctive, flying lower with slower wing-beats.

## STATUS AND DISTRIBUTION IN AUSTRALIA

The Painted Snipe is a rare bird, spread over a vast area of the continent (both wet and arid) and the comment by Lowe in 1963 (263) that much remains to be learnt of its status and distribution still holds true today.

It occurs over a wide area of south-eastern and eastern Australia, mostly east of a line from Eyre Peninsula, South Australia, to Mount Isa, Queensland (38). It has been recorded in Tasmania. Shallow freshwater wetlands, both permanent and ephemeral, particularly those with rushes and samphire, are its preferred habitat. Most *Atlas* project records come from the wetlands of south-eastern South Australia, northern central Victoria, central New South Wales (Macquarie and Namoi valleys), and the Fitzroy drainage system in central eastern Queensland. There was one *Atlas* record from Kununurra and there are historical records from other places in Western Australia prior to 1950 (38). There were only five breeding records during the *Atlas* project, all south of 28°S.

Few were counted during national counts, probably because of their cryptic behaviour and preference for wetlands with rank vegetation. Only 8 records involving 14 birds were submitted.

## MOVEMENTS IN AUSTRALIA

The Painted Snipe may be migratory, moving into the southern part of its range to breed in spring and summer (August–February) (27, 163, 195, 263). The *Atlas* project reporting rates in eastern Queensland, New South Wales and Victoria were higher in summer (0.5 per cent) than in winter (0.04 per cent) (38).

They flock after breeding in southern Australia, where groups of up to 25 birds have been seen. Soon after flocking they depart their breeding areas (263). However, a flock was observed near Melbourne between May and September 1951 (485).

It is most abundant in areas during floods, suggesting some opportunism, but much remains to be learnt about its movements.

## BREEDING

Most available information consists of anecdotal observations (263). The Painted Snipe breeds in shallow wetlands with a good growth of rushes or samphire. It nests on small islands, shores of swamps or banks of channels, usually among rank vegetation, but at times in the open (263). A platform of water weed and rushes is constructed in shallow water (276). Between 2 and 4 eggs are laid, and incubation probably takes between 21 and 24 days (263). The Painted Snipe is polyandrous, the male incubating the eggs.

It breeds between August and February in southern Australia. However, regular breeding occurs at Ayr, northern Queensland, in March and April (263).

## FEEDING

The Painted Snipe feeds in shallow water. No detailed account of its feeding behaviour or diet exists. It probably feeds on a variety of freshwater invertebrates. It has been recorded feeding on 'freshwater shells' and 'leaves, etc.' (62).

## CONSERVATION

The Australian race of the Painted Snipe is endemic, and so of considerable interest. It is one of our least-known shorebirds, and there is an urgent need for more research on it. Shallow swamps, its preferred habitat, have been destroyed at a faster rate in the more settled parts of Australia than other wetland types (79, 80). The impact of regulation of river flows for irrigation and flood mitigation on the availability of habitat is also a concern.

# PIED OYSTERCATCHER
## *Haematopus ostralegus*
(Plate 17)

**Other names** Black-and-white or White-breasted Oystercatcher; Redbill; Eugerie-bird; Seapie; Wong-bird; Oyk.

Almost world-wide in distribution: of 17 species or subspecies only 1, *longirostris*, occurs on the coasts of Australia, southern New Guinea and nearby islands (65).

## DESCRIPTION

A large, sturdy, black-and-white shorebird with long red bill and pink or red legs.

**Adult plumage** Head, neck and breast, entirely black; mantle, scapulars and wing-coverts, black; *back, rump and belly, white*; tail, black. Iris, red; *bill, long, straight, bright orange-red*; legs, pink, orange-red in fully mature (4-year-old) birds. Sexes similar.

**In flight** *Rump and back, white*; tips to greater coverts, white, showing as narrow white wing-bar on the inner halves of the wings.

**Voice** In flight it utters a high-pitched piping 'pleep-pleep'; repetitive 'tleepa tleepa tleepa tleepa ...'; during courtship, pairs give a rapidly repeated high-pitched whistle 'pipipipipipi ...'

**Juvenile plumage** Similar to adult, but black feathers have small buff spots on edges, which soon wear off. Bill, orange, variably tipped dark grey; eye, brown to dull red; legs, grey.

**Field notes** Found in scattered pairs or roosting flocks of up to a hundred birds on sandy beaches and intertidal mudflats; noisy displays and much chasing of rival pairs occur in the breeding months.

**Similar species** Sooty Oystercatcher. The Pied Oystercatcher can be distinguished by its slightly slimmer build and white belly, flanks, vent and inner wing-bar. Confusion is possible in bad light or at difficult angles of observation, when white parts are not obvious.

## STATUS AND DISTRIBUTION IN AUSTRALIA

The Pied Oystercatcher has an almost continuous distribution around the coast of Australia (38). The largest concentrations occur in large marine embayments in Victoria, Tasmania and South Australia. (See Figure 5.4, Table 5.2.) Its distribution in northern Australia is not fully documented as coverage is not complete, but concentrations have been reported from central and south-eastern Queensland, the Gulf of

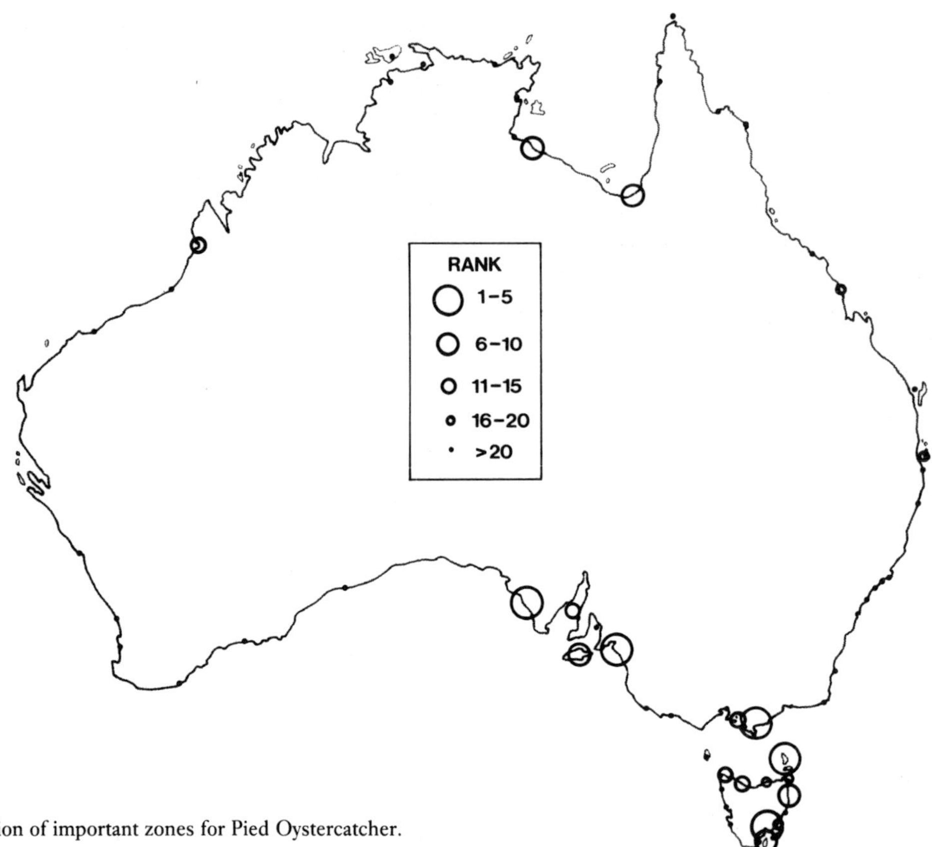



FIG. 5.4.   Location of important zones for Pied Oystercatcher.

TABLE 5.2. Ranked list of top twenty zones for Pied Oystercatcher. (maximum counts marked thus: *; other counts are averages) (120 or more birds)

| ZONE | NO. OF COUNTS | NO. OF INDIVIDUALS |
|---|---|---|
| Corner Inlet, Vic. | 10 | 870 |
| * The Coorong, SA | 1 | 630 |
| * Furneaux Islands, Tas. | 3 | 500 |
| Derwent estuary & Pittwater, Tas. | 9 | 480 |
| * W coast, Eyre Peninsula, SA | 8 | 450 |
| * Bruny Island, Tas. | 7 | 260 |
| * St Helen's area, Tas. | 4 | 250 |
| * Kangaroo Island, SA | 10 | 240 |
| * SE corner, Gulf of Carpentaria, Qld | 1 | 230 |
| * S coast, Gulf of Carpentaria, NT | + | 200 |
| * Spencer Gulf, SA | 9 | 200 |
| * Roebuck Bay, WA | 6 | 190 |
| * NW coast, Tas. | 4 | 180 |
| Westernport Bay, Vic. | 10 | 160 |
| * Central N coast, Tas. | 9 | 150 |
| * SE coast, Tas. | 9 | 150 |
| * Mackay area, Qld | 8 | 150 |
| * Moreton Bay, Qld | 10 | 140 |
| Tamar estuary, Tas. | 9 | 130 |
| * Cape Portland coast, Tas. | 10 | 120 |

(+ = aerial count)

Carpentaria, coastal Arnhem Land and Roebuck Bay near Broome in Western Australia.

The Pied Oystercatcher prefers the extensive intertidal mudflats of large marine embayments, but also occurs on sandy ocean beaches in lower numbers. It is absent from rocky coasts such as that of the Great Australian Bight (38). It forages mainly on intertidal mudflats and usually roosts at high tide on open sandy spits, where available. It sometimes moves into saltmarshes for shelter. It is almost entirely coastal, and only near Hobart does it leave the coast to forage in adjacent pasture when flooded.

## MOVEMENTS IN AUSTRALIA

The colour-banding of over 500 Pied Oystercatchers in Victoria and Tasmania shows that it frequently makes local movements within marine embayments and occasionally between them. The longest movement recorded is of a bird banded at Werribee Sewage Farm, Victoria, and resighted near Eddystone Point in north-eastern Tasmania (476).

Breeding adults are sedentary in the Hobart area, Tasmania (334), whereas an immature bird moves as much as 35 km in its first year. Many Pied Oystercatchers hold feeding territories for up to 6 months near where they are first banded (329). Flocks of non-breeding birds can occur all year round.

The Pied Oystercatcher is not migratory. Regular counts at Westernport Bay and Anderson's Inlet in Victoria, and round Hobart in Tasmania show that seasonal flocking occurs between May and August; they concentrate at traditional roost sites at this time then disperse to breed. Counts indicate that movements are irregular on ocean beaches. Little seasonal variation in numbers occurs at Corner Inlet, in Victoria.

## BREEDING

The Pied Oystercatcher's nest is a scrape in sand or shingle on beaches just above high-tide mark, in dunes, and occasionally in saltmarsh or earth. In the Hobart area, Tasmania, it sometimes breeds in pasture several hundred metres from beaches to escape disturbance (334). The usual clutch is 2, sometimes 1 or 3 and very rarely 4 (334, 385). In south-eastern Tasmania it regularly lays a second clutch, and occasionally a third, if earlier ones are destroyed (332). Incubation lasts for between 28 and 32 days and the young fledge 6 to 7 weeks after hatching and stay with the parents in the breeding territory for between 1 and 6 months (334). At Mortimer Bay in south-eastern Tasmania, the average number of young raised by a pair was 1 every 2 years (334). A pair stays together and breeds in the same place in successive years.

The Pied Oystercatcher has been recorded breeding between May and January; laying starts as early as May in northern Australia and as late as October in Tasmania (334).

Causes of nesting failure include flooding by extra-high tides, predation by ravens and disturbance by human activities, off-road vehicles and domestic animals; a wandering dog eating a young chick has been recorded (334). Other introduced and native predators probably take eggs and young.

## FEEDING

The Pied Oystercatcher forages at low tide on intertidal mudflats and has been observed eating marine worms, cockles and mussels (334). No detailed feeding studies have been undertaken in Australia.

## CONSERVATION

The Pied Oystercatcher is perhaps most vulnerable when breeding, and human disturbance can cause widespread nesting failure, especially on beaches that are popular for recreation (334). Non-breeding flocks also need undisturbed feeding and roosting areas. Identification of main breeding and non-breeding areas and sensitive management of these to reduce disturbance are necessary if numbers are to be maintained.

# SOOTY OYSTERCATCHER
## *Haematopus fuliginosus*
(Plate 17)

**Other names** Black Oystercatcher; Redbill; Black Redbill.

There are two races in Australia: *fuliginosus* and *opthalmicus* (65); *opthalmicus* has a more fleshy red orbital ring and a slightly longer bill.

## DESCRIPTION

A large sturdy black shorebird with a long red bill and pink legs.

**Adult plumage** Completely black plumage. Iris, red; bill, long, straight and red; legs, pink. Sexes similar.

**In flight** Completely black.

**Voice** High-pitched piping calls similar to the Pied Oystercatcher: 'tleepa tleepa tleepa tleepa ...'

**Juvenile plumage** Similar to adult plumage, but feathers have spotted buff edges that soon wear off. Bill, orange-red, variably tipped grey-green; legs, grey.

**Field notes** It roosts individually, in pairs or in small flocks and feeds on rocky shores. At times it occurs on mudflats and sandy ocean beaches, sometimes with Pied Oystercatchers.

**Similar species** Pied Oystercatcher. The Sooty Oystercatcher lacks white in the plumage and is slightly more robust.

## STATUS AND DISTRIBUTION IN AUSTRALIA

The race *fuliginosus* occurs from the Kimberleys, round the western, southern and eastern coasts to Cape York and *opthalmicus* from the Kimberleys, round the north coast, to Mackay, Queensland. Their ranges overlap along the Queensland coast, where Sooty Oystercatchers are rare (9). Sooty Oystercatchers have a very patchy distribution, and most occur at seven sites (see Figure 5.5, Table 5.3) consisting of rocky intertidal shoreline and sandy beaches with nearby intertidal mudflats near islands where they breed (102, 330, 385). Largest numbers of Sooty Oystercatchers occur in southern Australia. In south-eastern South Australia, where there are extensive rocky reefs but no islands, few Sooty Oystercatchers are reported. They occur only in ones and twos on a small number of islands in the Great Barrier Reef, where they are outnumbered by Pied Oystercatchers (78). There is a

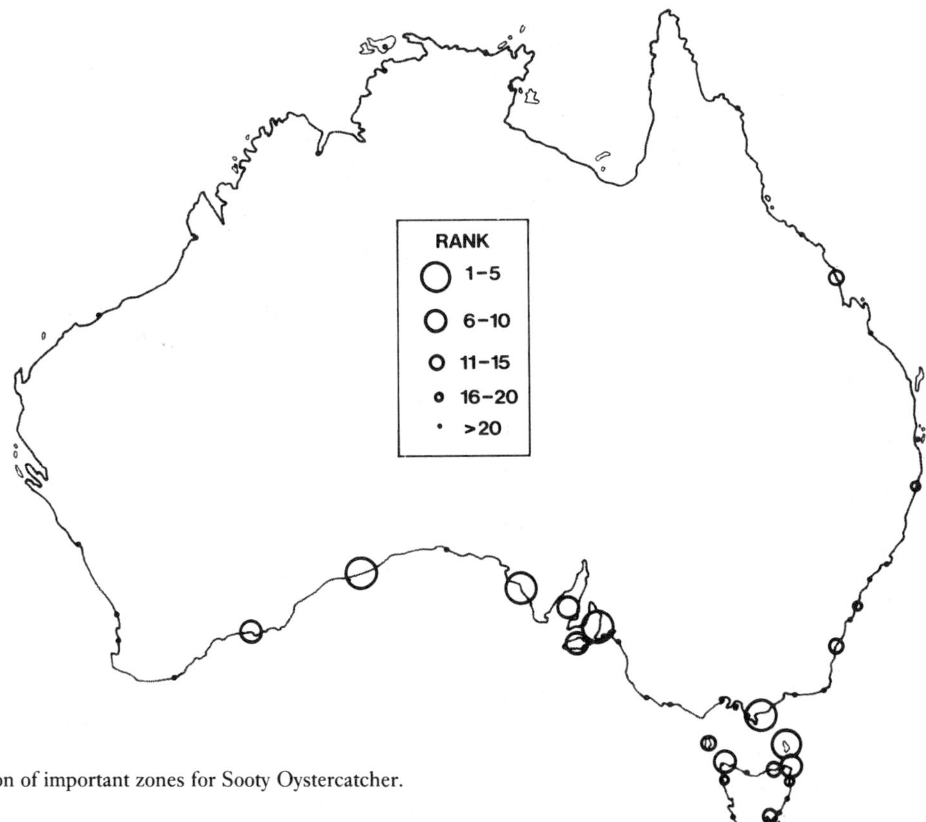


Fig. 5.5.   Location of important zones for Sooty Oystercatcher.

considerable gap in their distribution around the Gulf of Carpentaria, where there is little rocky shoreline (38).

The population of *opthalmicus* probably numbers less than 1000 individuals. The numbers of Sooty Oystercatchers in the Hobart area have declined by about 40 per cent since the 1960s (335).

## MOVEMENTS IN AUSTRALIA

Studies on the central Victorian coast, where Sooty Oystercatchers breed on islands between October and January (102, 385), show that numbers increase on mainland rocky shores through summer and on sandy ocean beaches in winter (70). These regular movements correspond to seasonal variations in the exposure of rocky intertidal feeding habitat, which is not exposed for as long during winter months (70). In the Hobart area, Tasmania, they move from islands to the Derwent estuary in winter (454). No long-distance movements have been reported, and they probably move locally, depending on food availability, between islands, rocky and sandy coast, and areas of intertidal mudflat.

## BREEDING

Most Sooty Oystercatchers breed on offshore islands and construct a simple scrape, often in sand, on a

TABLE 5.3. Ranked list of the top twenty zones for Sooty Oystercatcher. (all counts are maxima) (20 or more birds)

| ZONE | NO. OF COUNTS | NO. OF INDIVIDUALS |
|---|---|---|
| Furneaux Islands, Tas. | 3 | 480 |
| W coast, Eyre Peninsula, SA | 8 | 440 |
| Corner Inlet, Vic. | 10 | 270 |
| Nullarbor coast, WA | 8 | 140 |
| St Vincent Gulf, SA | 8 | 120 |
| Kangaroo Island, SA | 10 | 110 |
| Spencer Gulf, SA | 9 | 100 |
| NW coast, Tas. | 4 | 70 |
| Albany coast, WA | 9 | 60 |
| Cape Portland coast, Tas. | 10 | 50 |
| S coast, NSW | 10 | 50 |
| Tamar estuary, Tas. | 9 | 50 |
| Derwent estuary & Pittwater, Tas. | 9 | 40 |
| Mackay area, Qld | 8 | 40 |
| King Island, Tas. | 5 | 30 |
| St Helen's area, Tas. | 4 | 30 |
| Central NW coast, Tas. | 2 | 30 |
| Illawarra coast, NSW | 6 | 30 |
| Bruny Island, Tas. | 7 | 20 |
| N coast, NSW | 10 | 20 |

shingle beach, on the floor of a small cave, sometimes under tussocks of grass. Occasionally the scrape is made in soil on top of a rock. Usually the scrape is unlined, but occasionally it is lined with twigs, grass or leaves (385). The nest is usually close to the water's edge. Their use of islands and rock stacks makes them less vulnerable to disturbance while breeding than other coast-nesting shorebirds.

## FEEDING

The Sooty Oystercatcher feeds among rocks or walks along the backs of ocean beaches busily probing under washed-up seaweed. When foraging on intertidal mudflats, it often associates with the Pied Oyster-catcher.

The Sooty Oystercatcher has been observed to take a variety of common intertidal invertebrates of the mid and lower littoral, including sea-squirts, limpets, mussels and Chitons, as well as washed-up dead fish and, on sandy beaches, sandhoppers (*Orchestria* spp.) (69, 190).

Studies in Victoria show that it is able to obtain food more quickly from rocky shores than from sandy beaches, but during the winter months it leaves rocks to forage at the backs of sandy beaches among sea-weed. Food supplies are less available on rocky shores in winter due to reduced tidal range during the day (69). The young Sooty Oystercatcher must learn to prise off and break open its rocky shore prey and spends time feeding with its parents after fledging.

## CONSERVATION

The population of Sooty Oystercatchers is small and concentrated in seven main areas. The loss of a major habitat component (beach, rocky shore, mudflat or island) in one of these areas could significantly reduce a local population. Therefore co-ordinated reservation and management of these areas is desirable. More research is required to clarify the status of *opthalmicus* in northern Australia.

# MASKED LAPWING
## *Vanellus miles*
(Plate 1)

**Other names** Spur-winged Lapwing or Plover; Australian Spur-winged Plover; Masked, Spurwing or Wattled Plover; Alarmbird.

Two races occur in Australia: *miles* in the north and *novaehollandiae* in the south.

## DESCRIPTION

A large, heavily built plover with distinctive call and brown, black and white plumage.

**Adult plumage** of *novaehollandiae* Crown, nape, sides of neck and breast, black; mantle, scapulars, back and wing-coverts, tan. Underparts, white. *Yellow wattles cover front of head*; iris, yellow; fleshy yellow orbital ring around eye; bill, yellow; legs, red-brown; small spur on carpal joint of wing, yellow, tipped black, longer in males. Northern race birds (*miles*) lack black sides to the breast and neck, are slightly smaller than *novaehollandiae* and their face wattle is more extensive, extending behind the eye. Birds intermediate in plumage can occur. Sexes similar in plumage.

**In flight** Back, tan; rump, white; black subterminal bar to white tail; primaries black, contrasting with tan coverts.

**Voice** Loud, harsh 'kerr kee kee ki ki ki ki ki ...' Often utters a single 'ki ...'

**Juvenile plumage** Similar to adult plumage, but crown greyer; scapulars and wing-coverts, tipped black with buff subterminal bars; face wattle, smaller and duller yellow; bill, duller yellow to grey; legs, pinkish-grey.

**Field notes** Seen singly, in pairs or small flocks in a variety of open habitats from pasture to mudflat; aggressive and noisy when breeding.

**Similar species** Banded Lapwing. The Masked Lapwing is larger, duller brown on upperparts and lacks the broad black breast-bar, the red wattle in front of the eye and the white wing-bar.

## STATUS AND DISTRIBUTION IN AUSTRALIA

The Masked Lapwing is confined to Australia and islands north of Australia. The northern race (*miles*) occurs from the Kimberley region eastwards to Townsville and is common in lowland areas of Papua New Guinea and the Aru Islands (38, 184). The southern race (*novaehollandiae*) occurs in Tasmania, eastern South Australia, Victoria, New South Wales, southern Queensland and occasionally farther north, such as in the Port Moresby region of Papua New Guinea and the Solomon Islands (57, 184). The Masked Lapwing is also common in New Zealand, having first arrived naturally in the 1940s (230).

Masked Lapwings are widespread, occurring on pasture, playing fields, around the edges of wetlands, in saltmarsh and on intertidal mudflats. An accurate estimate of their population size is not possible, because many Masked Lapwings occur away from wetlands. Consequently, not all were counted. However, counts that were obtained totalled in the tens of thousands.

## MOVEMENTS IN AUSTRALIA

Regular counts show that Masked Lapwings form non-breeding flocks between November and August in most places, then disperse in late winter and early spring to breed. On the Queensland coast, numbers are fairly constant and low. In Victoria, numbers are higher in summer than winter. In the Darwin area, numbers are higher in the dry season (July), when birds concentrate on drying wetlands (88). No long-distance movements have been recorded (15). These observations suggest that Masked Lapwings move locally, dispersing from wetlands at the wettest time of year to breed, and concentrating at drier times.

## BREEDING

The nest consists of a shallow scrape in the ground, lined with small twigs, located on the edge of wetlands, in pasture, ploughed land or occasionally on playing fields, aerodromes or even the median strips of dual carriageway roads (385, 455). 3 or 4 eggs are laid, and incubation lasts for between 26 and 32 days (99, 455). A study in southern Victoria recorded 45 per cent of eggs hatching (99). Young leave the nest soon after emerging and may wander as far as 200 m from one another while the parents are still tending them (455). Young take between 5 and 8 weeks to fledge (455). In Victoria, nearly 20 per cent of young that hatched survived to fledge (99). At two locations in Victoria and one in Tasmania, between 4 and 9 per cent of eggs laid produced young that fledged (6, 99, 455).

In southern Victoria, the Masked Lapwing breeds in winter and spring, when soil fauna is most abundant and accessible for adults and young (99).

Causes of nesting failure include trampling by grazing stock and disturbance from roadworks (99, 455). Predation by introduced and native predators probably also contributes to nesting failure in some places.

## FEEDING

The Masked Lapwing feeds in a variety of habitats, from pasture to intertidal mudflats. On coastal mudflats it feeds at low tide and at high tide often roosts in saltmarsh.

There have been no studies of its diet when on coastal mudflats; but in pasture areas, beetle adults and larvae and earthworms are taken (99). A large range of insects and other arthropods and, less frequently, seeds and leafy material have also been recorded as food (62, 256).

## CONSERVATION

This species has adapted well to the changes wrought upon the environment by European settlement and is common and widespread in agricultural areas. It is in no danger of declining in numbers.

# BANDED LAPWING
## *Vanellus tricolor*
(Plate 1)

**Other names** Banded, Black-breasted, Brown, Flock or Plain Plover.
No races recognised.

## DESCRIPTION

A medium-to-large brown-and-white plover with a prominent black breast-bar.

**Adult plumage** Crown and nape, black; hindneck, tan; line behind eye, white and almost meeting below nape; sides of neck, black; mantle, scapulars, and wing-coverts, tan. Chin and throat, white; breast, black, continuous with black on sides of neck; belly, flanks and vent, white. Iris, yellow; orbital ring, fleshy yellow; *wattles on sides of forehead, red*; bill, yellow; legs and feet, dark grey-brown with pink 'knees'. Sexes similar.

**In flight** *Striking wing-pattern*: back, tan; rump, white; tail, white with black subterminal bar. Primaries, black; secondaries, black with white bases, broader nearer body, contrasting with tan wing-coverts and back; underwing, white, contrasting with black primaries.

**Voice** Softer than the Masked Lapwing; distinctive 'er-kill-kee, er-kill-kee . . .' repeated.

**Juvenile plumage** Similar to adults, but crown, nape, sides of neck and breast, mottled dark brown; feathers of upperparts, tan, edged buff; face wattle, smaller and duller.

**Field notes** Most frequently encountered in pairs and loose flocks inland in drier open habitats, such as pasture, ploughed fields and saltbush plains.

**Similar species** Masked Lapwing. The Banded Lapwing is smaller, has redder brown upperparts; black breast-bar, less extensive face wattle and white wing-bar.

## STATUS AND DISTRIBUTION IN AUSTRALIA

The Banded Lapwing occurs in drier habitats than the Masked, although it avoids the most arid parts of the continent and does not regularly occur north of 20°S (38). It also avoids *Acacia* scrub areas, except where these have become more open due to overgrazing in south-western Australia and western New South Wales. Observations of captive Banded Lapwings show that rank groundcover impedes movement and food-gathering (471). These observations and higher report-ing rates in the *Atlas* project in areas of agricultural land and saltbush plain show that they prefer open country in the southern semi-arid parts of the continent.

The Banded Lapwing occurs mainly away from wetlands and few were counted during national counts. It is generally less abundant than the Masked Lapwing where both species occur. In the Sydney area, it arrived in the 1940s (190), and in the south-west it has increased since European settlement (400). It has declined in parts of eastern South Australia; this has corresponded to a reduction in wheat farming on marginal land (39). In the Northam area of Western Australia, where improved pasture has caused a change to taller vegetation, it has also declined (299).

## MOVEMENTS IN AUSTRALIA

Records from the *Atlas of Victorian Birds* (27) show that the Banded Lapwing is most frequently reported in Victoria in winter and spring. This correlates with the timing of breeding records (385) and the average monthly rainfall at Horsham, in the heart of their Victorian range. This suggests that Banded Lapwings in Victoria disperse to breed when rainfall is highest. They may move into Victoria and other wetter areas of southern Australia to breed in winter and spring. But this probably involves only part of the population as they are recorded in Victoria all year round (27, 485).

Observations elsewhere in the continent suggest that the Banded Lapwing is nomadic, leaving areas subject to drought and returning when conditions become favourable (135, 195, 275).

## BREEDING

The Banded Lapwing generally breeds in territorial pairs, although loose colonies have been reported (135). It excavates a shallow scrape in the ground, usually lined with dry grass, plant stems or occasionally small stones. Sometimes it builds a small platform of grass on which to lay its eggs. It nests in pasture or mown grass (e.g. airfields), dry open plains and salt-bush, or ploughed fields (135, 385). 4 eggs is the usual clutch, occasionally more or less (190, 385). Incubation lasts for between 26 and 29 days, and in southern Victoria studies show that at least 31 per cent of all eggs hatched (99). The proportion of chicks that survived to fledge was 17.5 per cent and the number of young fledged was 5.5 per cent of all eggs laid (99). The young take about 60 days to fledge.

Throughout mainland Australia, the Banded Lapwing has been recorded breeding in late winter, spring and autumn (99, 135, 190, 385). In Tasmania, it breeds only in late winter and spring (385).

Reasons for nesting failure include trampling by grazing stock, destruction of nests by vehicles and excessive human disturbance (99, 385). Predation by introduced and native predators probably also contributes to nesting failure.

## FEEDING

The Banded Lapwing feeds predominantly from the surface of the ground (99). Its diet consists of worms, spiders, insects including ants, termites, beetles, cockroaches, grasshoppers, crickets, caterpillars and other grubs, as well as seeds and leaves of plants (61, 62, 99, 190, 256, 472).

## CONSERVATION

It is widespread in Australia and exploits a variety of open habitats, including agricultural land. There are no immediate or serious threats to its future survival.

# GREY PLOVER

## *Pluvialis squatarola*
(Plate 2)

**Other name**  Black-bellied Plover.
  No races recognised.

## DESCRIPTION

The largest of the migratory plovers, it has a stocky, big-headed appearance; mottled grey-and-white, but strikingly marked in breeding plumage.

**Adult non-breeding plumage**  Crown, nape, hindneck, mantle and scapulars, dark grey-brown feathers, fringed white, giving upperparts a generally mottled appearance; tertials and wing-coverts, dark grey-brown notched white, giving a speckled appearance. Supercilium, lores and ear-coverts, slightly paler than crown and streaked grey on white. Chin and throat, white; sides of neck and breast, white, blotched grey-brown; belly, flanks and vent, white. Iris, black; bill, short with bulge towards tip, black; legs, grey. Sexes similar.

**In flight**  Back and rump, dark grey-brown, feathers fringed white, but *uppertail-coverts and bases of tail feathers are white*; tail, white barred dark grey-brown. Broad white bases to primaries and secondaries form an obvious *white wing-bar; black axillaries* ('armpits') contrast with white underwing.

**Voice**  Plaintive whistles, most frequently 'tee-u-wee . . . tee-u-wee'.

**Breeding plumage**  Striking black, white and grey plumage. Crown and nape, black with pale grey fringes to feathers; hindneck, white, speckled black; feathers of mantle, black, fringed white; scapulars black, notched and fringed white, giving upperparts a *striking spangled appearance*. Supercilium, sides of neck and breast, white, contrasting with solid black lores, ear-coverts, throat, breast, belly and flanks; lower belly and vent, white.

**Juvenile plumage**  Like adult non-breeding, but feathers of upperparts heavily spotted and notched creamy-buff, rather than fringed pale. Breast and flanks are streaked (not blotched) grey-brown.

**Field notes**  Feeds in typical stop-start plover fashion; almost exclusively coastal. Roosts on sandy beaches and spits, often in flocks with knots and godwits. Birds in plumage transitional between breeding and non-breeding can be seen in August−October and February−April.

**Similar species**  Lesser Golden Plover; Great Knot. The Grey Plover is larger and paler, has greyer plumage, larger head, heavier bill and, in flight, distinctive black armpits compared with the Lesser Golden Plover. It often associates with the Great Knot, but is stockier with a shorter neck and bill and larger head and eye.

## BREEDING RANGE

The Grey Plover usually breeds north of 65°N in the tundra zone from the White Sea (W USSR) east to the Gulf of Anadyr (E Siberia), Alaska and arctic Canada, as far as western Baffin Island and south to Southampton Island (7, 85).

## ASIAN RANGE

In the non-breeding months it is found on all coasts of South America, Africa, southern Asia and Australia.

The Grey Plover is a regular migrant through Korea (162), Japan (344) and north-eastern China (255). In Japan, up to 2500 have been counted on southward (September) and 1285 on northward migration (488). It occurs in Taiwan (37), southern and south-eastern China (255), Hong Kong (54), Thailand (257), Malaysia (157, 305) and in small numbers in Vietnam (142, 489), the Philippines (171), Borneo (440) and parts of Indonesia (14, 128). In Micronesia, it is an uncommon visitor (18), and in the islands north-west

of Australia 'mainly transient' between September and November (487). A few are recorded on the coast of Papua New Guinea all year round. Single birds are occasionally reported from the Kermadec, Cook, Phoenix and Line islands (59, 60, 204, 312). In New Zealand, it is a regular but uncommon visitor (1983: 2 birds; 1984: 5 birds) (387, 388). During the non-breeding months there are few records of large numbers in Asia, except on migration, suggesting that most Asian Grey Plovers migrate to Australia. (See Figure 5.6.)

## STATUS AND DISTRIBUTION IN AUSTRALIA

The Grey Plover is more abundant on the western and southern coasts, where it is generally confined to a small number of large marine embayments. (See Figure 5.7, Table 5.4.) It occurs irregularly in small numbers on the east coast.

Its distribution contrasts with that of the Lesser Golden Plover, which occurs mostly on the east coast. In the south-east, where both occur in reasonable numbers, they tend to use different sites. For example, in Victoria, Corner Inlet holds almost exclusively Grey Plovers, and Shallow Inlet, only 15 km away, exclusively Lesser Golden Plovers. A similar separation occurs at sites on the South Australian coast: St Vincent and Spencer Gulfs, western Eyre Peninsula and Kangaroo Island hold mostly Grey Plovers, and the Murray mouth lakes and south-east coast mostly Lesser Golden Plovers.

It is almost exclusively coastal, foraging on intertidal mudflats and roosting at high tide on sandy beaches and spits.

## MOVEMENTS IN AUSTRALIA

In early September, the Grey Plover arrives in the Broome—Port Hedland area (where it stays until March) and at Darwin (where it does not stay). It reaches south-western Australia in October and November and coastal South Australia in November. Small numbers are present on the east coast in October. In southern New South Wales, an influx occurs in December. These results suggest that it probably arrives over a large part of the north-west and north coast in early September, continues migrating southwards in October and November, reaching maximum numbers in the south in December. There are few inland records of this species.

In March, temporary influxes occur in South Australia, on the east coast and at Darwin. In the Broome—Port Hedland area it departs by mid-March and does not appear to migrate northwards through this area. In the south-west, however, it does not depart until April. Birds departing northwards from Victoria

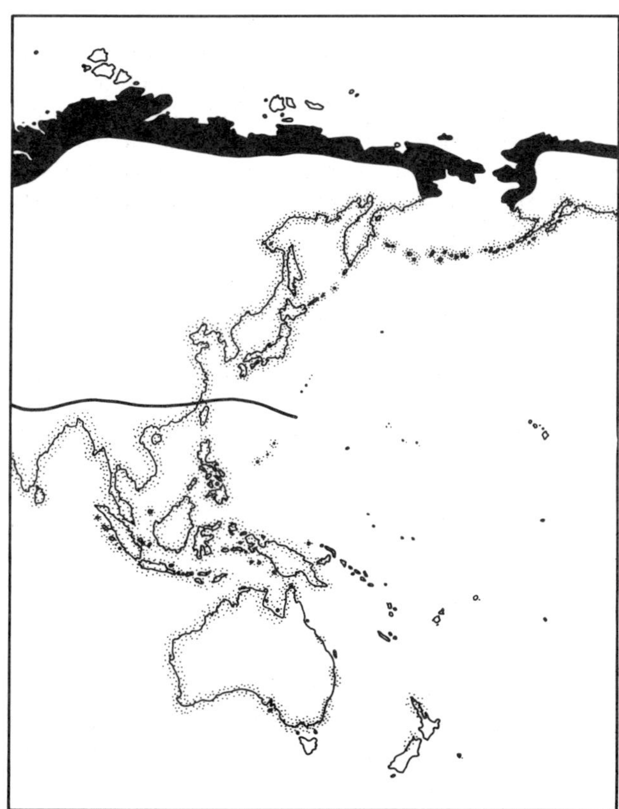

FIG. 5.6.   Breeding and non-breeding range of Grey Plover.

53

FIG. 5.7.  Location of important zones for Grey Plover.

TABLE 5.4.  Ranked listing of top twenty zones for Grey Plover. (maximum counts marked thus: *; other counts are averages) (20 or more birds)

| ZONE | NO. OF COUNTS | NO. OF INDIVIDUALS |
|---|---|---|
| * Eighty Mile Beach, WA | 5 | 1 650 |
| * SE corner, Gulf of Carpentaria, Qld | 1 | 1 550 |
| * Roebuck Bay, WA | 6 | 1 300 |
| * W coast, Eyre Peninsula, SA | 5 | 1 280 |
| * Spencer Gulf, SA | 5 | 740 |
| Corner Inlet, Vic. | 5 | 450 |
| St Vincent Gulf, SA | 4 | 390 |
| Port Phillip Bay, Vic. | 5 | 220 |
| * Albany coast, WA | 4 | 130 |
| * Kangaroo Island, SA | 5 | 120 |
| Swan coastal plain, WA | 5 | 100 |
| * Mackay area, Qld | 4 | 80 |
| Darwin area, NT | 4 | 80 |
| Peel Inlet, WA | 5 | 70 |
| * Esperance coast, WA | 5 | 50 |
| * Pilbara coast, WA | 3 | 40 |
| * Central W coast, WA | 4 | 30 |
| * NE coast, Arnhem Land, NT | 1 | 30 |
| * N coast, NSW | 5 | 20 |
| * Torres Strait, Qld | 3 | 20 |

fatten sufficiently to fly over the north coast of Australia to a stopover in South-East Asia (476). This, together with its late departure from the south-west, suggests that many southern birds do not touch down in northern Australia. However, some do pass through South Australia, the Darwin area and up the east coast during northward migration.

## FEEDING AND FOOD

Grey Plovers forage over intertidal mudflats at low tide. The birds spread well apart and detect food visually, pecking it from the surface of mudflats. There have been no detailed studies of the feeding behaviour of this species in Australia.

The stomach of a bird collected in South Australia contained small gastropod molluscs (256).

## CONSERVATION

The Grey Plover occurs in large numbers in only a handful of sites. Its continued survival depends on adequate preservation and management of these few sites. (See Table 5.4.)

Plate 1

Plate 2

Plate 3

Plate 4

PLATE 5

Plate 7

Plate 10

Plate 11

J N Davies

J.N.Davies

Plate 15

PLATE 16

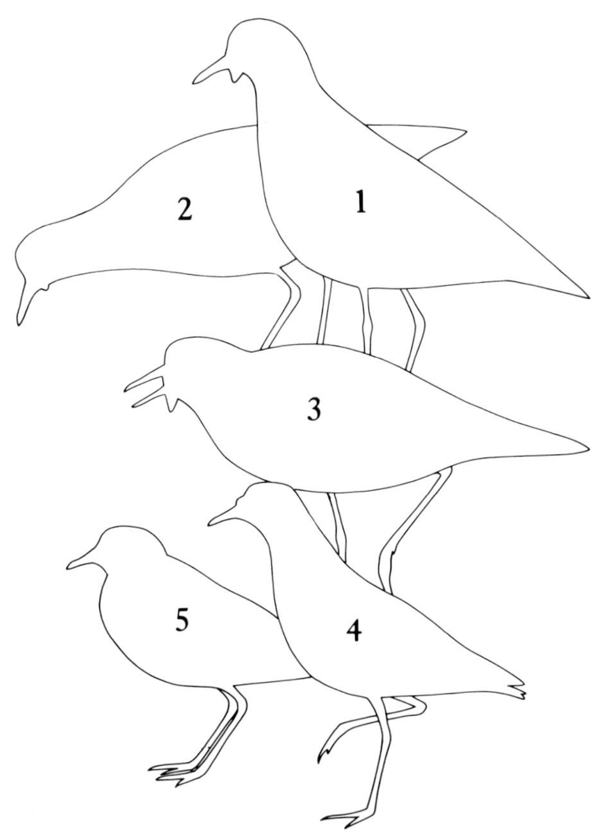

PLATE 1    1−3 Masked Lapwing, 1 adult *novaehollandiae*, 2 juvenile *novaehollandiae*, 3 adult *miles*; 4−5 Banded Lapwing, 4 adult, 5 juvenile.

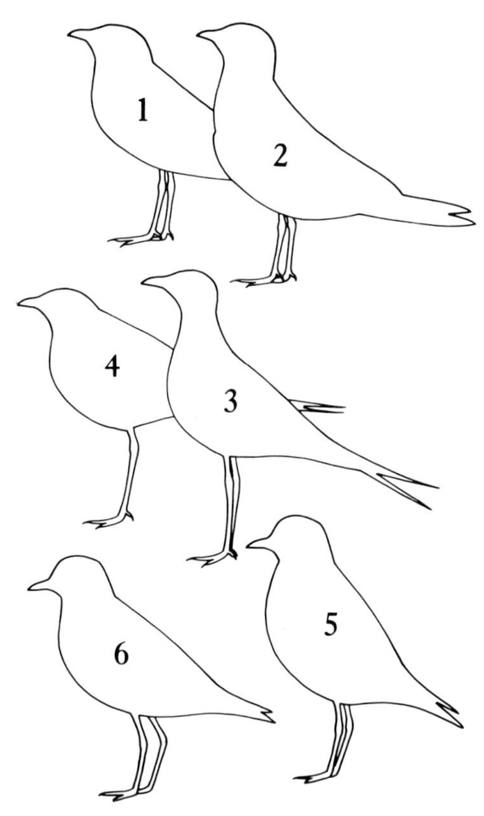

PLATE 3    1−2 Oriental Pratincole, 1 adult non-breeding, 2 adult breeding; 3−4 Australian Pratincole, 3 adult, 4 juvenile; 5−6 Inland Dotterel, 5 adult (breeding?), 6 juvenile (non-breeding?).

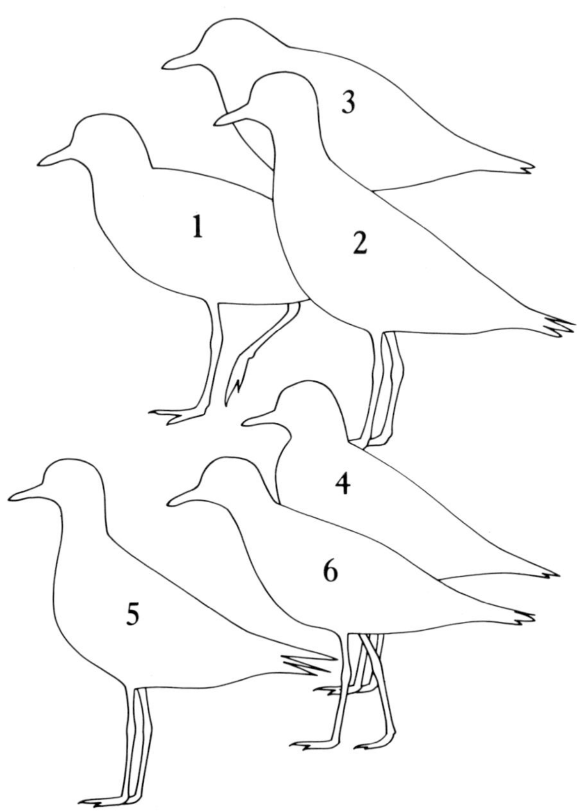

PLATE 2    1−3 Grey Plover, 1 adult non-breeding, 2 adult breeding, 3 juvenile; 4−6 Lesser Golden Plover, 4 adult non-breeding, 5 adult breeding, 6 juvenile.

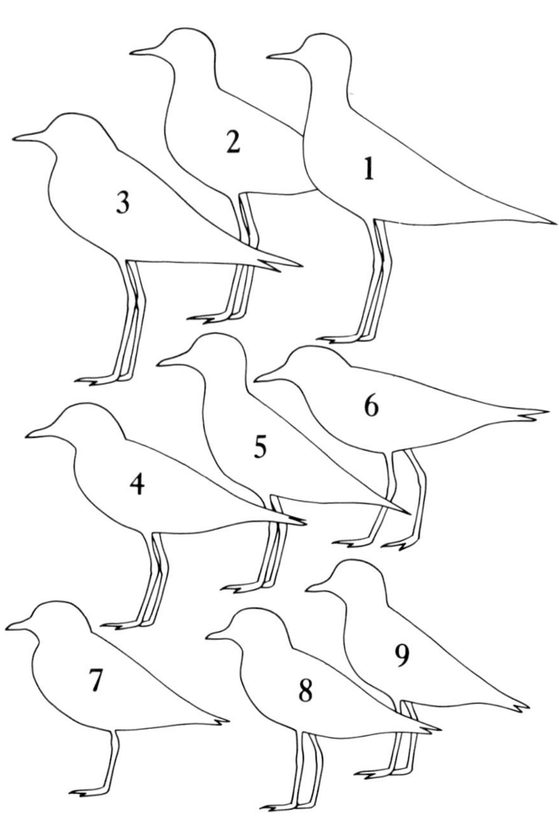

PLATE 4    1−3 Oriental Plover, 1 adult non-breeding, 2 adult breeding, 3 juvenile; 4−6 Large Sand Plover, 4 adult non-breeding, 5 adult breeding, 6 juvenile; 7−9 Mongolian Plover, 7 adult non-breeding, 8 adult breeding, 9 juvenile.

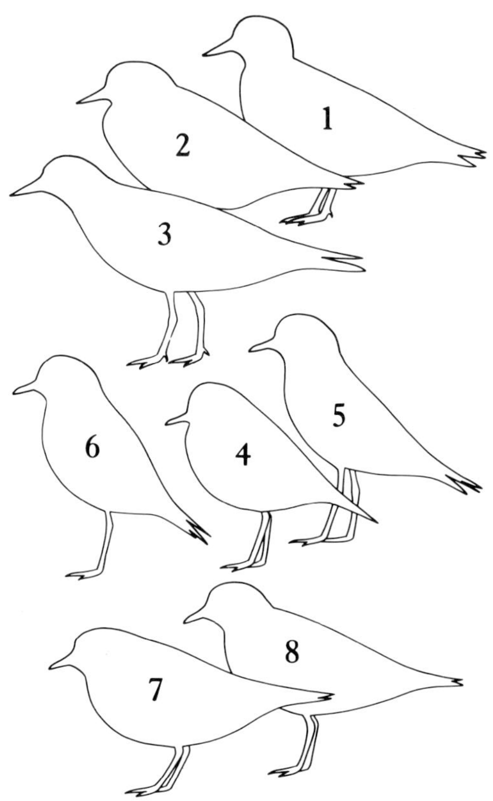

Plate 5    1−3 Ruddy Turnstone, 1 adult non-breeding, 2 adult breeding, 3 juvenile; 4−6 Double-banded Plover, 4 adult non-breeding, 5 adult breeding, 6 juvenile; 7−8 Hooded Plover, 7 adult, 8 juvenile.

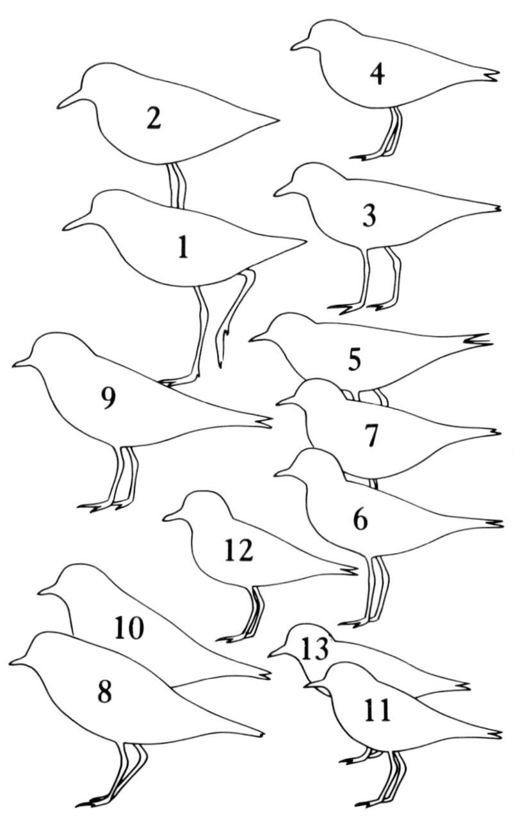

Plate 7    1−2 Red-kneed Dotterel, 1 adult, 2 juvenile; 3−4 Black-fronted Plover, 3 adult, 4 juvenile; 5−7 Little Ringed Plover, 5 adult non-breeding, 6 adult breeding, 7 juvenile; 8−10 Ringed Plover, 8 adult non-breeding, 9 adult breeding, 10 juvenile; 11−13 Red-capped Plover, 11 adult female, 12 adult male, 13 juvenile.

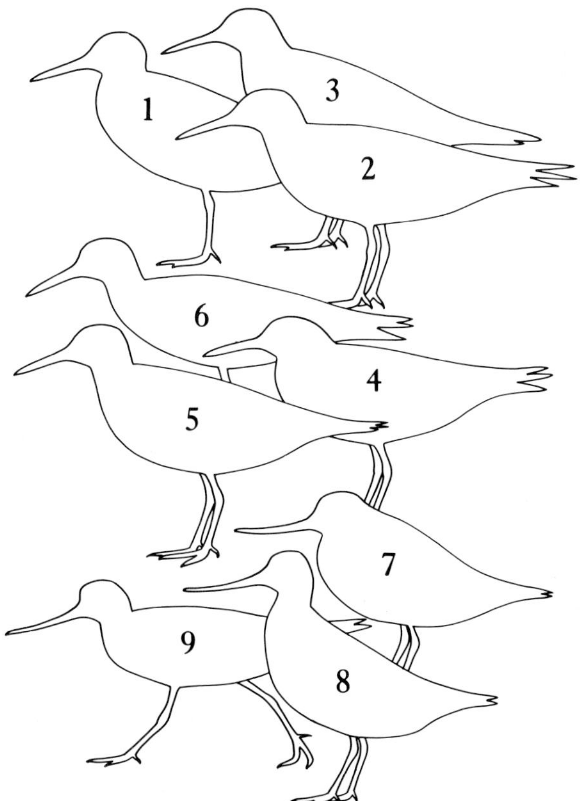

Plate 6    1−3 Wandering Tattler, 1 adult non-breeding, 2 adult breeding, 3 juvenile; 4−6 Grey-tailed Tattler, 4 adult non-breeding, 5 adult breeding, 6 juvenile; 7−9 Terek Sandpiper, 7 adult non-breeding, 8 adult breeding, 9 juvenile.

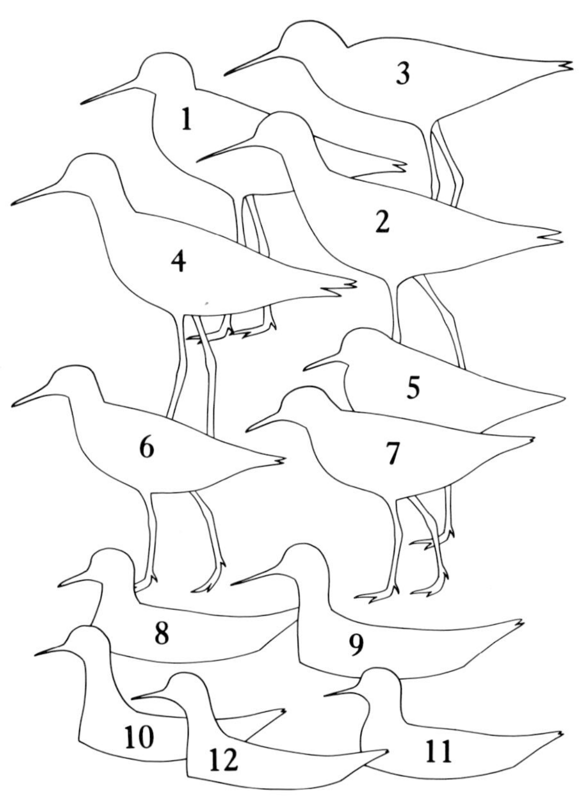

Plate 8    1−3 Marsh Sandpiper, 1 adult non-breeding, 2 adult breeding, 3 juvenile; 4 Lesser Yellowlegs, adult non-breeding, 5−7 Wood Sandpiper, 5 adult non-breeding, 6 adult breeding, 7 juvenile; 8 Grey Phalarope, adult non-breeding; 9 Wilson's Phalarope, adult non-breeding; 10−12 Red-necked Phalarope, 10 adult non-breeding, 11 adult breeding, 12 juvenile.

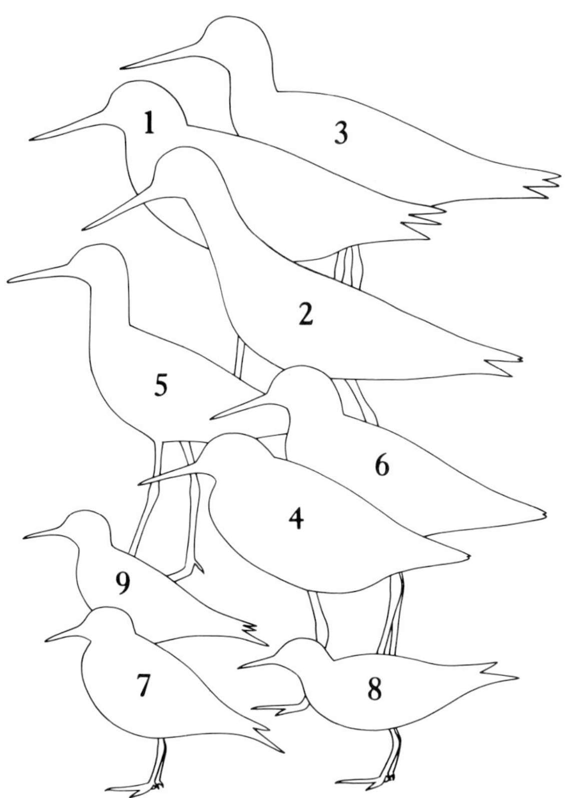

PLATE 9    1−3 Greenshank, 1 adult non-breeding, 2 adult breeding, 3 juvenile; 4−6 Redshank, 4 adult non-breeding, 5 adult breeding, 6 juvenile; 7−9 Common Sandpiper, 7 adult non-breeding, 8 adult breeding, 9 juvenile.

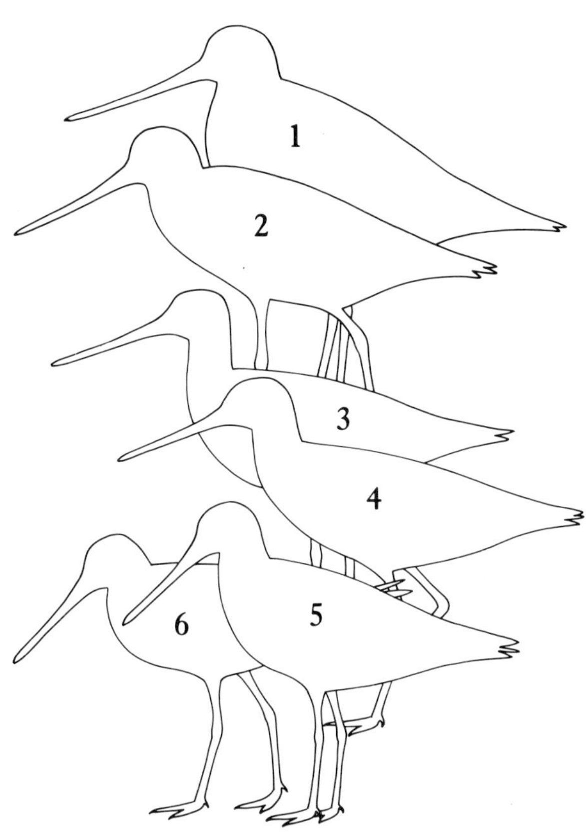

PLATE 11    1−2 Bar-tailed Godwit, 1 adult non-breeding, 2 juvenile; 3−4 Black-tailed Godwit, 3 adult non-breeding, 4 juvenile; 5−6 Asian Dowitcher, 5 adult non-breeding, 6 juvenile.

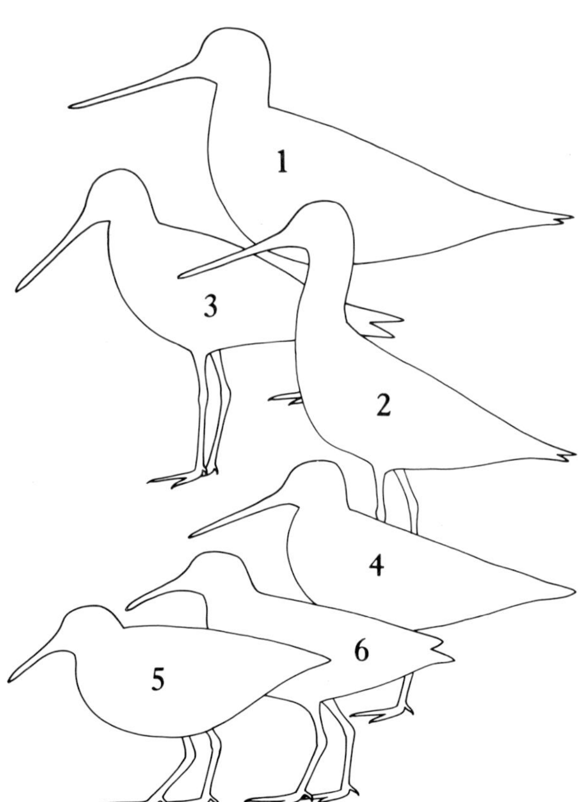

PLATE 10    1 Bar-tailed Godwit, adult breeding; 2 Black-tailed Godwit, adult breeding; 3 Asian Dowitcher, adult breeding; 4 Latham's Snipe, adult non-breeding; 5−6 Painted Snipe, 5 adult female, 6 adult male.

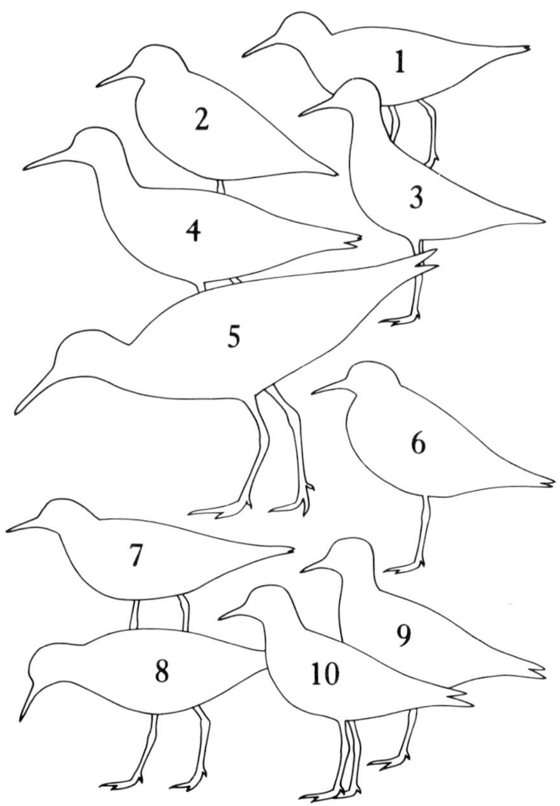

PLATE 12    1−3 Pectoral Sandpiper, 1 adult non-breeding, 2 adult breeding, 3 juvenile; 4−5 Ruff, 4 adult breeding (female or Reeve), 5 juvenile (male or Ruff); 6−8 Sharp-tailed Sandpiper, 6 adult non-breeding, 7 adult breeding, 8 juvenile, 9−10 Buff-breasted Sandpiper, 9 adult, 10 juvenile.

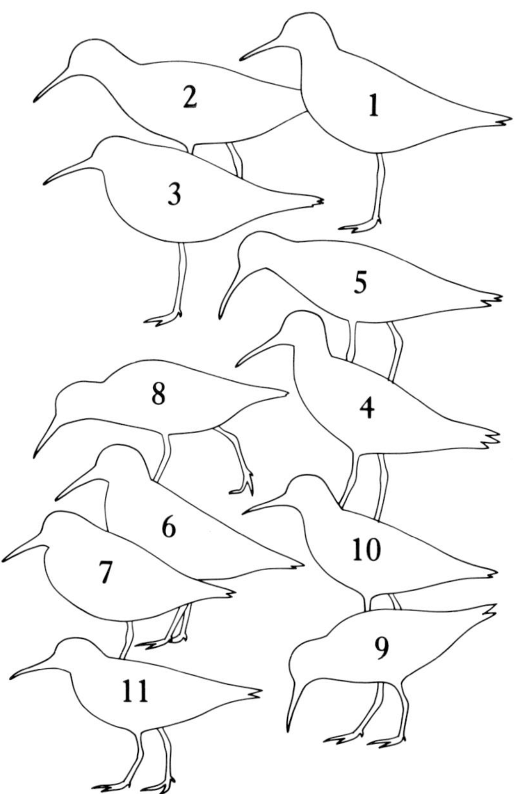

PLATE 13   1–2 Cox's Sandpiper, 1 adult non-breeding, 2 adult breeding; 3–5 Curlew Sandpiper, 3 adult non-breeding, 4 adult breeding, 5 juvenile; 6–8 Dunlin, 6 adult non-breeding, 7 adult breeding, 8 juvenile; 9–11 Broad-billed Sandpiper, 9 adult non-breeding, 10 adult breeding, 11 juvenile.

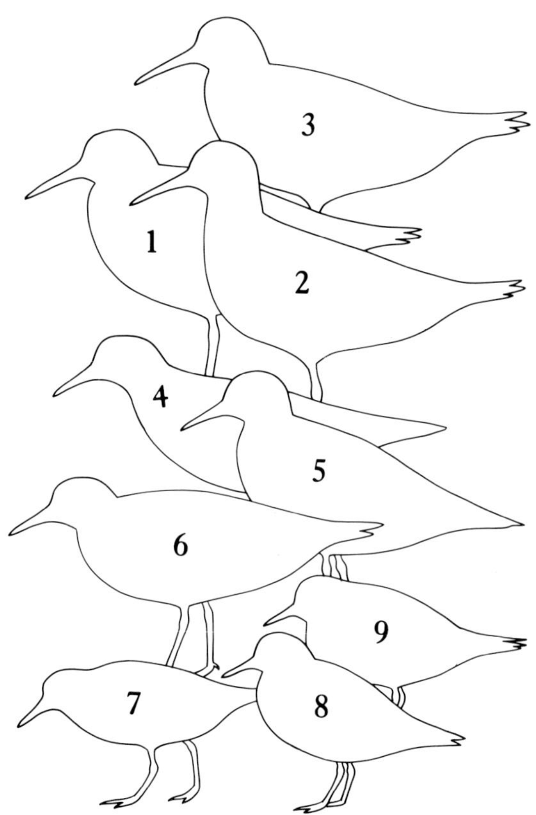

PLATE 15   1–3 Great Knot, 1 adult non-breeding, 2 adult breeding, 3 juvenile; 4–6 Red Knot, 4 adult non-breeding, 5 adult breeding, 6 juvenile; 7–9 Sanderling, 7 adult non-breeding, 8 adult breeding, 9 juvenile.

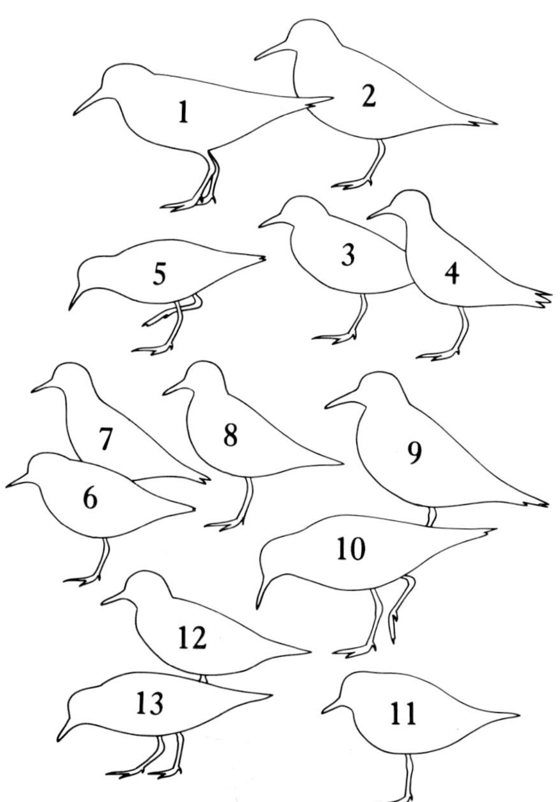

PLATE 14   1–2 Baird's Sandpiper, 1 adult non-breeding, 2 juvenile; 3–5 Long-toed Stint, 3 adult non-breeding, 4 adult breeding, 5 juvenile; 6–8 Little Stint, 6 adult non-breeding, 7 adult breeding, 8 juvenile; 9–10 White-rumped Sandpiper, 9 adult non-breeding plumage, 10 juvenile; 11–13 Red-necked Stint, 11 adult non-breeding, 12 adult breeding, 13 juvenile.

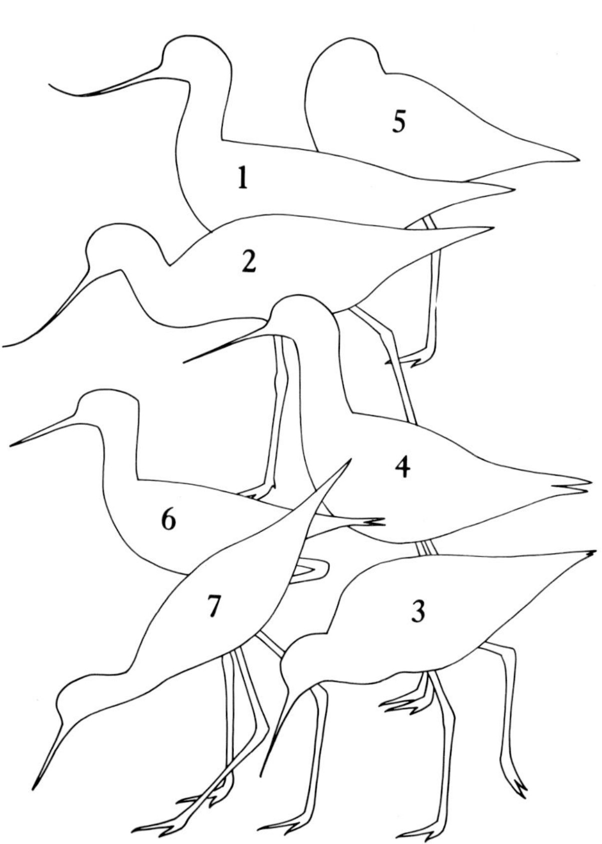

PLATE 16   1–2 Red-necked Avocet. 1 adult female, 2 adult male; 3–5 Banded Stilt, 3–4 adult, 5 juvenile; 6–7 Black-winged Stilt, 6 adult, 7 juvenile.

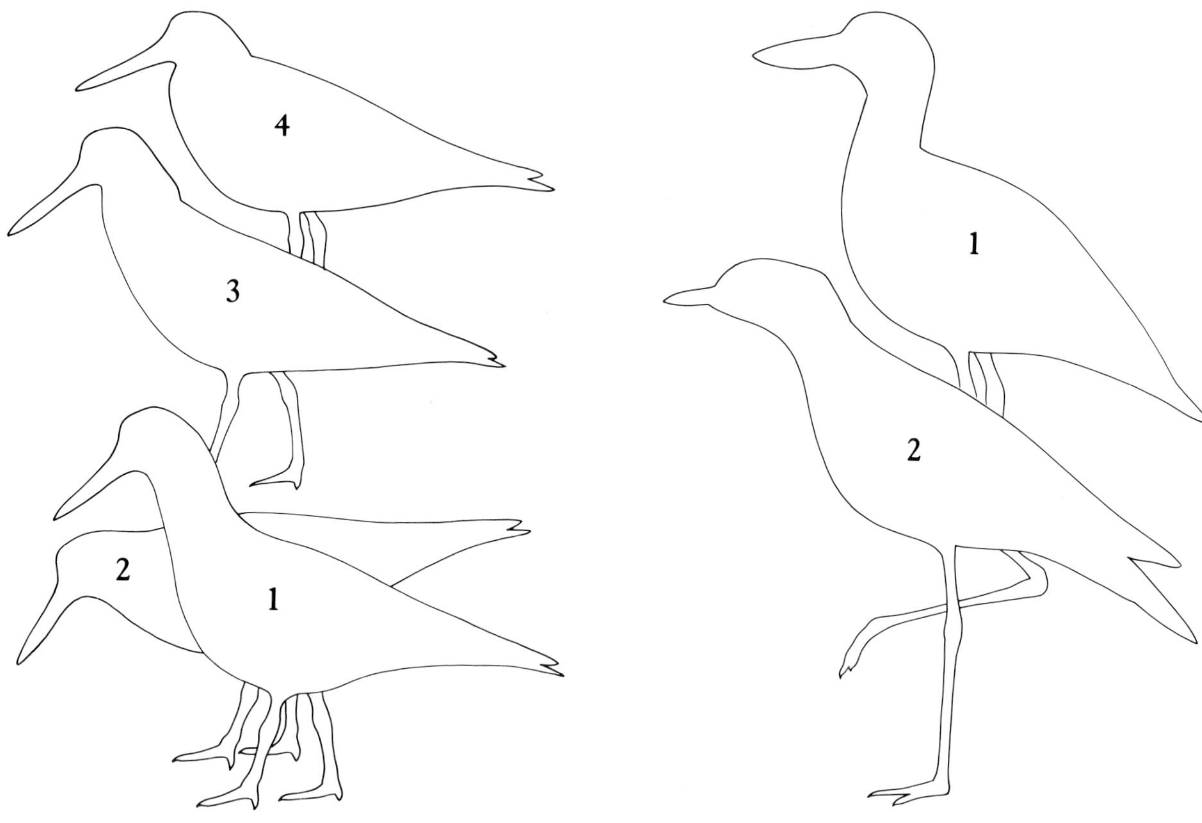

PLATE 17    1 – 2 Sooty Oystercatcher, 1 adult, 2 juvenile;
3 – 4 Pied Oystercatcher, 3 adult, 4 juvenile.

PLATE 18    1 Beach Thick-knee, adult; 2 Bush Thick-knee,
adult.

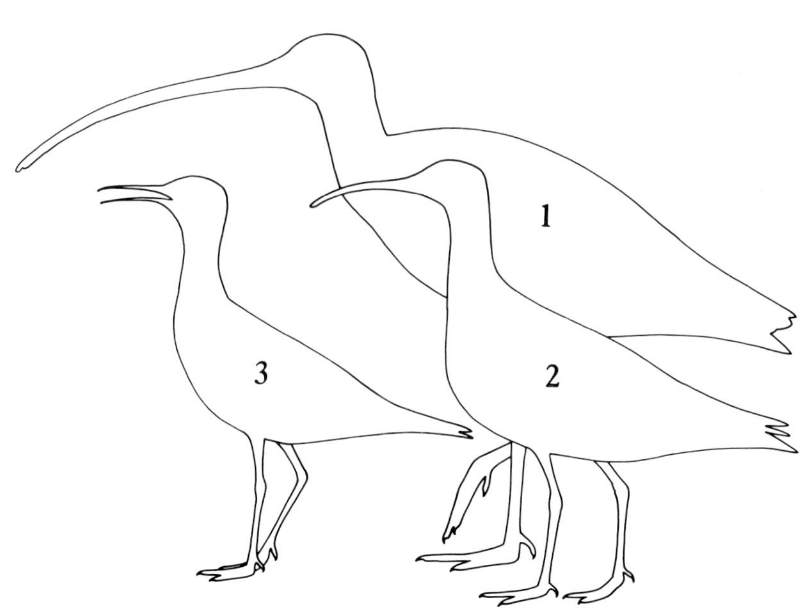

PLATE 19    1 Eastern Curlew, adult non-breeding,
2 Whimbrel, adult non-breeding, 3 Little Curlew, adult
non-breeding.

Plate 18

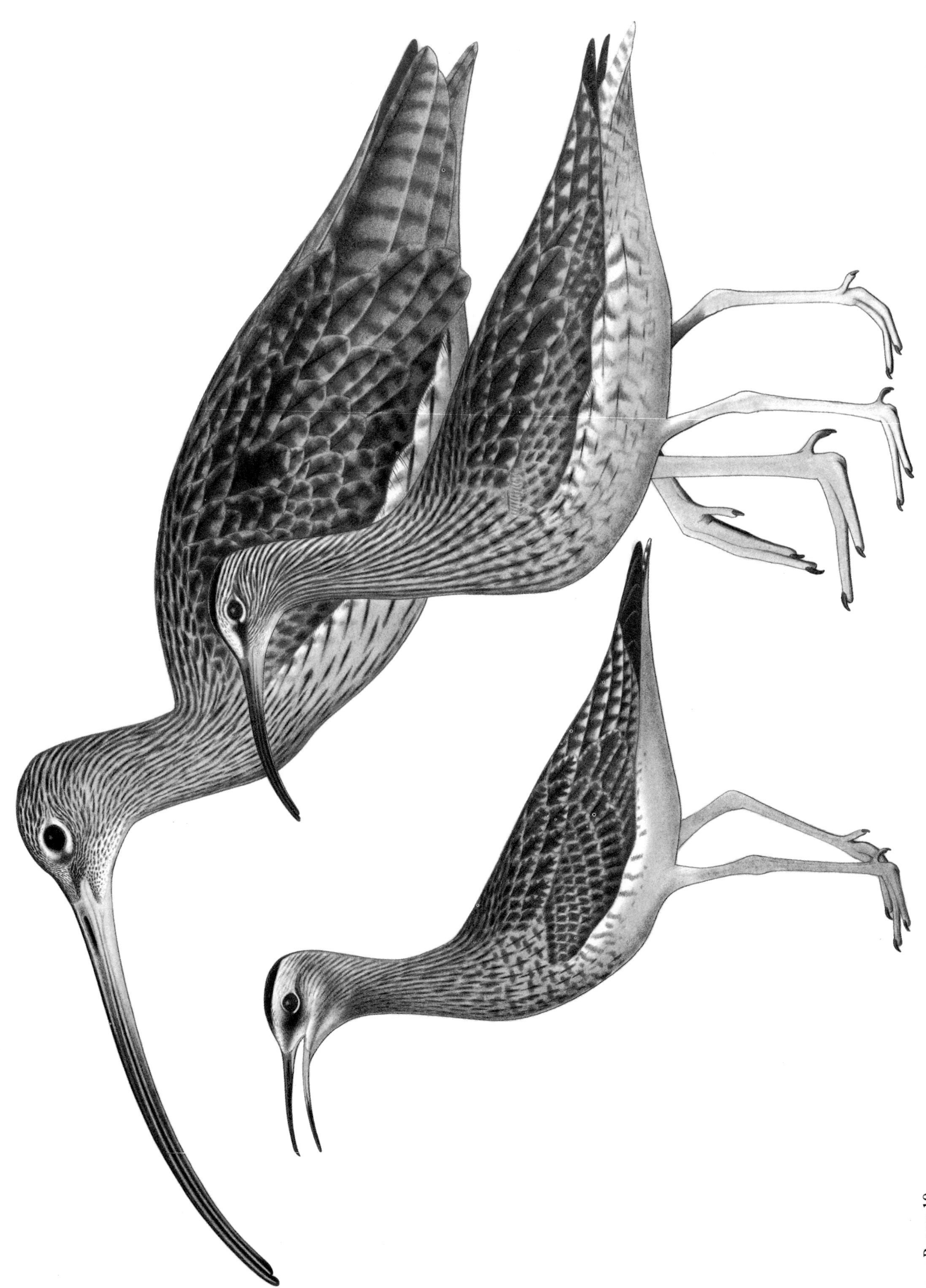

PLATE 10

# LESSER GOLDEN PLOVER
## *Pluvialis fulva*
(Plate 2)

**Other names** Eastern Golden, American Golden, Asiatic Golden or Pacific Plover.

Until recently this species was considered a race of *Pluvialis dominica*: subspecies *fulva*. It has been recommended that *P. dominica* be separated into two species: *P. fulva*, which migrates to Australia, Asia and the Pacific Islands, and *P. dominica*, which migrates to South America (68).

## DESCRIPTION

A large to medium-sized, golden-brown plover, with striking black, white and gold breeding plumage.

**Adult non-breeding plumage** Crown and nape, dark brown, streaked yellow; supercilium, lores and ear-coverts, golden-cream, sparingly streaked brown; mantle, dark brown, feathers edged yellow; scapulars, brown, notched and edged cream-yellow; wing-coverts, brown, notched cream and yellow. Underparts, generally cream with variable golden-brown blotching and streaking on breast and flanks; vent, white. Iris, black; bill, short and slender with slight bulge towards tip, black; legs and feet, grey. Sexes similar.

**In flight** Back and rump, brown, spotted yellow; wings and tail generally brown *without distinctive markings*; underparts paler; underwing-coverts, grey-brown.

**Voice** Whistles including 'tloo-ee' and hoarse trilling: 'chree chree chree …'

**Breeding Plumage** Upperparts similar to non-breeding plumage, although more *strikingly speckled brighter yellow and black*; forehead, supercilium and sides of neck and breast, white, contrasting with lores, chin, ear-coverts, throat, breast and belly, which are black; flanks and vent, white with large black blotches.

**Juvenile plumage** Generally paler and greyer with more heavily barred and streaked breast and flanks. Feathers of mantle, scapulars, tertials and lesser and median coverts, dark brown, notched cream.

**Field notes** Runs swiftly when feeding in typical stop-start plover fashion; often roosts in low saltmarsh herbage, where it is extremely well camouflaged; calls loudly when disturbed. Birds in plumage transitional between breeding and non-breeding can be seen in August–October and February–March.

**Similar species** Grey Plover; Oriental Plover. The Lesser Golden Plover is generally smaller, browner and darker than the Grey Plover and has a relatively smaller head, more slender bill and longer legs. It is slightly larger, more robust in body and bill, darker brown, more mottled and shorter of wings and legs than the Oriental Plover.

## BREEDING RANGE

Arctic and sub-arctic tundras beyond the tree limit are the favoured breeding habitat of the Lesser Golden Plover. It breeds from western Alaska, west through northern Siberia from the Chukotski Peninsula to the Yamal Peninsula (85). (*P. dominica* breeds in similar habitats in eastern Alaska and northern Canada, as far east as Baffin Island.) It breeds in drier habitats than the Grey Plover.

## ASIAN RANGE

The Lesser Golden Plover migrates southwards from Siberia and Alaska through most of Asia and the Pacific Islands.

Southward migration occurs between August and November in Korea (162), Mongolia, coastal and inland China, Taiwan (255), the Philippines (2) and Papua New Guinea (138, 278). Some Lesser Golden Plovers migrate across Asia to spend the non-breeding months in India, Sri Lanka, south-western Asia and eastern and southern Africa (85). In north-eastern China it is outnumbered by the Grey Plover (179), whereas elsewhere in Asia the Lesser Golden Plover is more abundant. A bird banded in Alaska in September has been recovered in Japan 6 weeks later (267). Major non-breeding areas include Taiwan (401), southern China (255), Vietnam (109, 489), Thailand (257), the Malay Peninsula (348), the Philippines (110), Borneo (440), the Pacific Islands (120, 231, 406), New Zealand (1983: 300 birds; 1984: 368 birds) (387, 388) and Australia. (See Figure 5.8.)

Counts of thousands in parts of Asia (e.g. 5000 estimated maximum in Serangoon estuary, Singapore (174)) suggest that Australia does not hold a large proportion of Asia's Lesser Golden Plovers. In Australia, it occurs mostly on the south-east coast and few have been observed elsewhere, suggesting that most birds that reach Australia are probably those that migrate through the Pacific Islands.

Northward migration occurs from mid-February to May, mostly in April. It arrives on the breeding grounds in late May and early June (367).

## STATUS AND DISTRIBUTION IN AUSTRALIA

The Lesser Golden Plover occurs in largest numbers on the coasts of southern Queensland, New South Wales, Victoria, Tasmania and south-eastern South Australia. Fewer occur on the west coast. Most of the

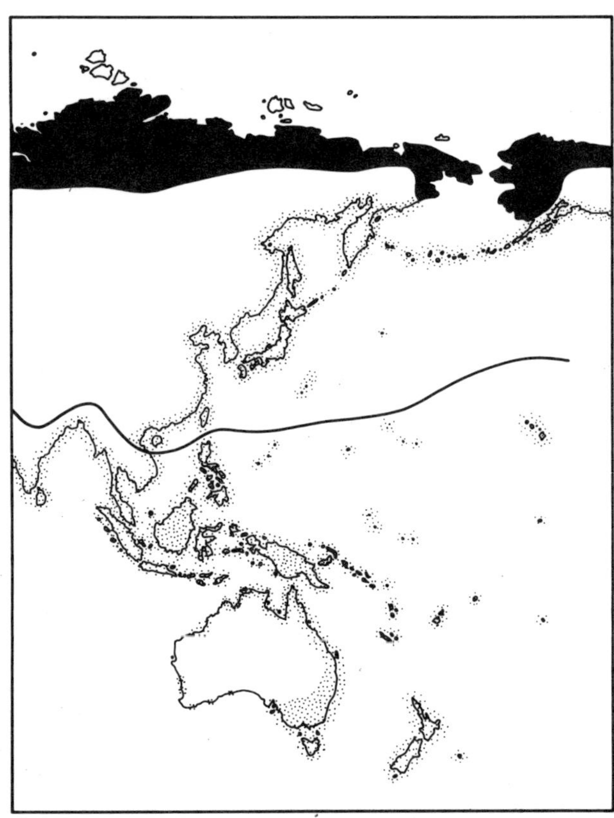

FIG. 5.8. Breeding and non-breeding range of Lesser Golden Plover.

TABLE 5.5. Ranked listing of top twenty zones for Lesser Golden Plover. (maximum counts are marked thus: *; other counts are averages) (50 or more birds)

| ZONE | NO. OF COUNTS | NO. OF INDIVIDUALS |
|---|---|---|
| * SE corner, Gulf of Carpentaria, Qld | 1 | 1 370 |
| * Eighty Mile Beach, WA | 5 | 440 |
| Hunter estuary, NSW | 4 | 410 |
| * N coast, NSW | 5 | 390 |
| * Moreton Bay, Qld | 5 | 360 |
| * The Coorong, SA | 1 | 290 |
| Port Phillip Bay, Vic. | 5 | 220 |
| * Botany Bay, NSW | 5 | 210 |
| SE coast, SA | 5 | 200 |
| Derwent estuary & Pittwater, Tas. | 4 | 180 |
| * Torres Strait, Qld | 3 | 120 |
| Shallow Inlet, Vic. | 5 | 90 |
| * Hervey Bay−Great Sandy Strait, Qld | 5 | 80 |
| Anderson's Inlet, Vic. | 4 | 70 |
| Darwin area, NT | 4 | 70 |
| * Port Stephens, NSW | 4 | 70 |
| Central coast, NSW | 3 | 60 |
| St Vincent Gulf, SA | 4 | 60 |
| Cape Portland coast, Tas. | 5 | 60 |
| Westernport Bay, Vic. | 5 | 50 |

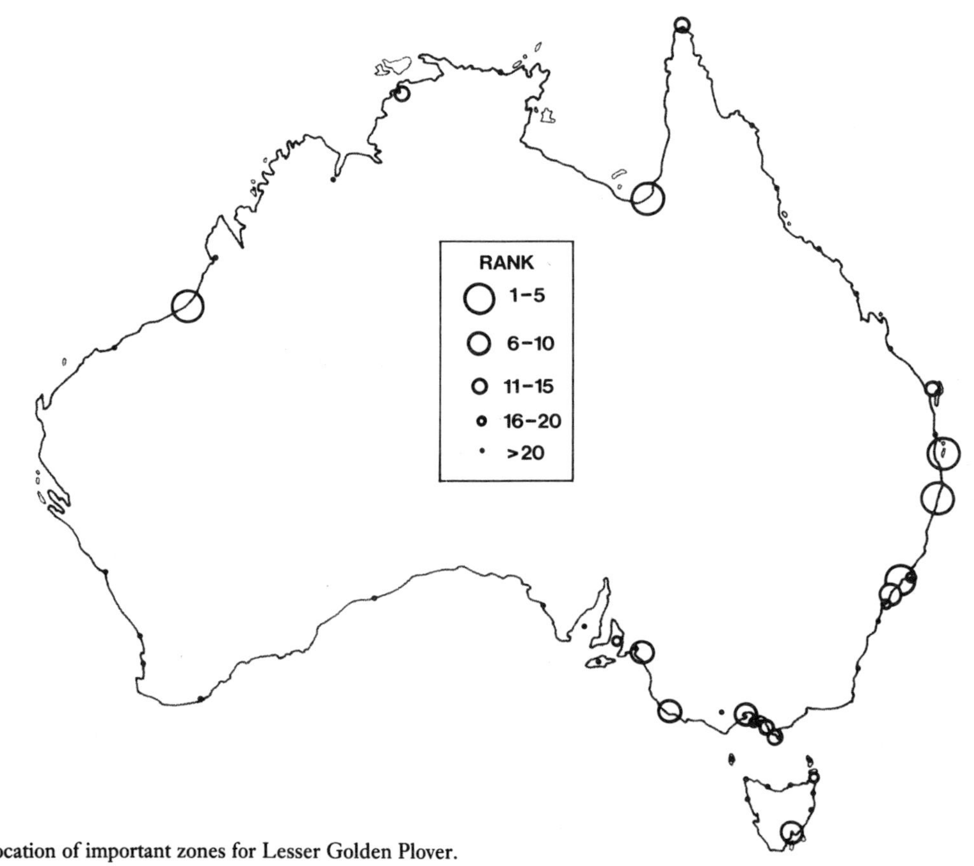



FIG. 5.9. Location of important zones for Lesser Golden Plover.

islands of the Great Barrier Reef were not covered, but observations suggest that it is as abundant there as on the mainland coast farther south (50). Largest numbers occur on the east coast, whereas most Grey Plovers occur in other parts of Australia (see comments for that species). (See Table 5.5, Figure 5.9.)

The Lesser Golden Plover occurs predominantly on coastal mudflats and near-coastal saltmarsh and pasture. At high tide it roosts in saltmarsh and short pasture, often feeding in these habitats as well. It also feeds on rocky shores, such as those near Sydney and in south-western Victoria and south-eastern South Australia, where it roosts at high tide on sandy beaches or rocks. It occurs on coral reef flats on the islands of the Great Barrier Reef.

## MOVEMENTS IN AUSTRALIA

The Lesser Golden Plover occurs during September and October on the east coast from Torres Strait, southwards to at least Mackay. It arrives in southern New South Wales, Victoria and Tasmania in September and in south-eastern South Australia in October and November. Small numbers move through inland south-eastern Australia between September and November. These results suggest that the Lesser Golden Plover moves southwards along the east coast between September and November, arriving later farther west along the south coast.

Numbers are fairly stable in south-eastern Australia from November to January, except in Victoria, where temporary influxes occur at some sites in December.

In late February and March, smaller numbers of birds move northwards up the east coast. No birds move through the inland. This suggests that most migrate northwards from south-eastern Australia without stopping in northern Australia. Birds that depart southern Victoria have fattened sufficiently to fly directly out of Australia (476).

## FEEDING AND FOOD

The Lesser Golden Plover forages individually on intertidal sand and mudflats, airfields, playing fields and rocky shores. It also forages on coral reef flats, where it locates its prey by sight (115). Its diet consists of molluscs and insects, including ants, beetles (adult and larvae) and caterpillars (61, 62, 256), as well as crabs and small crustacea (115).

## CONSERVATION

The Lesser Golden Plover is one of Australia's least abundant migratory shorebirds. It is more vulnerable than other species, as most sites holding larger numbers lie along the south-east coast of Australia, where the most people, industry and residential development occurs, so it is important that these areas are adequately protected and managed to ensure its survival.

# RED-KNEED DOTTEREL
*Erythrogonys cinctus*
(Plate 7)

**Other name**  Sandpiper.
No races recognised.

## DESCRIPTION

A small to medium-sized plover of freshwater wetlands with long legs and black, white and brown plumage.

**Adult plumage**  Crown, nape, hindneck, lores, supercilium and ear-coverts, black; mantle, scapulars and wing-coverts, dark brown. *Chin and throat, white; breast has thick black band continuing onto flanks, which are chestnut*; belly and vent, white. Iris, black; orbital ring, pink; bill, short, straight, pinkish-purple with black tip; legs, grey, pink at 'knees'. Sexes similar.

**In flight**  Back, brown; rump and tail, white with central dark brown line; tips of secondaries and inner primaries, white, showing as *white trailing edge to wing*; rest of wings, dark brown.

**Voice**  Various bubbling whistles when disturbed; slowly repeated 'chip chip'.

**Juvenile plumage**  Head and rest of upperparts, reddish-brown; chin, throat, breast, belly, flanks and vent, white; no black breast-bar; bill, dark brown.

**Field notes**  Behaves like most plovers, but more likely to wade and feed in shallow water; occurs mostly on freshwater wetlands; bobs head when approached, flying away with bubbling whistle call.

**Similar species**  Black-fronted Plover. The Red-kneed Dotterel is larger, has completely black crown, nape and hindneck, uniform dark brown upperparts and longer darker legs.

## STATUS AND DISTRIBUTION IN AUSTRALIA

The Red-kneed Dotterel is found mainly on freshwater wetlands, ephemeral and permanent, throughout most of Australia, but generally in lower numbers east of the Great Dividing Range. (See Figure 5.10, Table 5.6.)

It feeds mostly in shallow water or wet mud, generally avoiding dry shoreline. It lacks the nasal salt-secretion glands that most shorebirds have (292) and is rare on saline wetlands.

## MOVEMENTS IN AUSTRALIA

*Atlas* project records from Victoria (27) show that the Red-kneed Dotterel is recorded there most frequently in spring and summer, the usual breeding time (385).

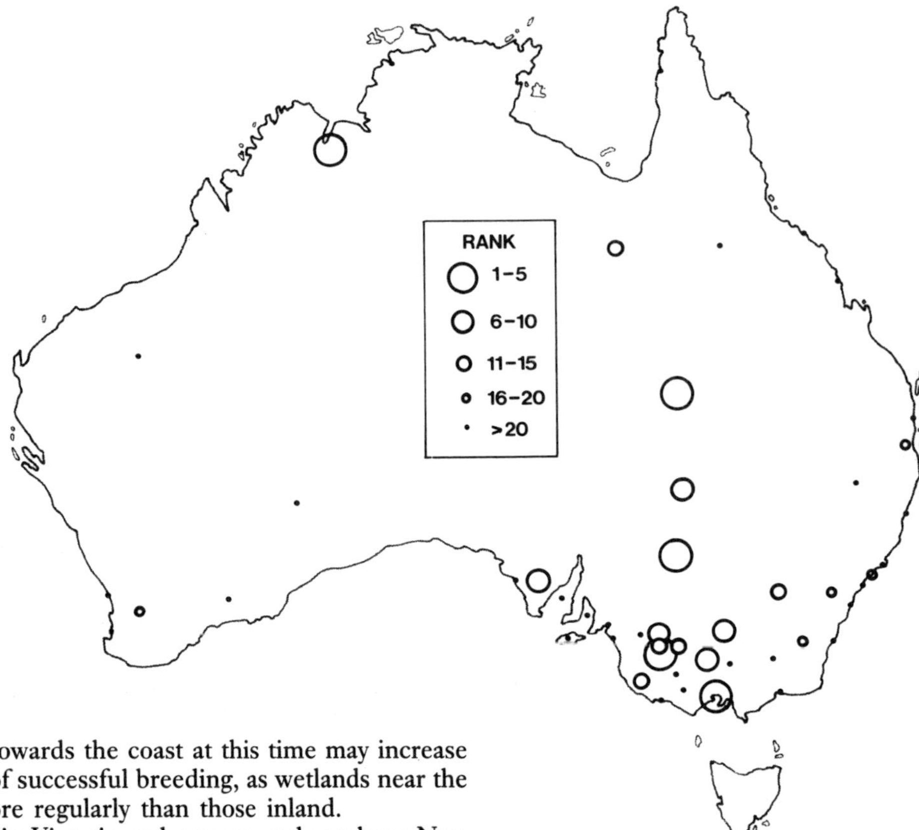



FIG. 5.10.   Location of important zones for Red-kneed Dotterel.

Movement towards the coast at this time may increase the chance of successful breeding, as wetlands near the coast fill more regularly than those inland.

Numbers in Victoria and eastern and southern New South Wales increased during the 1982−83 drought, but when it broke, and the Darling and Cooper basins flooded, the Red-kneed Dotterel virtually deserted this part of Australia. (See Figure 5.11.) By summer 1984, when the floods receded, large numbers reappeared in Victoria and southern New South Wales; many were young birds, suggesting that much breeding had occurred. In 1984 the Red-kneed Dotterel was counted in Darwin for the first time since counts began in 1981. At the same time a flock of 3000, including many juvenile birds, appeared on the Ord River in the Kimberleys.

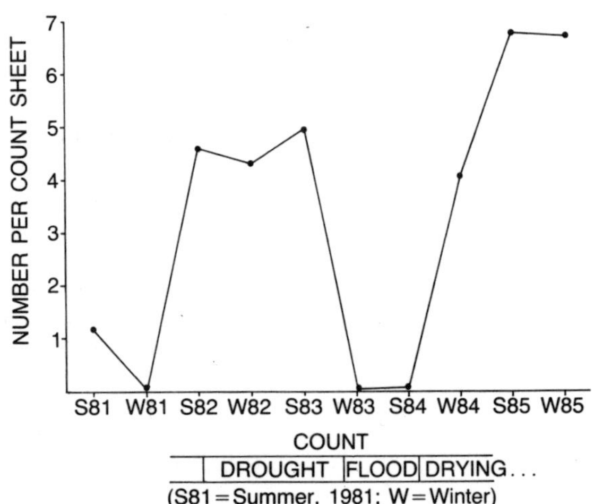



FIG. 5.11.   Numbers per count sheet for Red-kneed Dotterel. (Victoria and New South Wales)

TABLE 5.6.   Ranked listing of top twenty zones for Red-kneed Dotterel. (all counts are averages) (20 or more birds)

| ZONE | NO. OF COUNTS | NO. OF INDIVIDUALS |
|---|---|---|
| Kimberley region, WA | 6 | 750 |
| Lower Darling region, NSW | 6 | 240 |
| Port Phillip Bay, Vic. | 10 | 160 |
| Horsham area lakes, Vic. | 10 | 120 |
| Far SW lakes, Qld | 1 | 100 |
| Riverina, NSW | 7 | 90 |
| Mildura area, Vic. | 8 | 90 |
| Bendigo region, Vic. | 8 | 60 |
| Far NW lakes, NSW | 4 | 50 |
| Inland Eyre Peninsula, SA | 1 | 50 |
| Parkes region, NSW | 8 | 40 |
| NW lakes, Vic. | 3 | 30 |
| Mt Isa region, Qld | 8 | 30 |
| SE inland lakes, SA | 9 | 30 |
| Kerang−Swan Hill, Vic. | 10 | 30 |
| Brisbane area, Qld | 10 | 30 |
| Hunter estuary, NSW | 8 | 20 |
| Central tablelands, NSW | 7 | 20 |
| Southern tablelands, NSW | 3 | 20 |
| SW lakes, WA | 7 | 20 |

These observations suggest that the Red-kneed Dotterel moves into northern areas after prolific inland breeding, whereas wetlands in southern Australia are used as drought refuges and, in spring and summer, as reliable breeding areas.

## BREEDING

The Red-kneed Dotterel builds a saucerlike nest of plant stems on damp ground where it is susceptible to flooding if water-levels rise. The nest is often located under clumps of wetland plants (292), a nest site different to that of most of the small plovers, whose nests are usually placed in the open. The usual clutch is 4 eggs, sometimes 2 or 3, rarely 5, and they breed as individual pairs or in loose colonies, often within colonies of other breeding shorebirds. Breeding occurs between September and March, most records being from October to December (385).

There have been no detailed studies of its breeding biology, so the incubation and fledging periods and usual breeding success are not known.

Causes of nesting failure include human disturbance, trampling by domestic animals, flooding or drying of wetlands and predation by native and introduced predators (385).

## FEEDING

The Red-kneed Dotterel feeds by pecking and probing into sand and mud in shallow water. No detailed studies have been made of its feeding behaviour or diet. Small beetles and a midge larva were in the stomach of a bird found dead (292).

## CONSERVATION

Red-kneed Dotterels concentrate in largest numbers during droughts, so an adequate network of drought refuges in southern Australia is required for their long-term survival.

# HOODED PLOVER
## *Charadrius rubricollis*
(Plate 5)

**Other name**   Hooded Dotterel; Hoody.
   No races recognised.

## DESCRIPTION

A medium-sized sandy-brown plover with a black head, found on ocean beaches and, in Western Australia, on salt lakes.

**Adult plumage**   *Head, black; nape, white; hindneck and front of mantle, black,* continuous with broad black line on sides of breast; rest of mantle, scapulars and wing-coverts, pale sandy-brown. Breast, belly, flanks and vent, white. Iris, dark brown; orbital ring, red; bill, short, slender and red with black tip; legs, pink. Sexes similar.

**In flight**   Back, sandy-brown; rump and tail, white, with broad dark brown central line; greater coverts, dark grey; primaries and secondaries, dark grey with white bases showing as *prominent wing-bar.*

**Voice**   Quiet piping calls and deep 'chook-chook' call.

**Juvenile plumage**   Like adult plumage, but without black head, hindneck and front of mantle, which are sandy-brown; feathers of upperparts, edged white; bill, brown with slightly orange base; legs, pale orange.

**Field notes**   Seen in pairs and small parties on ocean beaches, where it often crouches next to washed-up seaweed and other debris; runs swiftly ahead of people. In south-western Australia, it is found in salt lakes, sometimes in flocks of hundreds.

**Similar species**   Sanderling. The juvenile Hooded Plover is distinguished from the Sanderling by having brown, more uniformly marked plumage and behaving in a plover-like fashion.

## STATUS AND DISTRIBUTION IN AUSTRALIA

Much of the suitable habitat for the Hooded Plover was not adequately covered during national counts, so additional surveys have been organised during October, when birds breed, to determine their population and distribution. The results presented in Figure 5.12 and Table 5.7 are a synthesis of surveys in Victoria, (1980, 1982, 1984), Tasmania (1982) (336), southern New South Wales (1980, 1982), South Australia (1982, 1983) (41) and Western Australia (1982), together with results of the national counts in some

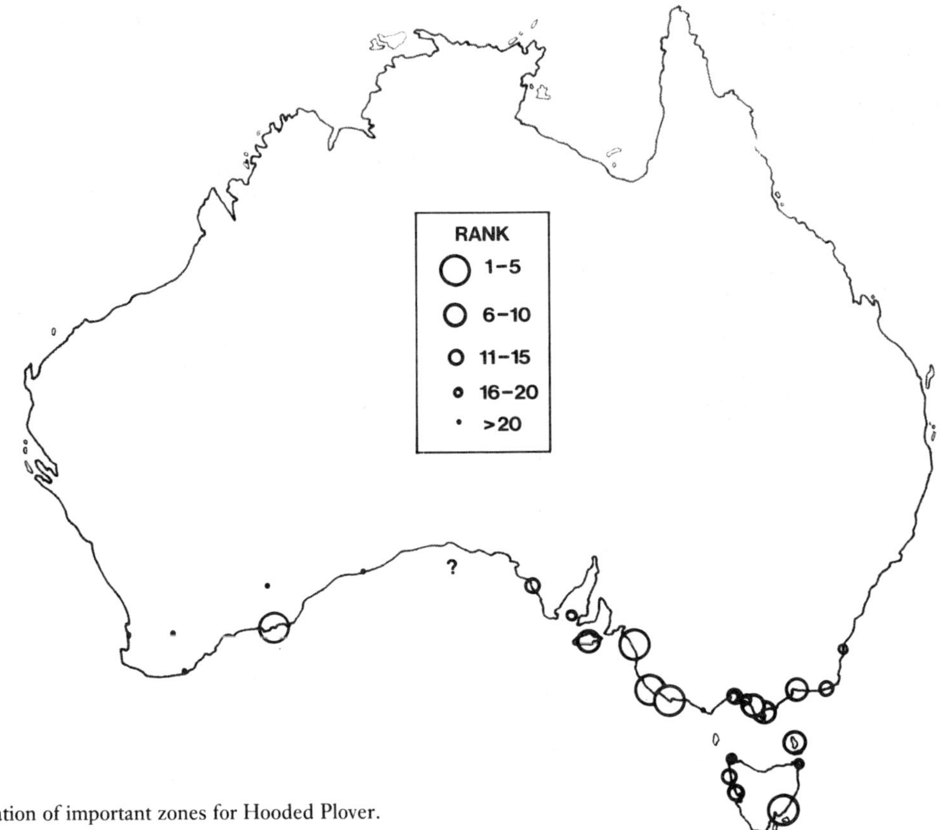


FIG. 5.12.   Location of important zones for Hooded Plover.

places and observations at other times (214) for areas not covered in these surveys.

The Hooded Plover occurs on sandy ocean beaches between Jervis Bay, New South Wales, and the Eyre Peninsula in South Australia; in Tasmania; and between Esperance and Perth in south-western Australia. In Victoria, highest densities occur on beaches with large amounts of beach-washed seaweed and, to a lesser extent, on those backed by extensive open dunes. Densities are lowest on narrow, steep beaches, where there are few or no dunes, and where human activities are most intensive. In the south-west, they also occur on inland salt lakes.

Hooded Plovers are not abundant, and the results of counts give an estimated minimum population of 1800. This is probably an underestimate, as ocean beaches in western South Australia and beaches and inland lakes in Western Australia have not been completely covered during the surveys. Another factor contributing to this is that counts were made while walking along beaches, and the Hooded Plover often shelters in sparsely vegetated dunes, where it can be difficult to detect (393). In eastern Australia, the species rarely congregates in flocks larger than 40 birds and then only in the nonbreeding season (March–July). It occurs in comparatively low densities on ocean beaches (average densities: Victoria, 1.0 per km; Tasmania, 1.7 per km; South Australia, 0.4 per km). In the west, densities are even lower (0.2 per km). Highest densities occur on beaches in southern Tasmania (up to 3.5 per km) and

TABLE 5.7. Ranked list of top twenty zones for Hooded Plover. (maximum counts marked thus: *; other counts are averages) (17 birds or more)

| ZONE | NO. OF COUNTS | NO. OF INDIVIDUALS |
| --- | --- | --- |
| * Esperance lakes, WA | 10 | 480 |
| SW coast, Vic. | 3 | 210 |
| * SE coast, Tas. | 1 | 190 |
| * SE coast, SA | 10 | 140 |
| * Murray mouth to Kingston, SA | 10 | 120 |
| Wilson's Promontory, Vic. | 3 | 80 |
| * Ninety Mile Beach, Vic. | 1 | 75 |
| * Kangaroo Island, SA | 10 | 75 |
| Venus Bay, Vic. | 3 | 65 |
| * Furneaux Islands, Tas. | 1 | 60 |
| * Ocean Beach, Strahan, Tas. | 1 | 60 |
| * W coast, Eyre Peninsula, SA | 10 | 40 |
| E Gippsland coast, Vic. | 8 | 30 |
| * Central W coast, Tas. | 1 | 30 |
| Mornington Peninsula, Vic. | 3 | 30 |
| * S coast, NSW | 5 | 30 |
| Phillip Island, Vic. | 3 | 20 |
| * NW coast, Tas. | 1 | 18 |
| * Cape Portland coast, Tas. | 1 | 17 |
| * Spencer Gulf, SA | 9 | 17 |

western Victoria (up to 4.4 per km). In the Esperance area of south-western Australia, up to 300 congregate on inland salt lakes in autumn and early winter (214).

In the 1930s and 1940s it occurred in New South Wales as far north as Port Stephens (254), but the species has since declined and is now considered endangered in that State (183).

## Movements in Australia

There is little detailed information on the movements of the Hooded Plover. On Venus Bay and Darby Beach in Victoria, numbers were consistently higher during winter (395), indicating movement into and out of these areas. Local movements to salt lakes immediately behind beaches occur in winter in the east. In the west, it moves from the coast to salt lakes some distance inland in winter (38).

## Breeding

The Hooded Plover excavates a shallow scrape in sand or fine gravel situated above the high-tide mark on ocean beaches or amongst dunes in which to lay its eggs. It is sometimes lined with shell fragments or occasionally with small twigs (385). The clutch is usually 3 eggs (58 per cent of records) but often only 2 (30 per cent) or, less frequently, 1 (12 per cent) (385). Usually only 1 or 2 eggs hatch, after approximately 30 days of incubation, and the downy young leave the nest within a day or two (385). Its incubation period is longer than that of other Australasian-breeding plovers, at about 30 days (333).

In south-eastern South Australia in 1982–83, 18 juvenile Hooded Plovers fledged from an estimated breeding population of 65 pairs, but this is only an approximate measure of breeding success (41). At Phillip Island in 1982, only 1 out of 12 breeding pairs successfully raised young (a single chick), the rest failing due to human disturbance (102). There are no detailed studies of its breeding biology.

Factors recorded as causing nesting failure include flooding by extra-high tides, trampling by stock, native animals and people, crushing by vehicles and predation of eggs by foxes (385, 393). Frequent disturbance of incubating adults leaves eggs exposed to predation by gulls, ravens and other scavengers (41).

## Feeding and Food

The Hooded Plover forages at all levels of the beach during all tide phases, but its feeding behaviour differs depending on the level of beach it is at, the phase of tide and daily and tidal invertebrate prey activity cycles (394).

Its diet includes insects, sandhoppers (*Orchestria* sp.), small bivalve molluscs, soldier crabs (*Mictyris*

*platycheles*) and amphipod crustacea. On rocky reefs it has been observed taking the small gastropod mollusc *Littorina unifasciata* (393).

## Conservation

The continued survival of the Hooded Plover depends on undisturbed beaches so that it can breed successfully. In South Australia, where off-road vehicles are allowed on beaches, densities average 0.4 per km, whereas in Victoria, where vehicles are not permitted on beaches, the density is 1.0 per km. More research is needed to ascertain the exact effects of such disturbance on the Hooded Plover.

The problems it encounters during breeding may lead to a decline in numbers, something that has already occurred in New South Wales, where the species is listed as endangered (183).

# MONGOLIAN PLOVER
## *Charadrius mongolus*
(Plate 4)

**Other name** Lesser Sand Plover; Mongolian Dotterel, Sand-dotterel or Sand Plover; Mong.

There are five races, which differ slightly in breeding plumage (see description). These fall into two broad groups: the 'atrifrons' group, comprising *pamirensis*, *atrifrons* and *schaeferi*, which have mostly black foreheads; and the 'mongolus' group, including *stegmanni* and *mongolus*, which have mostly white foreheads (370). Those of the race *mongolus* are most common in Australia, but there are a few confirmed records of *stegmanni* in the north. *C. m. schaeferi* may well occur in north-western Australia, but has not yet been recorded.

## DESCRIPTION

A small to medium-sized plover with grey-brown upperparts and white underparts; striking chestnut-and-black markings in breeding plumage.

**Adult non-breeding plumage** Forehead, white; crown, nape, hindneck, sides of neck, mantle, scapulars and wing-coverts, grey-brown; supercilium, white, thicker behind eye; *lores and ear-coverts, dark brown*. Chin and throat, white; incomplete brown breast-bar, thicker at sides; belly, flanks and vent, white. Iris, black; bill, slightly shorter than head with slight bulge towards tip, black; legs, dull olive-brown. Sexes similar.

**In flight** Back, grey-brown; white with central dark brown line; tail, grey-brown with narrow white tip; greater and secondary coverts, thinly tipped white giving impression of *narrow wing-bar*; primaries and secondaries, dark brown.

**Voice** Short chipping notes 'chip chip chreet . . .' or single 'chip'; trilling notes; whistling 'chew-eet'.

**Adult breeding plumage** *C. m. mongolus*: forehead, white with thin black upper edge and variable black central line (often absent); crown, brown; nape, pale chestnut, extending to behind eye; mantle, scapulars and wing-coverts, grey-brown; supercilium, white; lores and ear-coverts, black. Chin and throat, white, thinly edged black; *breast, flanks and upper belly, chestnut*; rest of underparts, white. Those of the race *C. m. stegmanni* have thicker black edges to the forehead and throat. Birds of the race *C. m. schaeferi* have a black forehead with a variable amount of white on its sides and lack the black edge to the throat. The variation between races is continuous, so certain identification requires careful observation and recording.

**Juvenile plumage** Similar to adult non-breeding plumage, but feathers of mantle, scapulars and wing-coverts, fringed buff.

**Field notes** Feeds in typical stop-start plover fashion. Birds in partial or full breeding plumage occur in August−October and February−May. It roosts in loose flocks on sandy beaches and spits and forages on intertidal mudflats and beaches; exclusively coastal.

**Similar species** Large Sand Plover; Double-banded Plover; Red-capped Plover. The Mongolian Plover can be distinguished from the Large Sand Plover by its shorter, more slender bill, slightly shorter legs, darker appearance round the face due to darker brown lores and ear-coverts and slightly less extensive white forehead. In breeding plumage it has a broader chestnut breastband. In southern and eastern Australia, it can be confused with the Double-banded Plover when both are in non-breeding plumage, but is generally paler, has a more clearly marked head and a stouter bill. It is conspicuously different in breeding plumage. It is larger than the Red-capped Plover.

## BREEDING RANGE

Five races of the Mongolian Plover are recognised (85). These are *mongolus* and *stegmanni*, which breed in north-eastern Siberia, *schaeferi* breeding in western China and southern Mongolia and *atrifrons* and *pamirensis* breeding in the Himalayas and mountain ranges to the west. (See Figure 5.13.) *C. m. mongolus* is the common race in Australia.

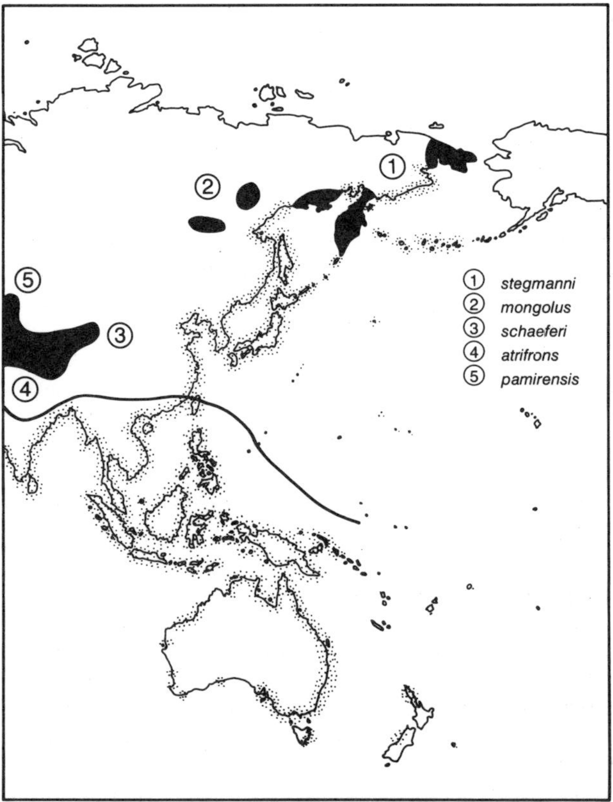


FIG. 5.13.    Breeding and non-breeding range of Mongolian Plover.

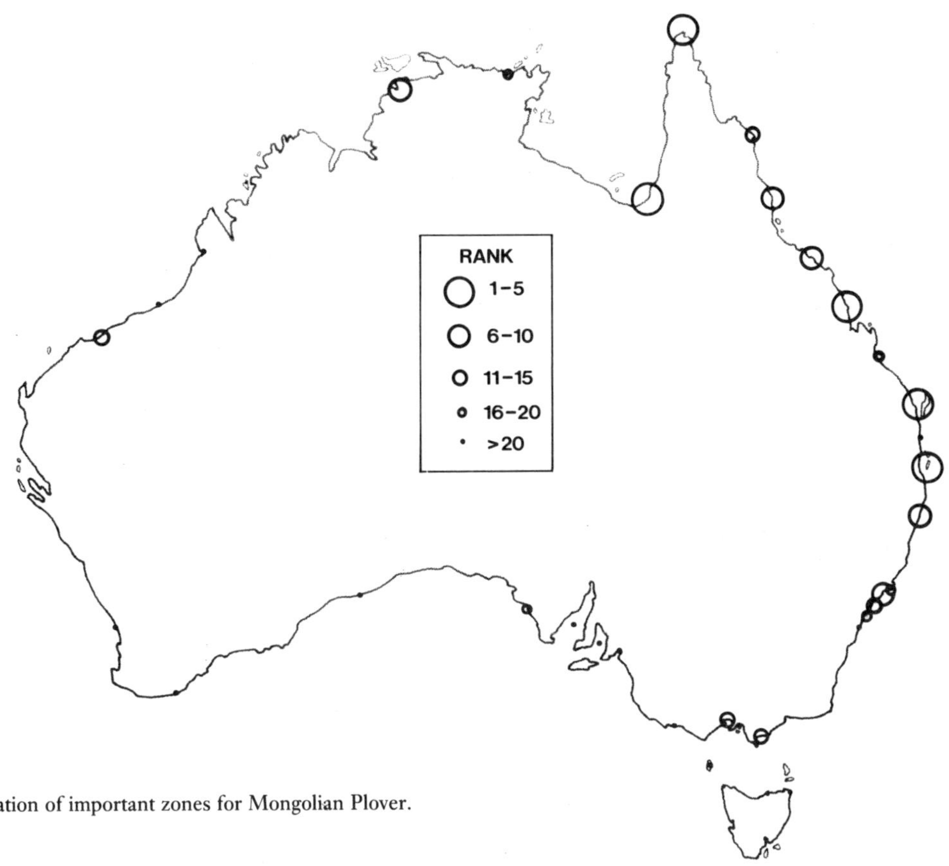



Fɪɢ. 5.14.   Location of important zones for Mongolian Plover.

# ASIAN RANGE

The Mongolian Plover occurs throughout Asia and eastern and southern Africa during the non-breeding months (see Figure 5.13), and the races have different non-breeding ranges. Those of *atrifrons* and *schaeferi* overlap in southeast Asia, and those of *mongolus* and *stegmanni* in southern China, the Philippines, Thailand, Malaysia, western Indonesia and possibly northern Australia (244).

**C. m. mongolus**  is a vagrant in Korea (162) and Japan (344), but common during migration on the Chinese coast (58,404), and in the Philippines (85) where it spends all the non-breeding months. It is not recorded in Borneo (440). Its status in eastern Indonesia is uncertain, in most cases the race there is not named (14, 128, 487), but *mongolus* has been listed for Sulawesi (449). It is one of the most common shorebirds in New Guinea (36, 184, 278). Birds occasionally reported in the South Pacific and New Zealand (134, 387, 388) probably belong to this race.

**C. m. stegmanni**  is common during migration in Korea (162) and in Japan (344), where 2500 to 3300 have been recorded on northward and 1300 to 2200 on southward migration (488). It spends the non-breeding months in Taiwan (56, 129), the Philippines, eastern Indonesia and Melanesia (85). It may occur regularly in northern Australia (244).

Tᴀʙʟᴇ 5.8.  Ranked list of top twenty zones for Mongolian Plover. (maximum counts marked thus: *; other counts are averages) (30 or more birds)

| ZONE | NO. OF COUNTS | NO. OF INDIVIDUALS |
|---|---|---|
| SE corner, Gulf of Carpentaria | 1 | 4 050 |
| * Moreton Bay, Qld | 5 | 1 770 |
| * Hervey Bay—Great Sandy Strait, Qld | 5 | 1 430 |
| * Mackay area, Qld | 4 | 1 090 |
| * Torres Strait, Qld | 3 | 700 |
| * N coast, NSW | 5 | 690 |
| Darwin area, NT | 4 | 610 |
| * Townsville area, Qld | 5 | 330 |
| * Cairns area, Qld | 4 | 150 |
| * Hunter estuary, NSW | 4 | 130 |
| Central coast, NSW | 3 | 130 |
| * Daintree—Melville, Qld | 2 | 70 |
| Corner Inlet, Vic. | 5 | 70 |
| Port Phillip Bay, Vic. | 5 | 60 |
| * Port Hedland Saltworks, WA | 12 | 60 |
| * Capricorn coast, Qld | 1 | 60 |
| * Port Stephens, NSW | 4 | 40 |
| * W coast, Eyre Peninsula, SA | 5 | 40 |
| * NE Arnhem Land, NT | 1 | 40 |
| Botany Bay, NSW | 5 | 30 |

**C. m. schaeferi** spends the non-breeding months in Malaysia, the Gulf of Thailand and western Indonesia (85). Although not yet confirmed, it may visit Australia, particularly the north-west (244).

**C. m. atrifrons** spends the non-breeding months in the Bay of Bengal, Malaysia, Thailand and western Indonesia (5, 85).

The inner Gulf of Thailand may be an important migratory stopover for the Mongolian Plover, as thousands occur there in March and April, mostly of the last two races (309). The Mongolian Plover is very common in southern Vietnam (489), and those of the 'mongolus' and 'atrifrons' groups have been seen in Hong Kong, the former group being more abundant (310).

## Status and Distribution in Australia

The Mongolian Plover is most numerous in the Gulf of Carpentaria and on the eastern Queensland and northern New South Wales coasts. (See Figure 5.14, Table 5.8.) Smaller numbers occur on the coast south to Victoria. It is rare in Tasmania, South Australia and Western Australia. The Mongolian Plover is exclusively coastal. There is one recent inland record, in north-eastern South Australia (38).

It feeds at low tide on intertidal mudflats and roosts on sandy beaches and promontories at high tide. Due to insufficient coverage of some parts of northern Australia and the difficulties of separating Mongolian from Large Sand Plovers, particularly in the north-west where the latter is abundant, the counts did not completely document the distribution and abundance of the Mongolian Plover.

## Movements in Australia

The Mongolian Plover arrives on and migrates southwards along the northern, eastern and south-eastern coasts of Australia in September and October. It continues to move along the east coast in November. Maximum numbers are attained in December at most places and remain stable until early February, although some fluctuations occur, possibly due to local movements.

In late February, numbers increase at some sites in northern Australia. At most other sites, influxes occur in March and April. This suggests that it moves short distances along the eastern and northern coasts before departing northwards from Australia. In the Broome — Port Hedland area it departs in late April (247).

## Feeding

The Mongolian Plover forages with the Large Sand Plover and other shorebirds in loose rather than tightly packed flocks, rarely along the water's edge, preferring wet intertidal mud. Nothing is known of the diet of the Mongolian Plover in Australia, but elsewhere in the non-breeding months it gleans crabs, molluscs, crustacea (crabs and amphipods) and worms from mudflats (85).

## Conservation

Important sites on the northern New South Wales and southern Queensland coasts are currently being extensively altered for recreational and real-estate developments. The survival of the Mongolian Plover at these sites depends on the maintenance of adequate areas of suitable feeding and roosting habitat. Elsewhere, human population density is lower and no significant areas are currently threatened.

# DOUBLE-BANDED PLOVER

## *Charadrius bicinctus*
(Plate 5)

**Other names** Banded Dotterel; Double-banded Dotterel; Double-bandy.
No races recognised.

## DESCRIPTION

A small to medium-sized brown-and-white plover with distinctive breeding plumage.

**Adult non-breeding plumage** Forehead, white, edged black or brown; crown, nape, hindneck, sides of neck, mantle, scapulars and wing-coverts, brown, feathers having narrow buff fringes; supercilium, white; lores and ear-coverts, grubby brown. Chin and throat, white; indistinct brown breast-bar; belly, flanks and vent, white. Iris, black; bill, short, slender and black; legs, dull olive-green. Slight difference between sexes: males show wider black edges to white forehead and slightly more distinct head markings.

**In flight** Back, brown; rump, white with broad black central line; tail, brown with narrow white tip; tips of greater coverts, white, showing as inconspicuous wing-bar.

**Voice** Short chirping 'chip chip'; various churring calls.

**Adult breeding plumage** Like non-breeding plumage, except for *two breast-bars, an upper black one and a lower chestnut one,* broader in the middle; head markings become more distinct; lores and ear-coverts, black, although sexual difference remains: males have heavier black markings on head and a more solid, darker chestnut lower breast-bar.

**Juvenile plumage** Similar to adult non-breeding plumage, except head markings less distinct; variable buff wash about head and breast; narrow white fringes to feathers of upperparts.

**Field notes** Forage in the usual plover-like fashion individually or in loose flocks on mudflats, pasture or saltmarsh; roost in loose flocks on sandy beaches, pasture or saltmarsh; call continuously in flight and utter quiet churring calls when feeding or roosting in flocks; show breeding plumage on arrival (February—April) and before departure (July—August).

**Similar species** Red-capped Plover; Mongolian Plover; Large Sand Plover. The Double-banded Plover is larger and more robust than the Red-capped Plover. It is slightly smaller and browner, and has a more slender bill than the Mongolian or Large Sand Plovers.

## BREEDING RANGE

The Double-banded Plover breeds in New Zealand, and many migrate across the Tasman Sea to Australia for the winter. It does not breed in Australia. It nests on sandy beaches throughout New Zealand, on pasture areas, on the gravel river-beds in the South Island and in a limited area inland on the North Island at lakes Rotorua and Tarawera (342). It arrives on its breeding grounds in July and August, and after the chicks have fledged in February it departs (342, 363).

## NON-BREEDING RANGE

After breeding, many migrate to the northern parts of the North Island of New Zealand and inland breeders move to the coast (342). Veitch (474) reported a flock of 3000 on pasture near the Firth of Thames, although no date was given. In March 1978 there were 500 on Kaipara Harbour, north-west of Auckland (475). Counts from Farewell Spit, on the northern end of the South Island, include up to 1620 on 11 March 1974 (124), and it is possible that this area could be used by inland-nesting birds from the South Island prior to trans-Tasman migration (381). A substantial proportion of the population moves to Australia, arriving in February and departing in July and August. (See Figure 5.15.) During winter in New Zealand, it uses pasture more readily than in Australia.

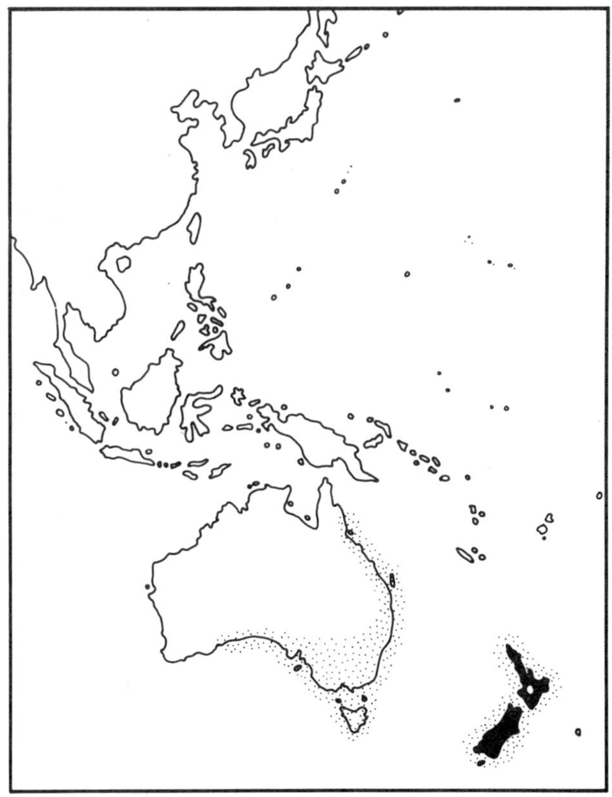

FIG. 5.15.   Breeding and non-breeding range of Double-banded Plover.

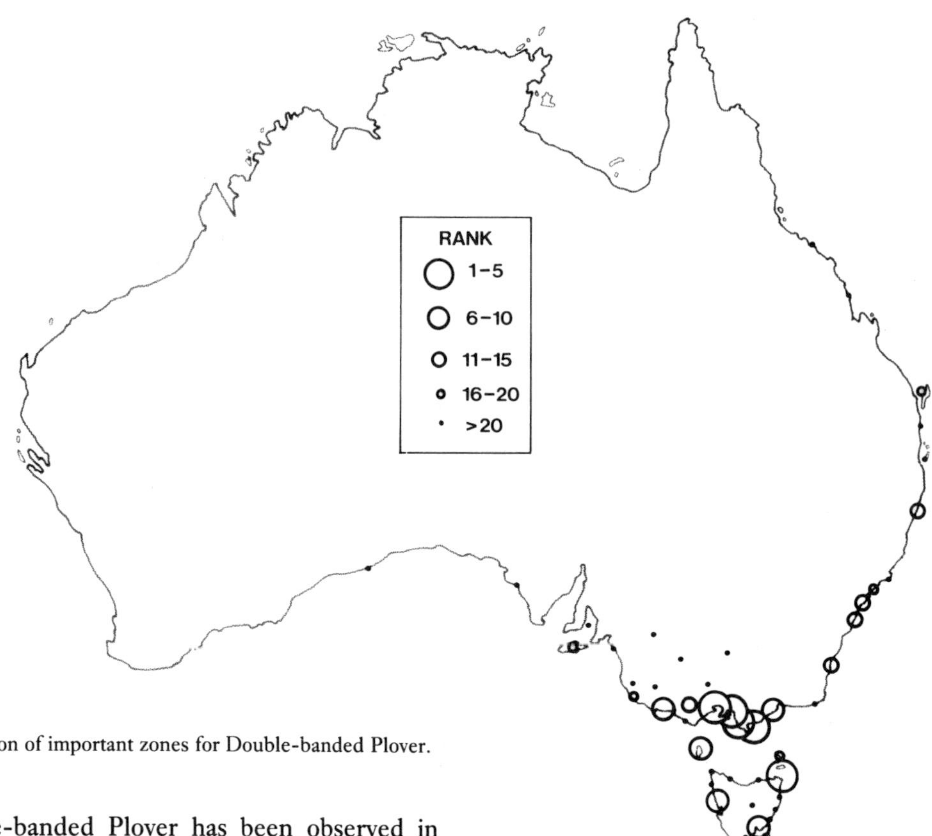



FIG. 5.16.   Location of important zones for Double-banded Plover.

The Double-banded Plover has been observed in small numbers in Fiji (405, 406) and on Lord Howe and Norfolk islands (342).

## STATUS AND DISTRIBUTION IN AUSTRALIA

The Double-banded Plover occurs in largest numbers in south-eastern Australia, especially on the Victorian coast, where over half those counted spend the winter. (See Figure 5.16, Table 5.9.) It regularly moves west as far as eastern Western Australia and north as far as Townsville in northern Queensland. There were *Atlas* project records in south-western Australia north as far as Perth (38). It also occurs at a number of inland salt lakes in Victoria and southern New South Wales.

The Double-banded Plover forages on intertidal mudflats, preferring places near grazed pasture or open saltmarsh, which are used for feeding and roosting at high tide. It occurs in small numbers on inland salt lakes, preferring those edged with low saltmarsh.

The maximum number counted in Australia was approximately 5600 in June 1985. At the same time, 6850 were counted in New Zealand (498). The world population is therefore at least 12 450.

## MOVEMENTS IN AUSTRALIA

The Double-banded Plover first arrives in Australia in early February. Many arrive in March, and in southern

TABLE 5.9. Ranked list of top twenty zones for Double-banded Plover. (maximum counts marked thus: *; other counts are averages) (50 or more birds)

| ZONE | NO. OF COUNTS | NO. OF INDIVIDUALS |
|---|---|---|
| * Port Phillip Bay, Vic. | 5 | 1 400 |
| Corner Inlet, Vic. | 5 | 730 |
| * Cape Portland coast, Tas. | 5 | 680 |
| * Anderson's Inlet, Vic. | 5 | 550 |
| * Westernport Bay, Vic. | 5 | 500 |
| * King Island, Tas. | 2 | 370 |
| * Derwent estuary & Pittwater, Tas. | 5 | 360 |
| * SW coast, Vic. | 5 | 300 |
| * Jack Smith's Lake, Vic. | 1 | 250 |
| * Ocean Beach, Strahan, Tas. | 1 | 170 |
| Western District lakes, Vic. | 4 | 170 |
| * N coast, NSW | 5 | 140 |
| * Illawarra coast, NSW | 3 | 140 |
| * S coast, NSW | 5 | 130 |
| Botany Bay, NSW | 5 | 130 |
| * Furneaux Islands, Tas. | 1 | 120 |
| * Hervey Bay – Great Sandy Strait, Qld | 4 | 100 |
| * Hunter estuary, NSW | 4 | 90 |
| SE coast, SA | 5 | 90 |
| * Kangaroo Island, SA | 4 | 50 |

New South Wales, Victoria and Tasmania largest numbers occur in April. Numbers then decline slightly by May and are stable until early July. These observations, together with those of small numbers in inland Victoria in April, suggest that it stops on the east coast in March and April before moving farther west. Some overfly inland Victoria, presumably to the South Australian coast.

It departs from Tasmania and South Australia in July, but in Victoria and southern New South Wales it is present until mid-August and sometimes as late as early September. Small numbers are seen in inland Victoria in August. In New South Wales and Victoria, temporary influxes occur in August, suggesting that birds from elsewhere in Australia use sites there as staging areas before their return migration to New Zealand.

## FEEDING

The Double-banded Plover forages at low tide on mudflats and at other times in saltmarsh and pasture. There have been no detailed studies of its feeding behaviour and diet. Four specimens from South Australia were recorded with the remains of insects, mostly beetles, in their stomachs (256).

## CONSERVATION

Australia holds a substantial proportion of the population of the Double-banded Plover. It is concentrated in a small number of coastal sites, which require adequate protection and management to ensure the species' survival.

# LARGE SAND PLOVER
## *Charadrius leschenaultii*
(Plate 4)

**Other names** Large Sand-Dotterel; Large-billed Dotterel; Greater Sand Plover.

Three races are recognised: *columbinus* breeds in the Middle East, *crassirostris* from the Caspian Sea eastwards to Kasakhstan and *leschenaultii* south of Lake Baikal in Siberia, Mongolia and northern China (85). Only the last comes to Australia (65).

## DESCRIPTION

A medium-sized plover with grey-brown upperparts, white underparts and robust bill; chestnut and black markings in breeding plumage.

**Adult non-breeding plumage** Forehead, white; crown, nape, hindneck, mantle, scapulars and wing-coverts, sandy grey-brown; supercilium, white; *lores, grey-brown, often incompletely marked; ear-coverts, grey-brown*. Chin and throat, white; breast, white with sandy grey-brown sides; belly, flanks and vent, white. Iris, black; bill, robust, as long as head and black; legs, dull greyish-green. Sexes similar.

**In flight** Back, sandy grey-brown; rump, white with central dark brown line; tail, sandy grey-brown with narrow white tip; greater and secondary coverts, tipped white showing as *conspicuous white wing-bar*.

**Voice** Trilling calls and double-noted whistle: 'chweep chweep'.

**Adult breeding plumage** Forehead, white with thick black upper edge and variable black central line; crown, sandy grey-brown; nape and sides of neck, pale chestnut, extending onto breast; hindneck, mantle, scapulars and wing-coverts, sandy grey-brown; lores, area round eye and ear-coverts, black. Chin and throat, white; *breast has broad chestnut band* extending onto flanks, sometimes with incomplete, thin, black upper edge; belly and vent, white. Males have slightly heavier markings round head.

**Juvenile plumage** Very similar to adult non-breeding plumage, but feathers of upperparts noticeably fringed white.

**Field notes** Often associate with Mongolian Plovers; roost in large flocks on beaches in north-western Australia; feed in stop-start plover fashion, spread out on intertidal mudflats; exclusively coastal; birds with varying amounts of breeding plumage can be seen in August−October and March−April.

**Similar species** Mongolian Plover; Double-banded Plover; Oriental Plover. The Large Sand Plover has paler markings on head, a longer, thicker bill, slightly longer legs and, in flight, a more noticeable white wing-bar than the Mongolian Plover. It is greyer and has a more robust bill than the Double-banded Plover or Oriental Plover and is conspicuously different in breeding plumage.

## BREEDING RANGE

*C. l. leschenaultii* breeds on dry stony plains in the northern Gobi Desert of Mongolia and north-western China, where its breeding range may overlap with that of *C. l. crassirostris*, which breeds from there west to the Caspian Sea (85, 129, 144, 237).

## ASIAN RANGE

When not breeding, the Large Sand Plover is found throughout coastal eastern and southern Africa, the Red Sea, Persian Gulf, Arabian Sea, Bay of Bengal, south-east Asia, New Guinea and Australia.

The Large Sand Plover is a vagrant in Korea (162) and uncommon in Japan on southward migration (488). It is scarce in north-eastern China (266), although it has been reported as common on the north-eastern coast of China (490). It occurs inland during southward migration in central China (58). In southern China and Hong Kong it is abundant on both migrations (late July–October; April and May) (54, 255), and it is one of the most abundant shorebirds in Vietnam (142, 489). It is common in Thailand (257), although considerably less numerous than the Mongolian Plover (309), and in Malaysia it is outnumbered by that species (348), although it has been described as fairly common there (157). In the Philippines it is widespread (171) and common (110); it considerably outnumbers the Mongolian Plover (2). It regularly migrates through western Micronesia (18). In Borneo, it probably outnumbers the Mongolian Plover (440), and at Brunei the largest numbers occur in late August and early September (477), at about the time they first arrive in large numbers in north-western Australia. Specimens of the Large Sand Plover in eastern Indonesia outnumber those of the Mongolian Plover 4 to 1, and most were collected in August and September during southward migration (487). It is very rare in the Pacific Islands, visits New Zealand almost annually, but in very small numbers (134, 387, 388), and occurs regularly in New Guinea and nearby islands, although it is not as common there as the Mongolian Plover (173, 184).

The extent of the non-breeding ranges of *C. l. leschenaultii* and *C. l. crassirostris* is unknown, but the birds of Borneo are *leschenaultii* (440). Most *leschenaultii* probably pass from their breeding grounds to southern China, the Philippines and Borneo to north-western Australia and New Guinea (Figure 5.17.) It spends December–February from Thailand and the Philippines southwards. Northward migration is most likely via a similar route.

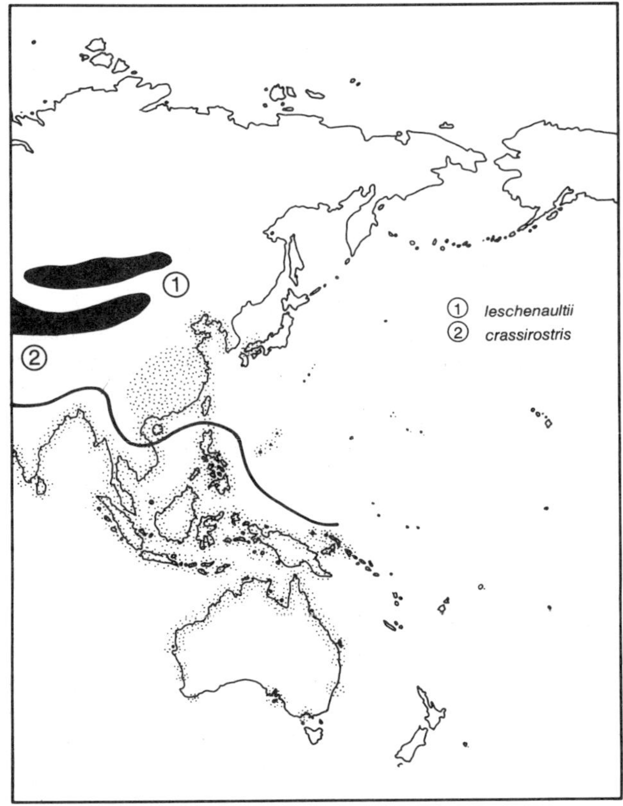


FIG. 5.17.   Breeding and non-breeding range of Large Sand Plover.

## STATUS AND DISTRIBUTION IN AUSTRALIA

The north-west coast of Australia, between Broome and Port Hedland, holds more Large Sand Plovers than anywhere else in Australia. It occurs regularly in small numbers in southern and eastern Australia, but it is mostly a bird of the northern coasts. In southern Australia, numbers are higher in the west. Unlike the Mongolian Plover, it is not abundant on the east Queensland coast, occurring in that State mostly in the Gulf of Carpentaria. It is exclusively coastal (38). (See Figure 5.18, Table 5.10.) It forages on extensive intertidal mudflats and roosts during high tide on sandy beaches and rocky points. The full extent of its distribution in the north is still unclear, as the Arnhem Land coast may also hold large numbers.

## MOVEMENTS IN AUSTRALIA

The Large Sand Plover arrives on the north-west coast and at Darwin in late August and September. Some leave the north-west by October and November, and there is a temporary influx at Darwin in October. Small numbers migrate southwards down the east coast and along the central southern coast between September and November. These observations suggest

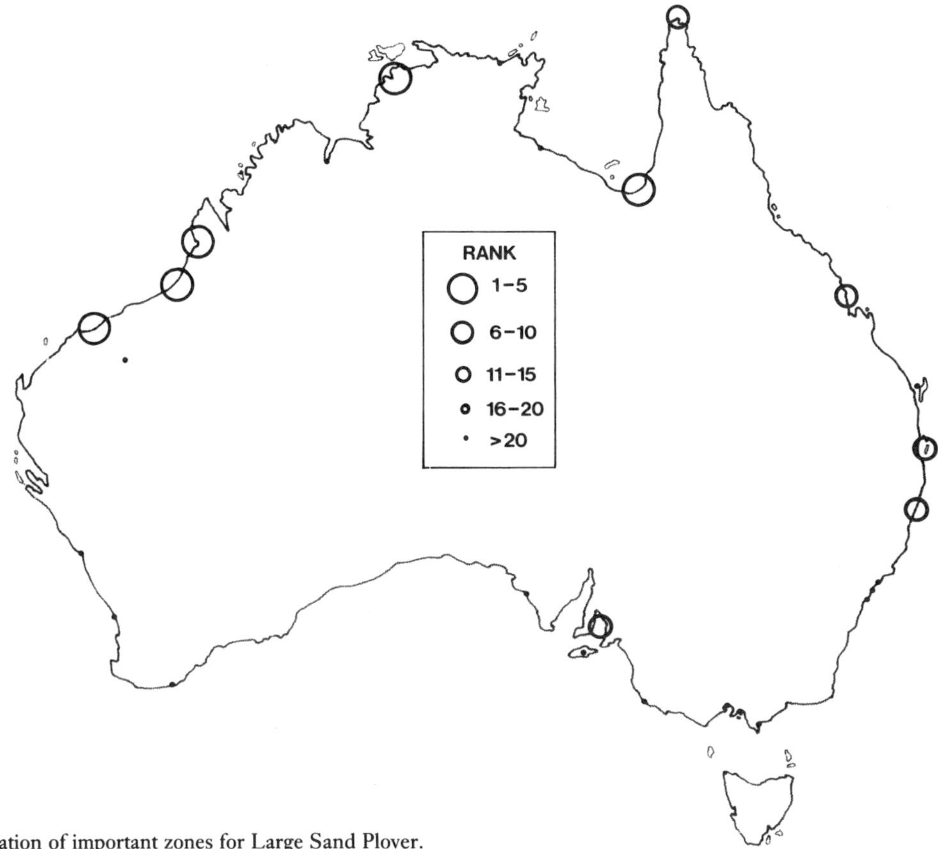



FIG. 5.18.    Location of important zones for Large Sand Plover.

that a small proportion of the population migrates eastwards and southwards from the north-west coast. However, large numbers stay in the north-west for the non-breeding months.

The Large Sand Plover migrates northwards through Darwin in March and along the eastern and central southern coasts in March and April. Most have departed from the north-west coast by mid-April.

## FEEDING

The Large Sand Plover feeds at low tide on wet mudflats rather than in shallow water. It detects prey visually, running short distances, stopping to look, running to peck, stopping again and continuing. Little is known of its diet in Australia. One specimen had eaten crabs (472), and they have been reported feeding on small crabs, marine worms and insects (e.g. beetles) in India (5).

## CONSERVATION

It is likely that the bulk of the population of the race *leschenaultii* comes to Australia, in which case the preservation and appropriate management of Roebuck Bay and Eighty Mile Beach is crucial for its survival.

TABLE 5.10.   Ranked list of zones that held 20 or more Large Sand Plovers. (maximum counts are marked thus: *; other counts are averages)

| ZONE | NO. OF COUNTS | NO. OF INDIVIDUALS |
|---|---|---|
| * Eighty Mile Beach, WA | 5 | 30 400 |
| * Roebuck Bay, WA | 6 | 29 900 |
| * SE Corner, Gulf of Carpentaria, Qld | 1 | 4 160 |
| Darwin area, NT | 4 | 660 |
| * Pilbara coast, WA | 3 | 230 |
| * Moreton Bay, Qld | 5 | 120 |
| Torres Strait, Qld | 4 | 90 |
| * N coast, NSW | 5 | 40 |
| Mackay area, Qld | 4 | 40 |
| * St Vincent Gulf, SA | 4 | 30 |
| * Central W coast, WA | 4 | 20 |
| Hervey Bay–Great Sandy Strait, Qld | 5 | 20 |
| Peel Inlet, WA | 5 | 20 |
| * Kimberley coast, WA | 1 | 20 |

# ORIENTAL PLOVER

## *Charadrius veredus*
(Plate 4)

**Other names**   Oriental or Eastern Dotterel; Eastern Sand Plover.

No races recognised. Some authors treat the Oriental Plover as a race of the Caspian Plover (*C. asiaticus*) (5).

## DESCRIPTION

A tall, slender, medium-sized plover, predominantly golden-brown in colour.

**Adult non-breeding plumage** *Upperparts, mostly uniform golden-brown*; forehead and lores, cream-brown; supercilium and sides of neck, cream with chestnut buff wash; ear-coverts, brown. Chin and throat, cream; breast, washed brown; belly, flanks and vent, cream. Iris, black; legs, pale orange; bill, short, slender and black. Sexes similar.

**In flight**   Uniform golden-brown upperparts; *no distinctive markings*.

**Adult breeding plumage**   *Head and neck, white*; rest of upperparts, golden-brown; breast, bright chestnut with thick black band on lower edge; belly, flanks and vent, white. Female lacks black breast-band and is browner on head.

**Juvenile plumage**   Very similar to adult, except all feathers of upperparts have chestnut-buff fringes, and the wing-coverts have paler, cream fringes. The head and sides of neck are more chestnut-buff than in the adult.

**Field notes**   Occurs on short open grassland as well as beaches, mostly in northern Australia; usually in loose flocks; feeds in typical stop-start plover fashion.

**Similar species**   Lesser Golden Plover; Large Sand Plover. The Oriental Plover has more uniform upperparts, is more slender, with a lighter bill, longer wings, which extend beyond the end of the tail when folded, and relatively longer paler legs than Lesser Golden or Large Sand Plovers.

## BREEDING RANGE

The Oriental Plover has a limited breeding range in the arid and mountain steppes of northern, eastern and western Mongolia and adjacent parts of eastern China (129, 232, 237, 255, 266, 473).

## ASIAN RANGE

The Oriental Plover is a vagrant in Korea (162) and Japan (344). It migrates southwards through eastern China, being very abundant in the Yangtze Valley in September (58, 255). It occurs in Hong Kong irregularly in September (54), and is not listed for Taiwan (401). It is rare in the Philippines (110), Thailand (257) and the Malay Peninsula (157, 305). There are a few records from Micronesia (127). In Borneo, it occurs regularly, but in small numbers, between September and November (161, 408, 440) and on Bali in September as an uncommon migrant (14). It occurs between September and November in the islands immediately north-west of Australia (487). It is rare in New Guinea (36, 184). Its presence in south-eastern China in September, the timing of its movement through the islands north-west of Australia (487), its arrival in north-western Australia in the first few days of September and its increase in numbers there until November suggest that it may migrate non-stop between China and Australia.

Most birds probably spend the non-breeding season in Australia; La Touche (255) describes the Oriental Plover as 'wintering' in the Philippines, whereas Hachisuka (171) records it from only two islands, and Delacour and Mayr (110) consider it rare there. (See Figure 5.19.)

As during southward migration, only small numbers occur between Australia and China during northward migration. It passes through south-eastern China in March and April (54,255), occurring as far north as Peking by 9 April (490). By late April it is in south-

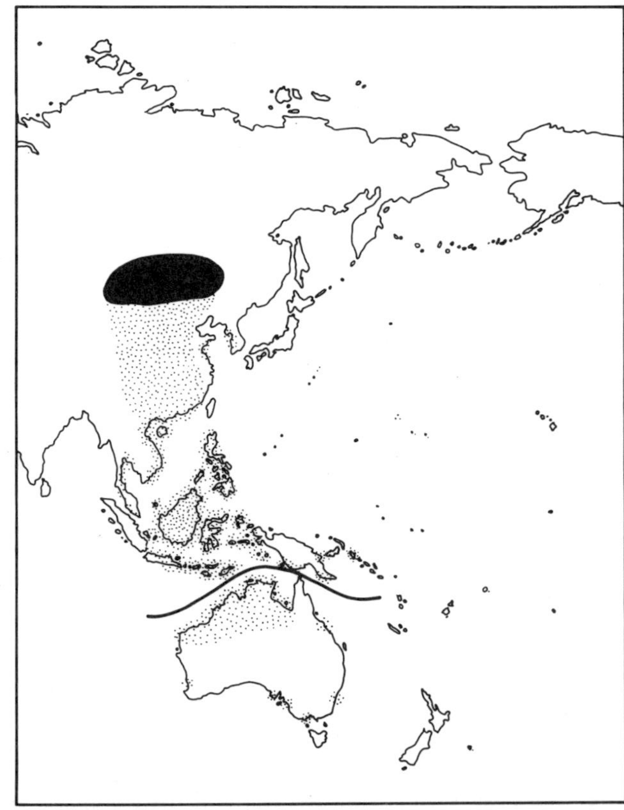

FIG. 5.19.   Breeding and non-breeding range of Oriental Plover.

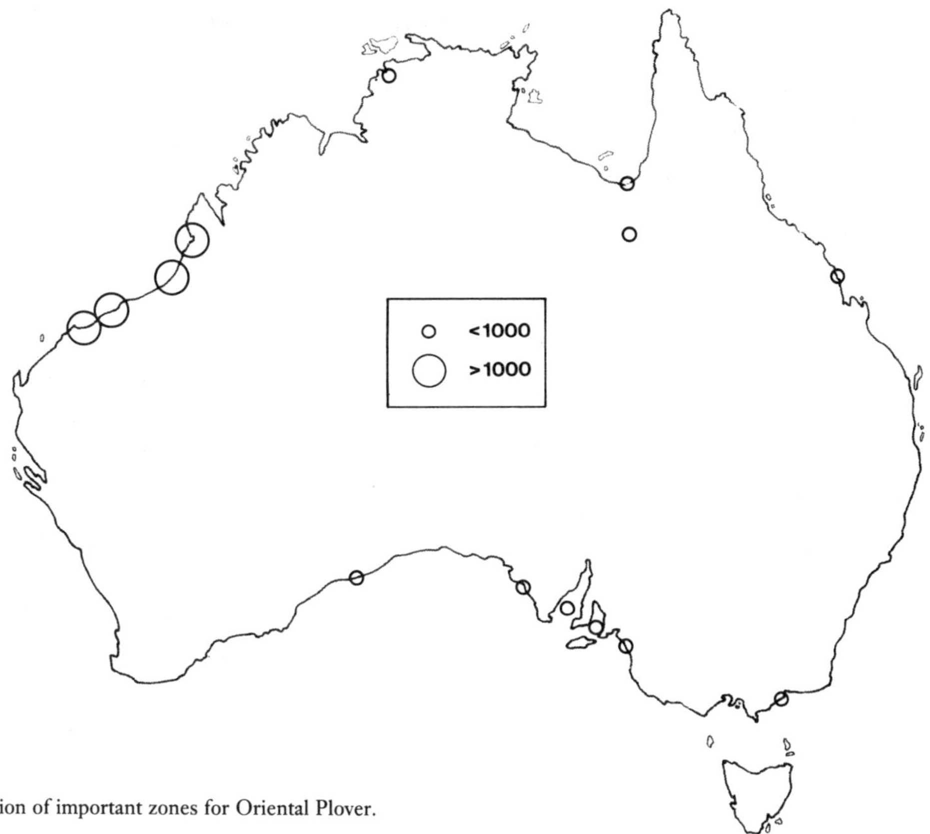


FIG. 5.20.   Location of important zones for Oriental Plover.

TABLE 5.11.  Ranked list of zones that held more than 1000 Oriental Plovers. (all counts are maxima)

| ZONE | NO. OF COUNTS | NO. OF INDIVIDUALS |
| --- | --- | --- |
| Port Hedland Saltworks, WA | 12 | 29 900 |
| Eighty Mile Beach, WA | 5 | 18 400 |
| Roebuck Bay, WA | 6 | 8 700 |
| Dampier Saltworks, WA | 5 | 1 830 |

central Mongolia, immediately south of the breeding grounds (232).

## STATUS AND DISTRIBUTION IN AUSTRALIA

Large numbers of Oriental Plovers occur in the Broome–Port Hedland area of Western Australia (max. 44 300), and smaller numbers occur in the Darwin area, where flocks usually number less than 200 (88). Counts did not adequately determine its distribution and abundance in Australia, as it spreads out over the vast grassy plains of northern Australia and most of these were not visited. Most *Atlas* project records were from northern Western Australia, the Northern Territory and far north-western Queensland (38). The north-west is probably the core of its non-breeding range. (See Figure 5.20, Table 5.11.) The

few Oriental Plovers seen during national counts were in coastal South Australia, the maximum being 178 on the west coast of the Eyre Peninsula. A record of 53 birds at Jack Smith's Lake in eastern Victoria in February 1983 is exceptional for that State.

The Oriental Plover inhabits the open grassy plains of northern Australia, but also occurs in recently burnt woodland (88). In the north-west, it feeds on the plains at night and roosts on the beaches with other shorebirds during the day.

## MOVEMENTS IN AUSTRALIA

The Oriental Plover arrives on the northern coast of Australia in the first week of September. It is present in Darwin until mid-October (88) and at Mount Isa until November. Maximum numbers in the Broome–Port Hedland area occur in October and November. Its whereabouts once the northern wet season starts in December and January are not known, but it probably moves inland to the extensive grassy plains of northern Australia. Small numbers occur in coastal South Australia between December and February each year, suggesting that some migrate southwards across the continent.

There are few records of Oriental Plovers in coastal northern Australia in March and April; many fewer occur in the Broome–Port Hedland area at this time. They appear to depart from their inland haunts directly to Asia.

## FEEDING

The Oriental Plover forages on open grassy plains, airfields and playing fields. At Eighty Mile Beach, Western Australia, it forages at night on plains and flies to roost with other shorebirds on the nearby beach during the day.

Its diet, based on gut contents of specimens, is mostly insects: termites, grasshoppers, crickets, bugs and beetles (300, 472).

## CONSERVATION

It is likely that most Oriental Plovers come to north-western Australia each year. This area is sparsely settled, and there are no immediate or serious threats to its continued survival.

# RED-CAPPED PLOVER
## *Charadrius ruficapillus*
(Plate 7)

**Other names** Red-capped Dotterel; Red-necked Plover; Sandlark; Sandpiper.
No races recognised.

## DESCRIPTION

A small, sandy-brown plover with white underparts.

**Adult plumage** Male: forehead, white, edged black; *crown and nape, chestnut*; hindneck, mantle, scapulars and wing-coverts, sandy-brown, thinly edged pale chestnut. Underparts, white; black mark on sides of breast at edge of nape. Iris, black; bill, short, slender and black, legs, black. Female paler with patchy chestnut cap and nape more thinly edged black.

**In flight** Back, sandy-brown; rump, white with black central line; pale bases to primaries form thin wing-bar.

**Voice** Whistle, 'drit drit' often uttered on the wing; various churring notes, and peeps.

**Juvenile plumage** Similar to adult female, but crown and nape are sandy-brown with very thin black edges; feathers of upperparts edged white.

**Field notes** Usually seen individually or in pairs, although they flock in the non-breeding months. They show the characteristic plover-like 'run-stop-peck-run . . .' behaviour, and tend to feed away from the edge of water. It can be encountered long distances from water.

**Similar species** Mongolian Plover; Double-banded Plover. The Red-capped Plover is smaller and has a more chestnut crown than these two species.

## STATUS AND DISTRIBUTION IN AUSTRALIA

The Red-capped Plover breeds throughout coastal and inland Australia. It does not occur outside Australia, except in New Zealand, where it is a vagrant (134). It is abundant on Eighty Mile Beach, Western Australia, and in the coastal and near-coastal wetlands of southern Australia. (See Figure 5.21, Table 5.12.) Some of the largest concentrations occur on inland salt lakes. The Red-capped Plover is very adaptable and occurs on a wide range of wetland types. In Tasmania it has been observed at an altitude of 1100 metres (143). In coastal areas, it feeds on intertidal mudflats and roosts on nearby sandy beaches, spits or saltmarsh, and generally avoids rocky shores.

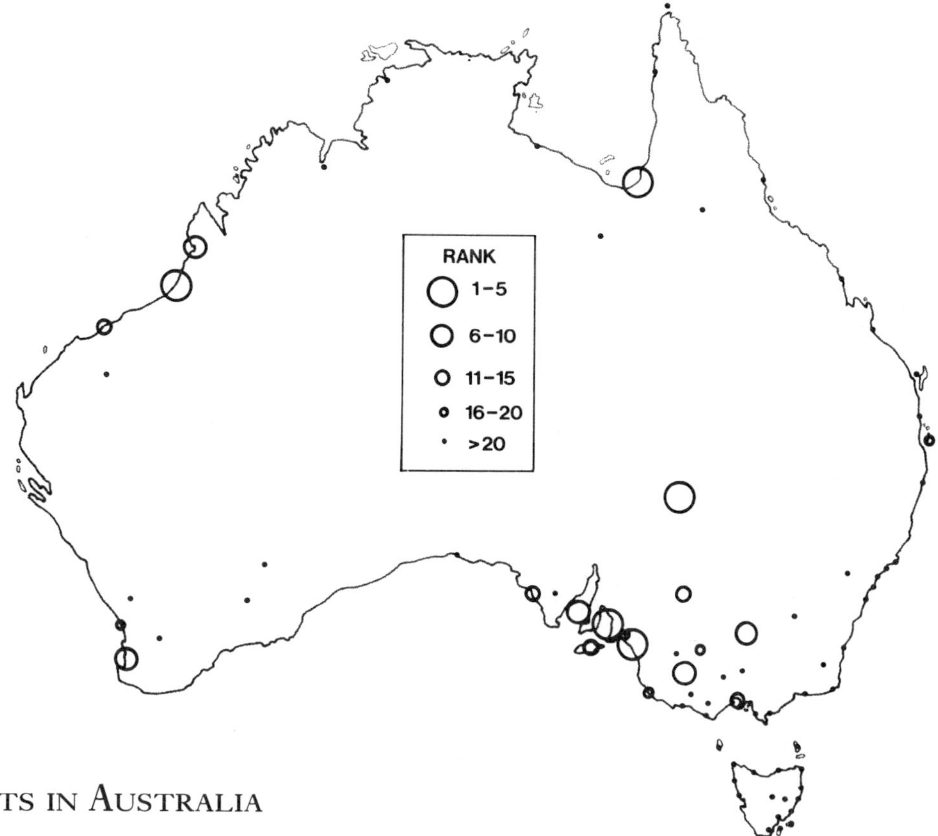



FIG. 5.21.   Location of important zones for Red-capped Plover.

## MOVEMENTS IN AUSTRALIA

Banding recoveries and counts at most coastal areas show that the Red-capped Plover does not usually undertake long-distance movements. On the coast, it spreads out to breed between August and January (385) and flocks at places with extensive intertidal mudflats in late summer.

In inland areas, longer movements of banded birds are recorded: e.g. 830 km in 4 weeks (15). On the south-western coast, more occur in summer than in winter (e.g. Peel Inlet and Swan coastal plain, mean counts: 2000 in summer, 130 in winter), suggesting movement to coastal areas for summer when little surface water remains inland. In Tasmania, it has been recorded moving onto inland salt lakes after rain fills them (168). In July 1984 a flock of more than 3500 was seen on Lake Buloke, Victoria, few having been present on previous counts. These observations show that the Red-capped Plover moves about in inland areas. Movements inland are likely to be more erratic than on the coast, as wetlands are not permanent.

In Darwin, more occur on saline plains behind the coast during the dry season (May–September) than during the wet season (October–May) (88). It may move to the coast during the wet season because these areas flood. On Eighty Mile Beach, Western Australia, more occur in the dry season (August–November counts: 4000 to 9000) than in the wet season (March–April counts: less than 1300) possibly because near-coastal saline plains are suitable only during the wet season; rainfall there is much less than at Darwin.

TABLE 5.12.  Ranked list of top twenty zones for Red-capped Plover. (maximum counts marked thus: *; other counts are averages) (350 or more birds)

| ZONE | NO. OF COUNTS | NO. OF INDIVIDUALS |
|---|---|---|
| * Eighty Mile Beach, WA | 5 | 9 600 |
| * Far NW lakes, NSW | 4 | 6 800 |
| * The Coorong, SA | 1 | 5 700 |
| * SE corner, Gulf of Carpentaria, Qld | 1 | 3 620 |
| * St Vincent Gulf, SA | 8 | 3 590 |
| * Roebuck Bay, WA | 6 | 3 300 |
| * Spencer Gulf, SA | 9 | 1 910 |
| * Peel Inlet, WA | 10 | 1 780 |
| * Riverina, NSW | 6 | 1 520 |
| Horsham area lakes, Vic. | 10 | 1 200 |
| * W coast, Eyre Peninsula, SA | 8 | 1 040 |
| * Kangaroo Island, SA | 10 | 810 |
| Port Phillip Bay, Vic. | 10 | 510 |
| Lower Darling region, NSW | 6 | 450 |
| * Moreton Bay, Qld | 10 | 450 |
| Kerang–Swan Hill, Vic. | 10 | 450 |
| * Murray mouth lakes, SA | 3 | 420 |
| Swan coastal plain, WA | 10 | 380 |
| SE coast, SA | 10 | 360 |
| * Central western lakes, Vic. | 7 | 350 |

## BREEDING

In coastal southern Australia, the Red-capped Plover breeds between July and March with most observations being between September and December (385). Breeding records from the Darwin area are between March and July. At inland sites it has been recorded breeding in all months of the year in response to unpredictable rainfall and flooding (136).

In coastal areas it breeds most frequently in isolated pairs, but at inland sites loose colonies may form. For example, Hobbs (196) documented a colony, at Fletcher's Lake in south-western New South Wales, of 43 nests spread along 800 metres of shoreline. In Tasmania, between 4 and 10 nests have been found in one hectare of saltmarsh (143).

The usual clutch of 2 eggs is laid in a small scrape in loose sand, shingle, gravel or occasionally short grass (385). In inland areas, the nest is situated near bushes or in the open. The amount of nest material varies, depending on the substrate; at Fletcher's Lake, nests on damp mud all consisted of substantial platforms of water-weed (196). In drier situations, such as on beaches, usually no nest material is used, but occasionally pieces of twig and grass or small stones are placed on the edge of the scrape.

Causes of nesting failure include flooding by high tides and rises in water-level at inland lakes due to flooding or rainfall (136), predation by foxes and scavenging birds (196), accidental or deliberate destruction by humans, or disturbance by wandering domestic animals. At Fletcher's Lake breeding success was quite low (10 per cent of eggs laid produced young that fledged) and fox predation contributed substantially to this (196). Replacement clutches are laid if the nest is destroyed early in the breeding season (143).

## FEEDING

The Red-capped Plover feeds on a wide variety of habitats, from intertidal mudflats, the edges of lakes and sandy beaches, to drier habitats, such as salt-marshes, pasture and gibber plain. It is a surface feeder, seldom walks into water, and locates its prey by sight (351).

In the Coorong, South Australia, 4 collected birds had eaten beetles and *Ruppia* seeds (351). At Lake Reeve, Victoria, it feeds on amphipod crustacea (*Parhyalella* sp.), gastropod molluscs (*Coxiella*), insect larvae (Diptera) and adult beetles of 4 different families (366).

## CONSERVATION

The Red-capped Plover is vulnerable when breeding. Disturbance of incubating adults leads to loss of unprotected clutches and broods from predation by gulls and scavenging birds. Successful breeding to prevent local extinction requires adequate areas of undisturbed habitat.

# BLACK-FRONTED PLOVER

## *Charadrius melanotus*
(Plate 7)

**Other names**   Black-fronted Dotterel; Guttersnipe; Sandpiper.
No races recognised.

## DESCRIPTION

A small, short-legged plover with streaky brown upperparts, black breast-band, red eye-ring, and red bill tipped black.

**Adult plumage**   Forehead, black; crown, streaked brown; black line from lores, through eye to nape; supercilium, white; mantle, brown; scapulars, chestnut; wing-coverts, elongated, brown with pale buff edges giving the appearance of heavy streaking. Chin, throat, belly and vent, white; thick black band on breast, broader at centre. Iris, black; eye-ring, red; bill, short, slender and red with black tip; legs, pinkish-orange. Sexes similar.

**In flight**   Back and rump, pale brown; outer tail feathers have white tips; rest of tail, dark brown; wings strongly marked: black primaries contrast with pale tips to greater coverts, which show as wing-bar; median coverts, streaky brown; lesser coverts, dark grey-brown; wings appear long and rounded in flight.

**Voice**   Single or repeated 'chip . . .', especially when in flight; quieter churring notes.

**Juvenile plumage**   Like adult plumage, but dark forehead and eye-line are washed-out brown; no breast-band; feathers of mantle, scapulars and wing-coverts, edged orange-buff; eye-ring and bill, brown.

**Field notes**   Found in pairs or small parties on small pools and farm dams, as well as large inland wetlands; when flushed, it calls often, landing with fluttering wings then running a short distance before stopping. When feeding, it shows the typical stop-start movement of plovers and bobs its head when approached closely.

**Similar species**   Red-kneed Dotterel; Red-necked Stint. The Black-fronted Plover is smaller, has shorter legs, streaky brown upperparts, thinner breast-band and a brown rather than black crown. The juvenile Black-fronted Plover is distinguished from the juvenile Red-kneed Dotterel by its smaller size, shorter legs and different head markings, and from the Red-necked Stint by its plover-like behaviour and different voice and head markings.

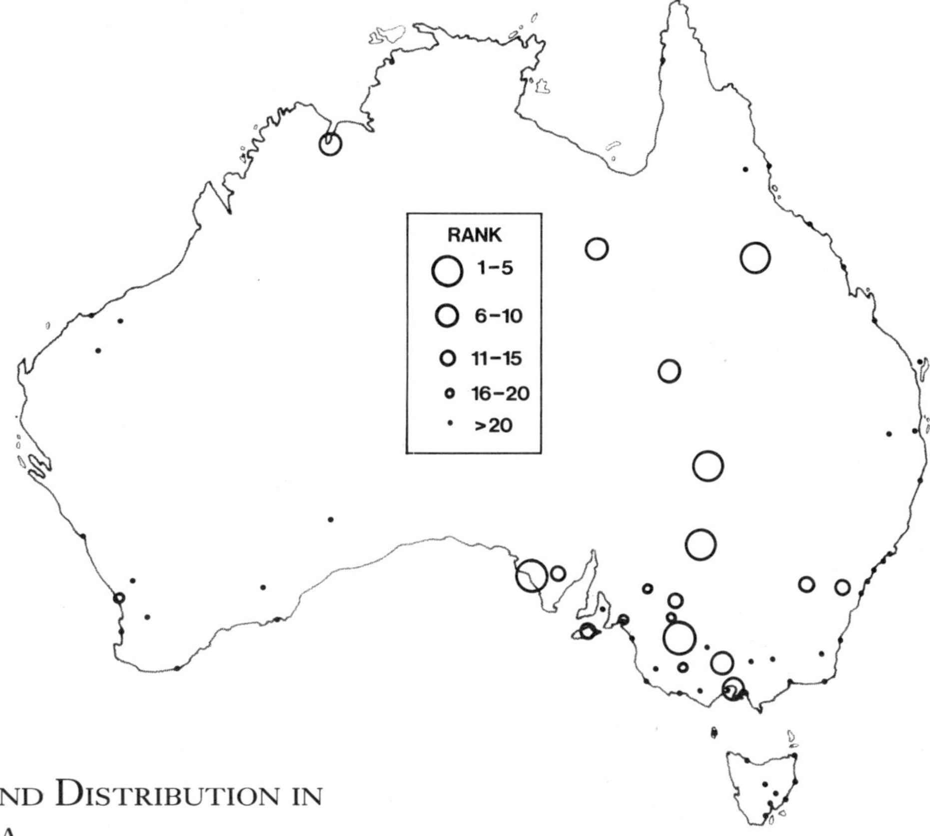


FIG. 5.22.   Location of important zones for Black-fronted Plover.

## STATUS AND DISTRIBUTION IN AUSTRALIA

The Black-fronted Plover is the most widespread of Australia's resident shorebirds. It occurs on wetlands in those parts of Australia subject to regular rainfall and on all the major inland drainage systems (38). It is absent from the most arid areas. (See Figure 5.22, Table 5.13.)

The counts did not cover all habitat suitable for Black-fronted Plovers, so the counts underestimated their abundance. The construction in the inland of farm dams, sewage treatment works and other artificial bodies of water may have greatly increased the amount of habitat available to this species (38). It generally avoids salt lakes.

## MOVEMENTS IN AUSTRALIA

Black-fronted Plovers flock between March and July and disperse to breed between August and February. This movement may be only local in scale. More occur on the Swan coastal plain in summer, possibly due to movement onto near-coastal wetlands when little surface water remains inland.

Figure 5.23 shows the number counted per sheet in Victoria and New South Wales during national counts. This does not show strong summer−winter differences. Before 1982, coverage of the inland was not good, hence the lower score. Numbers increased after the 1982−83 drought broke, due to expanded coverage and possibly because many young birds were present after successful breeding.

TABLE 5.13. Ranked listing of top twenty zones for Black-fronted Plover. (maximum counts marked thus: *; other counts are averages) (30 or more birds)

| ZONE | NO. OF COUNTS | NO. OF INDIVIDUALS |
|---|---|---|
| * Charters Towers region, Qld | 2 | 360 |
| * Far NW lakes, NSW | 4 | 200 |
| * Lower Darling region, NSW | 6 | 150 |
| * W coast, Eyre Peninsula, SA | 8 | 140 |
| Horsham area lakes, Vic. | 10 | 120 |
| * Far SW lakes, Qld | 1 | 120 |
| Port Phillip Bay, Vic. | 10 | 70 |
| * Kimberley region, WA | 6 | 60 |
| Bendigo region, Vic. | 8 | 60 |
| Mount Isa region, Qld | 8 | 50 |
| * Inland Eyre Peninsula, SA | 1 | 40 |
| Mildura area, Vic. | 8 | 40 |
| * Kangaroo Island, SA | 10 | 40 |
| Parkes region, NSW | 8 | 40 |
| Central tablelands, NSW | 7 | 40 |
| NW lakes, Vic. | 3 | 40 |
| Swan coastal plain, WA | 10 | 40 |
| * Murray lands, SA | 1 | 40 |
| * Murray mouth lakes, SA | 3 | 40 |
| Central western lakes, Vic. | 7 | 30 |

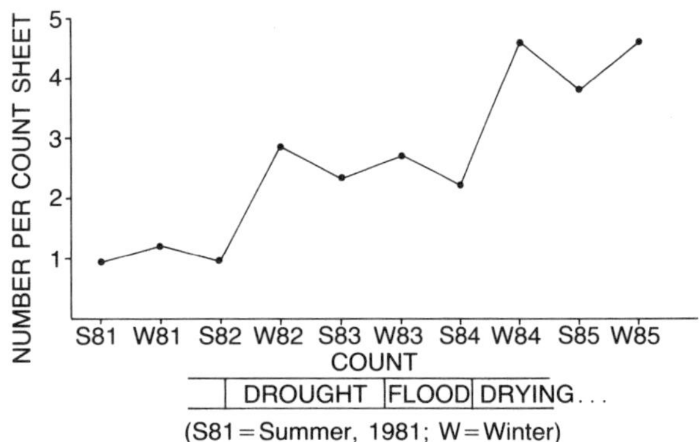



FIG. 5.23. Numbers per count sheet for Black-fronted Plover.
(Victoria and New South Wales)

## BREEDING

The Black-fronted Plover lays its usual clutch of 3 eggs in a shallow scrape located on open ground, often in gravel and not far from water (292, 385). They breed as individual pairs, and colonies have not been reported (385). It has been recorded as breeding between August and March, mostly from September to January (385). Incubation lasts for about 26 days (292), the young leave the nest within 2 or 3 days of hatching, and the adults tend them for a month or more (385). The exact period between hatching and fledging is not known. There have been no detailed studies of the breeding biology of the Black-fronted Plover.

Reasons for nesting failure include disturbance by human activities, predation by ravens and other native and introduced predators, and trampling by domestic animals (292, 385). They often nest on the gravel verges of roads; so nests, adults and young are often run over by cars.

## FEEDING

The Black-fronted Plover forages over wet and dry mud near water, pecking food from the surface. There have been no detailed studies of its feeding behaviour. It has been recorded as feeding on insects, including ants, beetles, flies and small caterpillars, as well as seeds (61, 62).

## CONSERVATION

The Black-fronted Plover is very widespread in Australia and occurs on a variety of wetlands, including man-made dams and sewage farms, indicating that it can adapt to changes in wetlands. There are no immediate or serious threats to its continued survival.

# INLAND DOTTEREL

## *Peltohyas australis*
(Plate 3)

**Other names**   Australian Dotterel; Desert Plover. No races recognised.

## DESCRIPTION

A medium-sized, cryptically marked plover found in desert areas of Australia.

**Adult plumage**   Upperparts, rich buff, heavily spotted black; ear-coverts, white, washed pale chestnut; black vertical line through eye, meeting on front of crown, separating white forehead from crown. Chin, throat and breast, white with black 'V' marking on breast, broader in middle and extending onto sides of neck and nape; broad black bar on belly separating white vent from chestnut sides of breast and flanks. Iris, black; bill, short and black; legs, buffish-grey. It may moult into a non-breeding plumage, the black markings becoming indistinct, giving the bird a paler appearance, but this requires confirmation (291).

**In flight**   Back, buff, spotted black; rump, paler, brownish; tail, brown with pale tip; no distinct markings on wing; primaries, dark brown, secondaries, dark brown with buff edges.

**Voice**   Generally fairly quiet, uttering plover-like notes, often when in flight.

**Juvenile plumage**   Black markings on upperparts are indistinct, giving them a paler appearance. The distinction between this plumage and the possible non-breeding plumage remains to be clarified.

**Field notes**   They live in small loose flocks in gibber and plains country of the inland, spreading out to feed individually at night, when they can often be seen in car headlight beams.

**Similar species**   Black-fronted Plover. The Inland Dotterel is quite distinctive, unlike any other shorebird in habitat preference; markings superficially resemble those of the Black-fronted Plover, but the Inland Dotterel can be easily distinguished as it is twice the size.

## STATUS AND DISTRIBUTION IN AUSTRALIA

Very few were recorded during counts, which did not cover its preferred habitat in the far inland. It occurs south of 20°S in inland desert and semi-arid regions of south-western and south-eastern Australia. Flocks of up to 400 birds have been observed (38).

The Inland Dotterel inhabits open ground with a

scattered covering of shrubs on which it feeds and uses to shelter from the desert sun. Clay pans, gravel flats and gibber plain are favoured (291), but it can venture onto ploughed wheatfields.

## MOVEMENTS IN AUSTRALIA

Very little is known of its movements. It possibly moves south in the spring and north in the summer, but it can move into and out of areas irregularly. Further study is necessary before a complete picture of its movements can emerge.

## FEEDING

The only study was done in north-western New South Wales, based on field observation and a small number of collected individuals (291). During the day it has been observed eating succulent vegetative tips of low shrubs, including *Tetragonia tetragonoides*, *Atriplex halocarpa*, *Bassia ventricosa* and *Babbagia acroptera*. It is thought these are taken primarily for their water content. It has been seen drinking. The Inland Dotterel is relatively inactive during the day, roosting in loose flocks. At night they spread out and feed more actively, predominantly on insects such as beetles, ants and occasionally grasshoppers (291, 295).

## BREEDING

The Inland Dotterel has been recorded breeding from October to February and in April and May, but it is not known if it breeds in winter, or if it breeds at the same time each year; it may breed in response to good rains (289, 291). The nest consists of a scrape in an open, stony position among short vegetation. The clutch is 3 eggs (289). The incubation and fledging periods are not known. Young are fed by their parents until after they can fly (289).

Causes of nesting failure have not been documented.

## CONSERVATION

The Inland Dotterel mostly inhabits the desert areas of Australia where human settlement is sparse, consequently there are probably no immediate or serious threats to its survival.

# BLACK-WINGED STILT

## *Himantopus himantopus*
(Plate 16)

**Other names**  Pied or White-headed Stilt; Longshanks; Stiltbird.

The race in Australia is *H. h. leucocephalus*, which also occurs from the Philippines to Java and New Guinea.

## DESCRIPTION

A tall, slender, large shorebird; black and white with exceptionally long pink legs, which trail in flight.

**Adult plumage**  Head, white; *nape and hindneck, black*; mantle, white; scapulars, back and wings, black. Underparts, white. Iris, black; bill, long, needle-like, straight and black; *legs, very long pink*. Sexes similar.

**In flight**  Very distinctive: rump and tail, white; wings, pointed and black; *long legs trail well beyond end of tail*.

**Voice**  A repeated yapping call, higher pitched in juveniles.

**Juvenile plumage**  Like adults, but lack black nape and hindneck; have a grey smudgy head, nape and hindneck; wing-coverts, scapulars, and feathers of back, black with brown edges; legs, grey; adult plumage is attained within a year.

**Field notes**  They occur singly, in pairs or in small flocks of up to a hundred birds, more commonly on freshwater wetlands than the Banded Stilt or Red-necked Avocet.

**Similar species**  Banded Stilt; Red-necked Avocet. The Black-winged Stilt is longer-legged and taller than the Banded Stilt, with a less extensive white mantle and a black hindneck. It seldom swims. It is less robust than the Red-necked Avocet and lacks the chestnut head and neck, upturned bill and blue-grey legs of that species.

## STATUS AND DISTRIBUTION IN AUSTRALIA

The Black-winged Stilt is found throughout mainland Australia wherever wetland habitat exists, mostly in areas of regular rainfall (38). It is a vagrant in Tasmania. Largest numbers occur in coastal Western Australia. (See Figure 5.24, Table 5.14.) It occurs on a variety of coastal and inland wetlands, generally only in small numbers on intertidal mudflats.

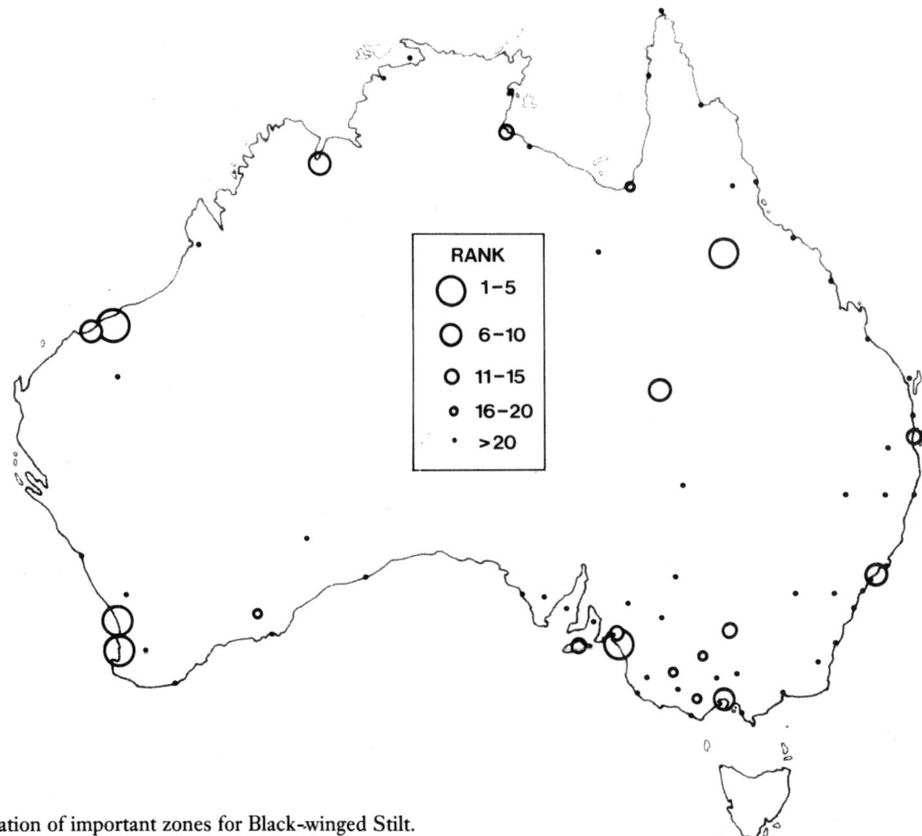


Fig. 5.24.   Location of important zones for Black-winged Stilt.

TABLE 5.14. Ranked listing of top twenty zones for Black-winged Stilt. (maximum counts marked thus: *; other counts are averages) (230 or more birds)

| ZONE | NO. OF COUNTS | NO. OF INDIVIDUALS |
|---|---|---|
| Swan coastal plain, WA | 10 | 1 680 |
| * Port Hedland Saltworks, WA | 12 | 1 230 |
| * The Coorong, SA | 1 | 990 |
| Charters Towers area, Qld | 2 | 940 |
| Peel Inlet, WA | 10 | 850 |
| Kimberley region, WA | 6 | 690 |
| Hunter estuary, NSW | 8 | 550 |
| * Far SW lakes, Qld | 1 | 510 |
| Port Phillip Bay, Vic. | 10 | 510 |
| * Pilbara coast, WA | 3 | 440 |
| Riverina, NSW | 7 | 410 |
| * Murray mouth lakes, SA | 3 | 400 |
| Brisbane area, Qld | 10 | 370 |
| * Kangaroo Island, SA | 10 | 350 |
| * SW corner, Gulf of Carpentaria, NT | + | 340 |
| Horsham area lakes, Vic. | 10 | 330 |
| Kerang–Swan Hill, Vic. | 10 | 300 |
| SE lakes, WA | 9 | 270 |
| * SE corner, Gulf of Carpentaria, Qld | 1 | 230 |
| Western District lakes, Vic. | 9 | 230 |

(+ = aerial count)

## MOVEMENTS IN AUSTRALIA

In Victoria, the Black-winged Stilt shows seasonal movements, being more abundant in summer (average count: summer 2545; winter 673). *Atlas* project records support this; reporting rates in spring and summer were significantly higher than in autumn and winter (27). Winter rains fill many Victorian wetlands, so that by spring they are suitable for breeding. The movement into Victoria coincides with the breeding season (385).

On the Swan coastal plain, Western Australia, numbers were highest in summer 1984, after a dry spring and summer inland, and lowest 5 months later, after a wet autumn inland (based on Kalgoorlie rainfall figures). These changes indicate that movements to and from the coast occur, most likely correlated with changes in inland wetland availability.

At Darwin, Northern Territory, numbers of Black-winged Stilts average ten times higher during the dry season (July) than during the wet season (February). This is possibly due to local movement into and out of the numerous wet-season swamps behind the coast.

There were considerable increases in numbers in Victoria and New South Wales between 1983 and 1985 due to prolific breeding in spring 1983 and 1984 after the 1982–83 drought broke. (See Figure 5.25.) Movements of adults and young from areas farther

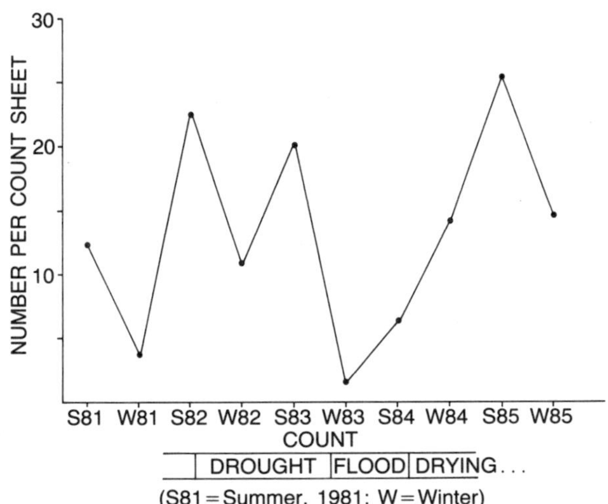


FIG. 5.25. Numbers per count sheet for Black-winged Stilt. (Victoria and New South Wales)

inland that flooded after the drought to those nearer the coast also may have supplemented numbers.

These results show that Black-winged Stilt movements correlate with seasonal and year-to-year fluctuations in rainfall and wetland availability.

## BREEDING

The Black-winged Stilt builds a substantial nest platform of twigs, dry grass herbage or water-weed, usually placed on a small island in shallow water, on tussocks of vegetation or at times floating, but anchored to aquatic plants (385). They nest as isolated pairs or in colonies normally of no more than 25 pairs, but at times colonies of up to a hundred pairs have been observed (385).

The clutch usually consists of 4 eggs, sometimes 2 or 3 and, rarely, 5; and 2, sometimes 3, young usually fledge (385). The breeding season extends from August to April, most records being for October and November (385).

## FEEDING

The Black-winged Stilt feeds by wading in shallow water and mud, and walking on drying mud and sand, pecking and jabbing at small prey items. It has been recorded feeding on brine shrimps (*Artemia* sp.) and brine fly larvae (*Ephydrella* sp.) (217). Stomach and crop contents from specimens indicate that it also takes small shellfish, insects (such as dragonfly larvae and water beetles) and diatoms (61, 62, 300).

## CONSERVATION

The Black-winged Stilt inhabits a variety of wetlands, including artificial ones, such as saltworks and sewage farms. There are no immediate or serious threats to its future survival.

# BANDED STILT

*Cladorhynchus leucocephalus*
(Plate 16)

**Other names**  Bishop or Rottnest Snipe. No races recognised.

## DESCRIPTION

A large, slender, long-necked, long-legged, brown-and-white shorebird, often found in large flocks on salt lakes.

**Adult plumage**  *Head, neck and mantle, white*; scapulars and wing-coverts, sooty black. *Breast has a chestnut-brown band*, broadening at its centre and trailing backwards onto belly. Iris, black; bill, long, thin, straight and black; legs, pink. Sexes similar.

**In flight**  Back, rump and tail, white; inner primaries and secondaries, sooty black, tipped white, showing as *white trailing edges to wings*; legs trail well beyond tail.

**Voice**  A feeble barking sound: 'chuk-uk, chuk-uk . . .'

**Juvenile plumage**  Like adult, but lacks chestnut breast-band; wings browner; some individuals show a grey wash on the head. Individuals are often seen moulting into adult plumage, in which case the breast-band is very patchy; legs, grey. (NB. Many individuals lack the breast-band despite no evidence of recent breeding. A vagrant individual observed in Tasmania lost the breast-band then regained it 5 months later (180). More research is required to elucidate the plumages and moults of the Banded Stilt.)

**Field notes**  The Banded Stilt can be very conspicuous on salt lakes, where it feeds and roosts in tight flocks, often numbering thousands, in shallow water. It has partially webbed feet and can swim well, often far from the shore.

**Similar species**  Black-winged Stilt; Red-necked Avocet. The Banded Stilt can be distinguished by its completely white head, white back, and slightly shorter legs and neck.

## STATUS AND DISTRIBUTION IN AUSTRALIA

Most Banded Stilts in eastern Australia are confined to a small number of sites: the Coorong, St Vincent and Spencer gulfs and the large inland salt lakes (when they hold water) in South Australia, and the salt lakes of western and north-western Victoria. (See Figure 5.26, Table 5.15.) In Western Australia, the southern coastal lakes round Esperance hold large numbers, and there are records of hundreds of thousands on large inland salt lakes in that State (44).

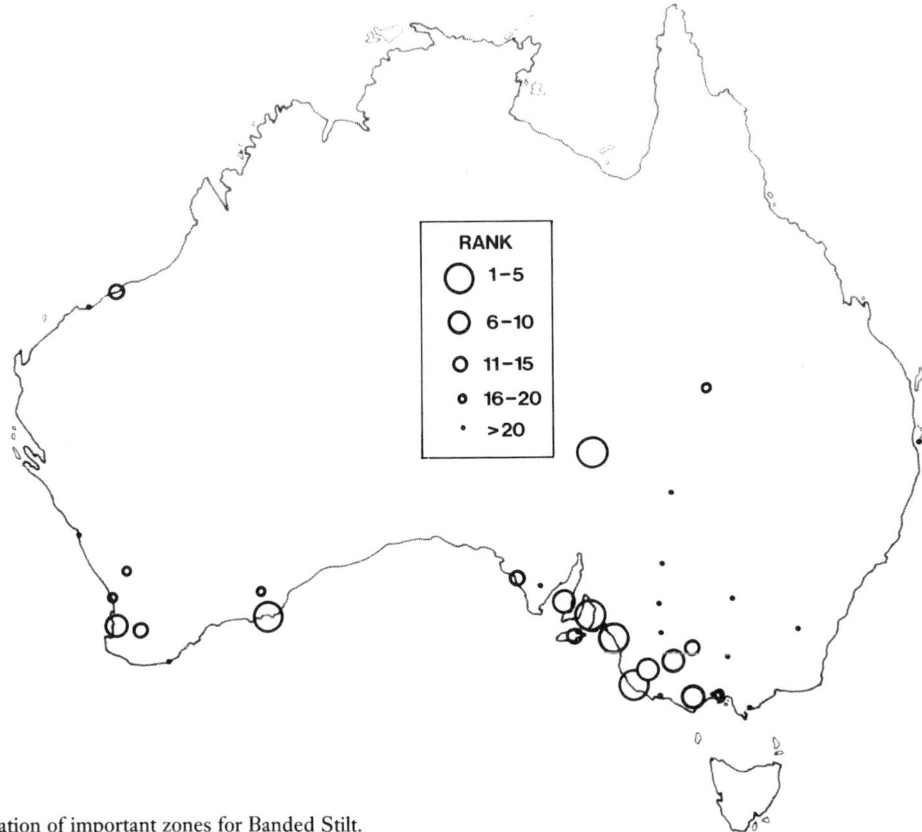


Fig. 5.26. Location of important zones for Banded Stilt.

At the St Kilda saltfields in South Australia, the Banded Stilt occurs in higher densities and is the dominant shorebird on ponds of higher salinity (89). Only in the northern half of Spencer Gulf in South Australia do large numbers of Banded Stilts inhabit intertidal mudflats. This area is an incursion of the sea into the arid zone, and high evaporation rates frequently increase water salinity to twice that of the open sea.

Banded Stilts are the most abundant of Australia's resident waders. The maximum number counted, 133 697, probably underestimates their abundance, as they are highly mobile birds. Indeed, at Lake Barlee in Western Australia, a deserted breeding colony was discovered in September 1980 holding an estimated 179 000 nests (44).

## Movements in Australia

Numbers are highest at St Kilda saltfields in South Australia in late spring, summer and early autumn, suggesting seasonal movements to that site. On inland wetlands, however, numbers fluctuate unpredictably and often dramatically. For example, in January 1983, 24 000 appeared on Lake Wyn Wyn, in Victoria; by April all had left. Near by, at Lake Mitre, in October 1985, an estimated 50 000 were feeding in shallow water, but 6 days later most had departed (3). At salt lakes near Mildura, in Victoria, a flock of over 1000

Table 5.15. Ranked listing of top twenty zones for Banded Stilt. (maximum counts marked thus: *; other counts are averages) (210 or more birds)

| ZONE | NO. OF COUNTS | NO. OF INDIVIDUALS |
|---|---|---|
| * The Coorong, SA | 1 | 77 000 |
| * Lake Eyre, SA | + | 30 000 |
| St Vincent Gulf, SA | 8 | 6 100 |
| SE coastal lakes, SA | 10 | 5 900 |
| Esperance coast, WA | 8 | 2 380 |
| * SE inland lakes, SA | 9 | 2 190 |
| Western District lakes, Vic. | 9 | 2 050 |
| Spencer Gulf, SA | 9 | 1 900 |
| Peel Inlet, WA | 10 | 1 660 |
| Horsham area lakes, Vic. | 10 | 1 560 |
| * Kangaroo Island, SA | 10 | 1 510 |
| W coast, Eyre Peninsula, SA | 8 | 1 340 |
| * SW lakes, WA | 7 | 1 100 |
| Kerang–Swan Hill, Vic. | 10 | 980 |
| * Port Hedland Saltworks, WA | 12 | 800 |
| Port Phillip Bay, Vic. | 10 | 600 |
| * Far SW lakes, Qld | 1 | 500 |
| * SE lakes, WA | 9 | 400 |
| Swan coastal plain, WA | 10 | 240 |
| * Mid-W lakes, WA | 2 | 210 |
| (+ = aerial count) | | |

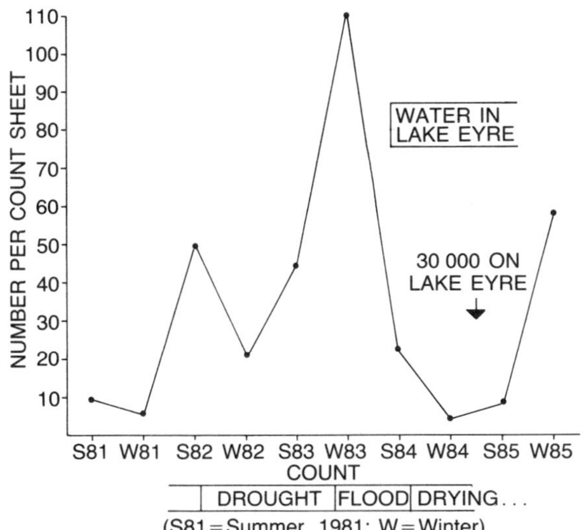


FIG. 5.27.　Numbers per count sheet for Banded Stilt. (Victoria)

disappeared after a heavy fall of rain. During the 1982–83 drought in Victoria, numbers increased, reached a maximum *after* the drought broke, then declined dramatically once Lake Eyre filled in February 1984. (See Figure 5.27.) An aerial survey of Lake Eyre in mid-July 1984 revealed 30 000 there (251).

There is no consistent pattern to Banded Stilt movements. It is nomadic, departing from some areas immediately after rain, from others despite no apparent changes, and arriving elsewhere when conditions are suitable.

## BREEDING

Published breeding records of the Banded Stilt show that it breeds in any month; the filling of salt lakes far inland does not necessarily trigger breeding. Breeding attempts have been listed (38), but few have been described in detail, probably because it breeds in such remote areas.

Between 1971 and 1980 it bred sporadically on lakes Barlee, Ballard and Goongarrie in the eastern gold-fields of Western Australia (38, 236). The only documented breeding records in the eastern part of their range are in South Australia at Lake Callabonna in 1930 and 1936 (38, 277), but many remote salt lakes there are yet to be searched.

It breeds in colonies numbering many thousand nests on low islands in salt lakes (44, 236). Clutches consist of between 1 and 5 eggs, usually 3 or 4 (236). Little else is known of its breeding biology, but breeding attempts often fail if lakes dry quickly (44, 277).

## FEEDING AND FOOD

The Banded Stilt forages on salt lakes, either by wading in shallow water or swimming, often some distance from the shore, using its partially webbed feet.

Salt lakes have a low diversity of potential invertebrate prey, but it is often in great abundance (26). The Banded Stilt feeds on various crustacea and molluscs, including the small abundant gastropod *Coxiella*, brine shrimps (*Artemia* and *Parartemia*), insect larvae (notably, brine-fly larvae, *Ephydrella*), occasionally the isopod *Haloniscus*, various Ostracods and water beetles and bugs (225). Tadpole shrimps (*Triops* and *Lepidurus*) also possibly form part of the Banded Stilt's diet (225). In the Coorong, it feeds on small fish (*Craterocephalus* sp.), beetles (*Clivinia* sp.), the small salt-tolerant gastropod *Coxiella striata* and the seeds and turions (small root stocks) of the salt-tolerant plant *Ruppia maritima* (351). When very large numbers of Banded Stilts were in the Coorong in 1981 and 1982, brine shrimps (*Parartemia* sp.) were not recorded, so it is not dependent on them (351). At Lake Eyre in 1984, it was observed swimming or wading, pecking floating food items, which it located visually; the most likely food was the floating pupae of chironomid midges *Tanitarsus* sp. (251). At Price saltfields, South Australia, it has been observed taking insects from the water surface while swimming (31).

## CONSERVATION

Large numbers of Banded Stilts use all Australian commercial saltworks west of the Great Dividing Range, indicating that they adapt to human-induced changes to wetlands. The increasing salinity of many wetlands in south-western Australia and in the Murray River basin in eastern Australia may make them more attractive to the Banded Stilt.

# RED-NECKED AVOCET

## *Recurvirostra novaehollandiae*
(Plate 16)

*Other names*  Cobbler; Cobbler's Awl; Painted Lady; Scooper; Trumpeter; Yelper.
 No races recognised.

## DESCRIPTION

A large black-and-white shorebird, mostly of inland wetlands.

**Adult plumage**  *Head and neck, dark chestnut* with a varying amount of white or grey around eye and base of bill; rest of body, white except for black primaries, outer secondaries, outer upper wing-coverts and scapulars. Iris, black; bill, *upturned at tip* (more so in female) and black; legs and feet, blue-grey and partially webbed.

**In flight**  Back, rump and tail, white; primaries, outer secondaries, outer wing-coverts and scapulars, black; *rest of wing, white*; long legs trail well beyond tail; distinctive *'floppy' flight*.

**Voice**  A quiet, nasal 'toot toot', often uttered while flying.

**Juvenile Plumage**  Like adult, but head and neck paler chestnut, at times appearing quite washed out; more pale grey or white around eye and base of bill; scapulars, grey.

**Field notes**  The Red-necked Avocet is most often encountered on brackish and salt lakes, often in flocks numbering hundreds; readily swims, feeding by upturning its body, as some ducks do; often scythes bill from side to side when feeding over mud or in shallow water; often associates with the two stilts.

**Similar species**  Black-winged Stilt; Banded Stilt. The Red-necked Avocet is easily distinguished from the two stilts by its upturned bill and chestnut head and neck.

## STATUS AND DISTRIBUTION IN AUSTRALIA

The Red-necked Avocet occurs in largest numbers in the southern half of the continent, including coastal areas between Melbourne and Perth. (See Figure 5.28, Table 5.16.) In the far inland, it occurs in large numbers when conditions are suitable. It is rarely recorded in Tasmania, and occurs irregularly in small numbers in the Northern Territory. Small numbers occur in eastern Queensland south of Townsville, coastal New South Wales and north-western Australia.

The Red-necked Avocet prefers salt and brackish lakes in inland Australia. Smaller numbers visit intertidal mudflats on the coast. It also inhabits all commercial saltworks west of the Great Dividing Range.

National counts did not adequately cover the more remote habitat of the Red-necked Avocet. In September 1984 an aerial survey found 95 000 on Lake Eyre (251), when there were very few elsewhere in eastern Australia. This concentration probably represented the bulk of the species' population in eastern Australia. Their population level probably varies considerably, depending on the availability of large inland wetlands for prolific breeding.

## MOVEMENTS IN AUSTRALIA

More Red-necked Avocets occur on the Swan coastal plain in summer (average count: 559 in summer; 26 in winter). This suggests that it moves out of the inland in summer. Victorian *Atlas* project records (27) show that they are more frequently reported in spring, summer and autumn, when breeding (385). During the 1982–83 drought, numbers increased in Victoria and southern New South Wales. (See Figure 5.29.) They declined when the drought broke in mid-1983. Birds moved to wetlands far inland; in September 1984, 95 000 were seen on Lake Eyre. Numbers increased substantially again in Victoria and southern New South Wales in late 1984, after Lake Eyre started to dry out. At the same time, they appeared on coastal wetlands in New South Wales and Queensland for the first time in some years. These observations indicate that Red-necked Avocets undergo seasonal movements inland in winter, and that this pattern can be interrupted by mass movement out of drought-affected areas or into flooded inland areas.

## BREEDING

The Red-necked Avocet excavates a shallow scrape in gravel, dry mud or sand and lines it with twigs and dry vegetation. The clutch is usually 4 eggs, sometimes 3 (385). It nests in the open or near low herbage, often in colonies on islands in temporary lakes.

The breeding season extends from August to February (385).

## FEEDING

The Red-necked Avocet has a characteristic feeding action; it sweeps its upturned bill from side to side through water, often when up-ending while swimming. It takes prey from the surface or top layer of mud. It also forages in this manner in soupy mud at the edge of water or by pecking the surface of mud or water when wading or swimming. It is mostly a tactile feeder (251).

At Lake Eyre it feeds on the larvae of non-biting midges (Chironomidae; *Tanitarsus* sp.) taken from the surface of submerged mud (251). It is inclined to feed in shallow water, where it can reach the bottom and in wet mud at the edge of water. At Port Hedland Saltworks it feeds on brine shrimps (*Parartemia* sp.) by jabbing in shallow water with a 100 per cent success

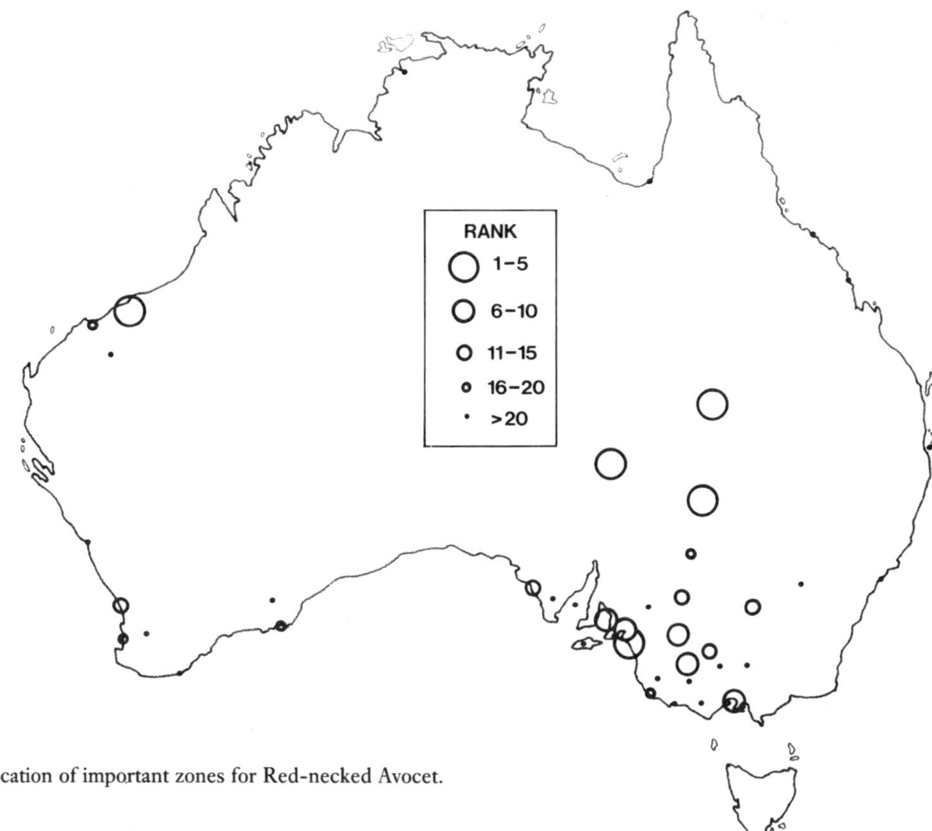



Fig. 5.28. Location of important zones for Red-necked Avocet.



Fig. 5.29. Numbers per count sheet for Red-necked Avocet.
(Victoria and New South Wales)

rate (this study). It has been observed feeding on small fish (*Craterocephalus* sp.) at the Coorong (351).

## CONSERVATION

The Red-necked Avocet is widespread in Australia and capable of breeding in response to temporarily favourable conditions. Provided that an adequate network of inland and near-coastal (drought refuge) wetlands is maintained, its survival will be assured.

TABLE 5.16. Ranked listing of top twenty zones for Red-necked Avocet. (maximum counts marked thus: *; other counts are averages) (100 or more birds)

| ZONE | NO. OF COUNTS | NO. OF INDIVIDUALS |
|---|---|---|
| * Lake Eyre, SA | + | 95 000 |
| * Far NW lakes, NSW | 4 | 6 850 |
| * The Coorong, SA | 1 | 5 400 |
| * Port Hedland Saltworks, WA | 12 | 2 700 |
| * Far SW lakes, Qld | 1 | 1 000 |
| Port Phillip Bay, Vic. | 10 | 980 |
| St Vincent Gulf, SA | 8 | 740 |
| Horsham area lakes, Vic. | 10 | 600 |
| Murray mouth lakes, SA | 3 | 580 |
| * NW lakes, Vic. | 3 | 560 |
| Mildura area, Vic. | 8 | 480 |
| Riverina, NSW | 6 | 420 |
| Swan coastal plain, WA | 10 | 370 |
| Kerang − Swan Hill, Vic. | 10 | 310 |
| * W coast, Eyre Peninsula, SA | 8 | 310 |
| Peel Inlet, WA | 10 | 300 |
| * Pilbara coast, WA | 6 | 300 |
| Lower Darling region, NSW | 7 | 170 |
| SE coast, SA | 10 | 120 |
| Esperance coast, WA | 8 | 100 |

(+ = aerial count)

# RUDDY TURNSTONE

## *Arenaria interpres*
(Plate 5)

**Other names**  Turnstone; Eastern Turnstone; Sea-Dotterel; Calico-bird; Beachbird.

Two races are recognised: *interpres*, breeding throughout the Arctic, and *morinella*, breeding in arctic Canada north to 74°N (82). Birds of the former race occur in Australia. The race *cinclus* described as breeding in eastern Siberia and Alaska and migrating to Australia (65), is not recognised by most authors, who consider it synonomous with *interpres*.

## DESCRIPTION

A small to medium-sized, stocky, short-legged, dark brown shorebird with conspicuous red legs and striking flight pattern.

**Adult non-breeding plumage**  Crown, nape, hindneck, mantle, scapulars and wing-coverts, dark brown edged paler brown, giving upperparts a generally mottled appearance. Rest of head and throat, mottled dark brown; *breast shows black bib, narrower in the centre;* flanks, belly and vent, white. Iris, black; bill, stout, pointed and black; *legs and feet bright orange-red.* Sexes similar.

**In flight**  Shows conspicuous black-and-white markings: back and rump, white; uppertail-coverts, dark brown; *tail, white with broad black subterminal band;* dark brown primaries and secondaries have white bases that show as an *obvious wing-bar.*

**Voice**  Hoarse, high-pitched whistles; descending trill in flight or when disturbed: 'kikikikiki . . .'

**Adult breeding plumage**  The sexes are slightly different. Male: bold black-and-white patterning on head and breast; crown, nape and hindneck, white, finely streaked brown; mantle, scapulars and wing-coverts, rich chestnut, streaked black; broad black centres to outer scapulars show as black line above folded wing. Flanks, belly and vent, white; soft parts same as in non-breeding birds. Female: like male, but duller; less distinctive black-and-white markings on head and breast; heavier streaking on crown, nape and hindneck.

**Juvenile plumage**  Like adult non-breeding plumage, but duller; mantle, scapulars and wing-coverts edged buff.

**Field notes**  Usually feeds individually or in small parties on mudflats, rocky reefs, sandy beaches and among seaweed; almost exclusively coastal. Feeds by flicking over debris with powerful bill to obtain prey; on some islands it is quite tame.

**Similar species**  This shorebird is quite distinctive: its stocky shape, dark markings, red legs and bold flight pattern are characteristic.

## BREEDING RANGE

The Ruddy Turnstone breeds on Arctic coasts no more than a few kilometres from the sea and has a circumpolar range. It also breeds on the coast of the Baltic Sea (65). The nest is placed on gravelly open tundra or shingle shores, occasionally in rank vegetation (85).

## ASIAN RANGE

The Ruddy Turnstone occurs on the coasts of all continents in the non-breeding months.

The Ruddy Turnstone that migrates to Australia comes from breeding grounds in eastern Siberia and possibly Alaska. Its migration takes it along the east coast of Asia and across the Pacific Ocean. (See Figure 5.30.)

Birds passing through the Pribilof Islands in the Bering Sea breed in far eastern Siberia and on St Lawrence Island (466). Colour-marking studies showed that these birds migrated to the Pacific Islands from Hawaii to the Marshall Islands in Micronesia.

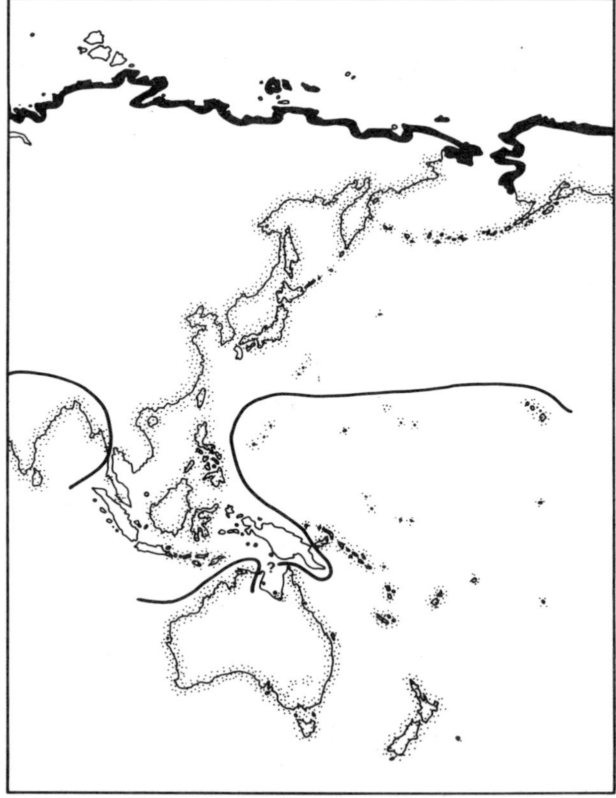

FIG. 5.30.  Breeding and non-breeding range of Ruddy Turnstone.

Small numbers of colour-marked birds from the Pribilofs have been seen elsewhere in the south-western Pacific Ocean: Kiribati, Tuvalu, Samoa, Tonga, Fiji, the Solomons, Vanuatu and New Caledonia. Marked birds also have been recorded on the North Island of New Zealand, and two were seen on the east coast of Australia; at Bowen, Queensland, and Sydney, New South Wales. This suggests that the Ruddy Turnstones of eastern Australia migrate across the Pacific. The same birds probably migrate northwards through Japan (466). Recoveries of birds banded in Japan support this notion (12, 494, 495). Pribilof-marked birds were not recorded in Japan during southward migration, when another population may be passing through on its way to the Philippines (267), where they are not uncommon (171).

Ruddy Turnstones are more abundant in Japan during northward migration (488), whereas in Korea they are numerous during southward migration. They are common during both migrations in eastern China (255), but more abundant during northward migration at Hong Kong (306). They are not common on the Malay Peninsula (348) or Thailand (257), and occur only in small numbers on migration in Vietnam (142, 489), at Brunei (477) and in Sabah (161). They are uncommon in Java (206) and have been recorded from Timor in October and November (285). They are recorded throughout the Pacific Islands, where some remain during the breeding months (216, 220).

Few Ruddy Turnstones occur in New Guinea (36, 138, 184) or in the Gulf of Carpentaria, whereas there are many thousands in eastern and western Australia. This suggests that there are two separate migration streams into Australia: birds that move along the eastern coast of Asia and through the Philippines, coming to north-western Australia; and those that fly across the Pacific Ocean, coming to eastern Australia.

Tens of thousands of Ruddy Turnstones migrate through the Pribilofs (466) and between 4000 and 5000 occur in New Zealand each year (387, 388). Only 2000 to 3000 birds occur in eastern Australia, so it is likely that only a small proportion of the trans-Pacific population reaches Australia. The relative importance of Turnstones in Western Australia to the Asian population is unknown, as there are no records of numbers in China, the Philippines or the islands north-west of Australia.

## Status and Distribution in Australia

Ruddy Turnstones occur on most of the Australian coast. (See Figure 5.31, Table 5.17.) They are widespread in the islands of the Great Barrier Reef (228), but no comprehensive counts have been made. They number in the hundreds on Norfolk and Lord Howe islands. There are few inland records of Ruddy Turnstones.

In the north it prefers places that have wide mudflats, such as the Broome−Port Hedland area in Western Australia, the Darwin area, and the Mackay area and Moreton Bay in Queensland. In the south, by contrast, it is found on rocky coastline, such as that of the western Eyre Peninsula, southern Spencer Gulf and Kangaroo Island in South Australia, the Esperance and Perth regions of southern Western Australia, south-western Victoria, King Island in Tasmania, and central New South Wales. In the south it is generally absent from large embayments with extensive mudflats.

## Movements in Australia

The Ruddy Turnstone first arrives in northern and southern Western Australia and in Darwin in August. On the east coast, it arrives in September and continues migrating southwards until October in Queensland and into November in New South Wales. It arrives in South Australia and Victoria in September and reaches maximum numbers in November or December. There are inland records in south-eastern Australia in October and November. These observations show that it arrives in the north and west earlier than in the south and east. It migrates down the east coast, and some probably fly across the continent to the south coast.

Large numbers stay in the north-west over the non-breeding months. In the south-east, numbers are stable from December to February at most sites, although some movement occurs during this period.

In southern Australia, numbers of Ruddy Turnstones remain high at some sites well into April. Influxes occur in March and April in coastal South Australia and southern and central New South Wales. Ruddy Turnstones banded in Victoria fatten sufficiently before northward departure to fly directly to south-east Asia or nearby Pacific Islands (476). No influxes occur in north-eastern and northern Australia in March and April, and there are no inland records at this time. This suggests that they move short distances along the south coast of Australia before departing, and that they do not touch down in northern Australia. In the third week of April, there is, however, an influx of Ruddy Turnstones at Roebuck Bay, near Broome; banding shows that these birds are different from those that pass southwards through the bay in August and September (247). An explanation for this awaits further research.

## Feeding

The Ruddy Turnstone forages on sandy ocean beaches, rocky shores and intertidal mudflats. It often searches strandline debris for prey, flicking over seaweed, stones and shells and probing under rocks. On islands in the Great Barrier Reef it has been observed

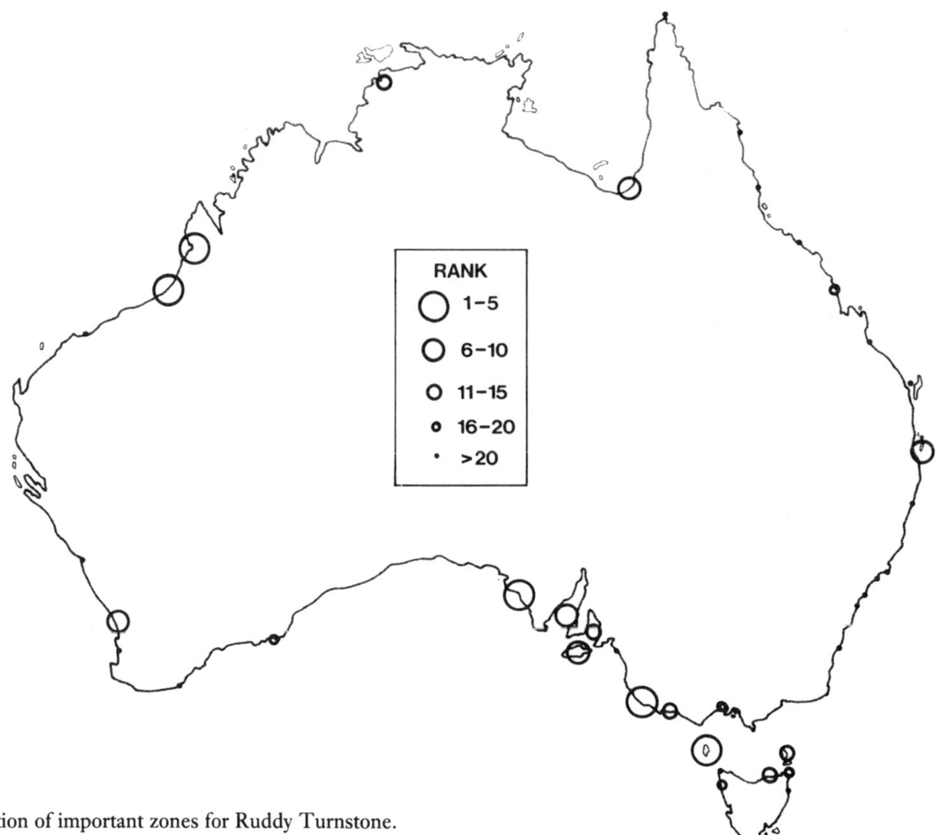


FIG. 5.31.   Location of important zones for Ruddy Turnstone.

TABLE 5.17.  Ranked listing of top twenty zones for Ruddy Turnstone. (maximum counts marked thus: *; other counts are averages) (100 or more birds)

| ZONE | NO. OF COUNTS | NO. OF INDIVIDUALS |
|---|---|---|
| * Roebuck Bay, WA | 6 | 2 060 |
| SE coast, SA | 5 | 900 |
| King Island, Tas. | 3 | 830 |
| * W coast, Eyre Peninsula, SA | 5 | 780 |
| * Eighty Mile Beach, WA | 5 | 740 |
| * Spencer Gulf, SA | 5 | 470 |
| * Kangaroo Island, SA | 5 | 440 |
| * Swan coastal plain, SA | 5 | 280 |
| * Moreton Bay, Qld | 5 | 230 |
| * SE corner, Gulf of Carpentaria, Qld | 1 | 230 |
| SW coast, Vic. | 5 | 210 |
| * St Vincent Gulf, SA | 4 | 180 |
| Darwin area, NT | 4 | 170 |
| Tamar estuary, Tas. | 4 | 140 |
| Furneaux Islands, Tas. | 3 | 130 |
| * Cape Portland coast, Tas. | 5 | 130 |
| * Esperance coast, WA | 4 | 120 |
| * Mackay area, Qld | 4 | 120 |
| Port Phillip Bay, Vic. | 5 | 110 |
| * Central NW coast, Tas. | 1 | 100 |

foraging in this way on the crests of reefs at low tide (115).

In Victoria, it has been recorded feeding on exposed rocky reef, where it takes mostly small gastropod molluscs and some barnacles, and on sandy beach where it feeds on sandhoppers (*Orchestria* sp.) obtained by foraging in beach-washed seaweed (392). The Ruddy Turnstone is opportunistic in its choice of prey, as demonstrated by predation on Sooty Tern eggs at Michaelmas Reef (77) and its curiosity for bathers at Lord Howe Island, where it will take bread, chocolate and other unhealthy morsels thrown to it.

## CONSERVATION

Most Ruddy Turnstones in Australia are confined to a small number of sites around the coast of Australia. The reservation and proper management of these sites is essential to the survival of the Ruddy Turnstone.

# EASTERN CURLEW

## *Numenius madagascariensis*
(Plate 19)

**Other names**   Far Eastern, Australian or Sea Curlew.
No races recognised.

## DESCRIPTION

It is the largest of Australian shorebirds; generally
mottled brown and buff, with an exceptionally long
down-curved bill and long neck and legs.

**Adult non-breeding plumage**   Head, hindneck,
sides of neck, mantle, scapulars and wing-coverts, light
brown, heavily speckled darker brown. Chin, throat,
foreneck, sides of neck and breast, pale buff, heavily
streaked dark brown; supercilium, paler; lower breast
and flanks, off-white, lightly streaked brown; belly,
pale buff; vent, pale buff with fine brown streaks. Iris,
dark brown; legs, blue-grey; *bill, exceptionally long,
down-turned*, black, grading to grey at base of upper
mandible and pink at base of lower mandible. Sexes
similar in plumage, but females are noticeably larger
than males.

**In flight**   *Back and rump, brown*, with darker speckling; tail, buff, heavily barred dark brown; outer primaries and greater coverts, dark brown; inner primaries and secondaries, buff, heavily barred brown; no
wing-bar; underwings white, heavily barred dark
brown; fly with slow wing-beats.

**Voice**   Hoarse, repeated, ascending 'cheer-loo,
cheer-loo ...', most often uttered in flight; also, a
hoarse 'cheer ...'

**Adult breeding plumage**   Not very different from
non-breeding plumage; markings more accentuated;
generally more buff.

**Juvenile plumage**   Like adult plumage, but underparts more finely streaked.

**Field notes**   It occurs solitarily or in loose parties
when feeding, often flying in long lines to roost, calling
as it goes. It roosts in mangroves or trees, but more
often on sandy spits or in low saltmarsh. In northern
Australia, it is often encountered as a solitary bird
along mangrove-fringed beaches or tidal rivers.

**Similar species**   Whimbrel; Eurasian Curlew. The
Eastern Curlew has a longer bill and a dark rump,
lacks a distinctive supercilium and lateral crown stripes
and is larger than the Whimbrel. It is separated from
the rare Eurasian Curlew by barred, not white, underwing, and dark, not white, rump.

## BREEDING RANGE

The Eastern Curlew breeds in eastern Siberia, from
the Kamchatka Peninsula to trans Baikalia, northern
Mongolia, northern Manchuria and the far south-east
of Siberia (7, 171).

It builds a nest lined with dry grass and twigs in
swampy ground close to areas where wild berries grow
(338). When nesting, it is subject to predation by foxes.
Draining of wetlands in far eastern USSR has led to
its disappearance from parts of its range over the last
20 years (338).

## ASIAN RANGE

The Eastern Curlew is common on migration in Korea
(162) and over 600 were counted on the Nakdong
estuary in September 1984 (364). In Japan, up to 255
have been counted on northward and 1377 on southward migrations (488). It is uncommon during migration in eastern China (58, 129) and Hong Kong
(54). It is rarely reported in Thailand (308), Malaysia
and the Philippines (110), but is common on migration
in Borneo (440). It occasionally strays to Fiji, Western
Samoa (120) and New Zealand in small numbers (387,
388). Its non-breeding range includes southern Korea
(scarce) and Taiwan (129), but most probably come to
Australia (171). (See Figure 5.32.)

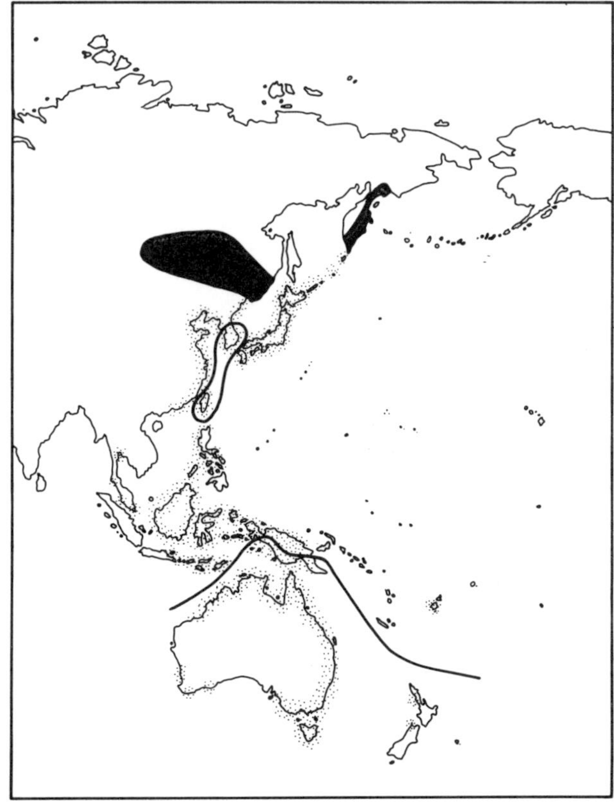

FIG. 5.32.   Breeding and non-breeding range of Eastern Curlew.

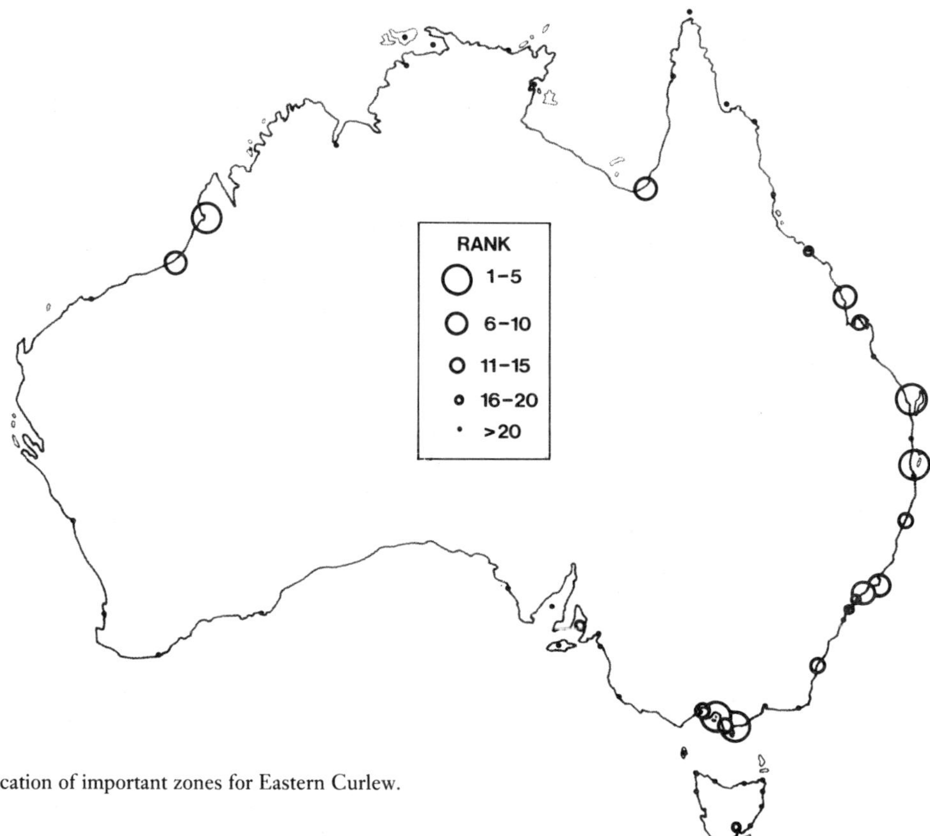



Fig. 5.33.   Location of important zones for Eastern Curlew.

Table 5.18.   Ranked listing of top twenty zones for Eastern Curlew. (maximum counts marked thus: *; other counts are averages) (100 or more birds)

| ZONE | NO. OF COUNTS | NO. OF INDIVIDUALS |
|---|---|---|
| * Hervey Bay – Great Sandy Strait | 5 | 1 970 |
| Corner Inlet, Vic. | 5 | 1 740 |
| * Moreton Bay, Qld | 5 | 1 080 |
| Westernport Bay, Vic. | 5 | 1 020 |
| * Roebuck Bay, WA | 6 | 960 |
| * Mackay area, Qld | + | 710 |
| * SE corner, Gulf of Carpentaria, Qld | 1 | 690 |
| Port Stephens, NSW | 4 | 530 |
| Hunter estuary, NSW | 4 | 490 |
| * Eighty Mile Beach, WA | 5 | 480 |
| * Broad Sound & Shoalwater Bay, Qld | + | 430 |
| * N coast, NSW | 5 | 390 |
| * S coast, NSW | 5 | 350 |
| Anderson's Inlet, Vic. | 4 | 230 |
| Port Phillip Bay, Vic. | 5 | 160 |
| Derwent estuary & Pittwater, Tas. | 4 | 150 |
| St Vincent Gulf, SA | 4 | 130 |
| Botany Bay, NSW | 5 | 120 |
| * Central coast, NSW | 3 | 100 |
| * Townsville area, Qld | 5 | 100 |

(+ = aerial count)

# Status and Distribution in Australia

Eastern Curlews are most abundant at a small number of large marine embayments on the eastern and south-eastern coasts of Australia and at Roebuck Bay in the north-west. (See Figure 5.33, Table 5.18.) There are very few in south-western Australia. It is exclusively coastal.

It occurs mostly on intertidal mudflats, particularly those with extensive seagrass meadows (Zosteraceae). It is rarely seen feeding on near-coastal, non-tidal wetlands, although it occasionally roosts in them at high tide. Saltmarsh behind mangroves or sandy beaches and spits are its preferred roosting habitat.

It has declined in parts of south-eastern Australia, especially in St Vincent Gulf, South Australia, and in the Hobart area, Tasmania (63). The reasons for this decline are not clear, but hunting and habitat destruction on the breeding grounds may be responsible (338).

# Movements in Australia

The Eastern Curlew arrives in north-western and eastern Australia as early as late July. Those in the north-west move on by October. Influxes occur on the east coast from Torres Strait southwards, in September, October and November. Most arrive in Victoria in November. Small numbers migrate westwards along the Victorian coast as early as August. This indicates that it migrates through the north-west and down the east coast to southern Australia.

Numbers are stable in most sites in south-eastern Australia between December and February, indicating that little movement occurs at this time.

Influxes occur in March and April, along the coast of New South Wales. No influxes occur at this time north of Brisbane or in north-western Australia. This suggests that the Eastern Curlew moves northwards along the east coast before departure, then overflies northern Australia.

Influxes of Eastern Curlews occur on the east coast of Queensland in May and June. July numbers in northern New South Wales and Queensland are between 30 and 60 per cent of February numbers, whereas in southern New South Wales, Victoria and Tasmania, they are usually less than 10 per cent of February numbers. This suggests that non-breeding birds move north along the east coast after the adults have departed.

## FEEDING

Like most coastal shorebirds, the Eastern Curlew forages at low tide on intertidal mudflats and roosts at high tide. Each individual can hold a feeding territory, but in some circumstances they feed in loose flocks. (See Chapter 3.)

At Westernport Bay, Victoria, it feeds on ghost shrimps (*Calianassa* sp.) and other crustacea (101), obtained by probing its bill down burrows.

## CONSERVATION

The Eastern Curlew is particularly sensitive to human disturbance when both feeding and roosting. It is often the first species in a high-tide roost to take flight if disturbed, moving to an alternative roosting site, often many kilometres away (264).

In the past the species was frequently shot for food (349), and in Tasmania this is thought to be a reason for its decline (63). Curlews are large, easy targets, and considerable numbers are probably shot in Asia for food (338). This probably accounts for its exceptional wariness.

Most Eastern Curlews counted were confined to ten sites in south-eastern Australia. Those sites that are not yet reserved should be protected and managed to ensure the continued existence of suitable, undisturbed feeding and roosting habitat.

# WHIMBREL
## *Numenius phaeopus*
(Plate 19)

**Other names**   Jack or Little Curlew; Mayfowl; Shipmate.

There are four races of the Whimbrel: *hudsonicus*, *phaeopus*, *alboaxillaris* and *variegatus* (85). Only the last visits Australia.

## DESCRIPTION

A smaller version of the Eastern Curlew with shorter bill and legs, a white rump and more distinct head markings.

**Adult non-breeding plumage**   *Crown, striped dark brown and white*: dark brown lateral crown stripes and whitish median crown stripe. Nape, hindneck, mantle, scapulars and lesser and median wing-coverts, brown with heavy black speckling. Chin and ear-coverts pale buff; supercilium, white; dark brown eye-stripe; throat and breast, buffish-white with heavy brown streaking; flanks, white with brown barring; belly and vent, white. Iris, dark brown; bill, long, *down-curved*, black at tip, grading to grey at base of upper mandible and pink at base of lower mandible; legs, bright blue-grey. Sexes similar.

**In flight**   *Back and rump, white*; uppertail-coverts, white, sparsely barred dark brown; tail, buffish-white, barred dark brown; outer primaries, black to dark brown; inner primaries and secondaries, barred dark brown on white; primary coverts, black to dark brown; secondary coverts, barred dark brown and buff; no wing-bar.

**Voice**   A rapid, hoarse trill, often uttered when disturbed; 'che-che-che-che-che-che-che ...', descending in scale.

**Adult breeding plumage**   Like non-breeding plumage, but generally more buff.

**Juvenile plumage**   Not unlike adult plumage, but head darker and markings generally less distinct; coverts and scapulars, more speckled buff.

**Field notes**   It is seen most often in northern Australia; lives singly or in small parties along mangrove-fringed tidal rivers and shorelines or on coastal islands; roosts in trees, mangroves or on sandy or rocky beaches and forages among quite tall mangroves as well as on open mudflats.

**Similar species**   Eastern Curlew; Little Curlew; Eurasian Curlew; Bar-tailed Godwit; Black-tailed Godwit. The Whimbrel is distinguished in flight by its white rump and back, and when standing, by its shorter bill and legs and striped crown. It is larger than the

Little Curlew, which lacks the white rump and back, and has a shorter bill. It is smaller than the rare Eurasian Curlew and has a down-curved bill, unlike the similar-sized Bar-tailed Godwit and Black-tailed Godwit, which have straight bills.

## BREEDING RANGE

Members of the race *variegatus* that visit Australia breed from possibly as far west as the south-eastern Taimyr Peninsula eastwards to north-eastern Siberia (85). The limits of its breeding range are not completely known. It is known to nest near the coast, but is not dependent on it. Three other races breed from the Taimyr Peninsula to Iceland (*phaeopus*), the lower Volga River south and east of the Urals (*alboaxillaris*) and in Alaska and Canada (*hudsonicus*).

## ASIAN RANGE

The different races occur in different areas during the non-breeding months. Those of the race *variegatus* occur from the Burmese side of the Bay of Bengal, through south-east Asia to the Pacific Islands, Australia and New Zealand; of *hudsonicus* in southern North America and Central and South America, as well as the eastern Pacific Islands and occasionally New Zealand; of *phaeopus* in western and southern Africa as well as western Asia; and of *alboaxillaris* in the western Indian Ocean (85). The last two races have been considered by some authors to be synonymous (85). The eastern limit of the non-breeding range of *phaeopus* is not clear, and specimens from the Indian subcontinent, Thailand and Indonesia may be of both *phaeopus* and *variegatus* (5, 85, 382).

Those of the race *variegatus* occur in Korea and Japan only on migration, being more common on northward migration (162, 488) (maximum in Japan: 6800 in April and 1200 in September). In China and Hong Kong, they are common on migration from late July to September (until October in the south) and in April and May (54, 255). They migrate through Taiwan, but some occasionally stay for most of the non-breeding months (401). In the Philippines they are common and widespread (110, 171). There have been two recoveries of Whimbrels banded in the Philippines; one in north-eastern Siberia, presumably on the breeding grounds, and one on the Soviet island of Sakhalin in September, while on southward migration (267). They are common visitors to Thailand (257), and have been described as 'a [non-breeding] visitor in impressive numbers' in southern Vietnam (489). They are widespread on the Malay Peninsula, numbering in the hundreds in places (174, 348). In coastal Borneo, they are common (440). White (487) documents 74 specimens from the islands immediately north-west of Australia, most collected in August and September, suggesting that they are mainly transient, although present for most of the non-breeding months. In New

Guinea, they are widespread, being recorded from north-western Irian Jaya (207), the Solomon Islands (176) and round Port Moresby (138). Up to 1100 have been seen in south-eastern Irian Jaya (36). A bird banded in Japan during northward migration in 1969 was recovered on the Solomon Islands in November 1974 (11).

In the Pacific, it occurs regularly on most islands of western Micronesia (18), Fiji (405, 406), the Kermadecs (312) and the Cook Islands (24). Small numbers visit New Zealand (387, 388). Numbers in New Zealand are small compared to those on the east coast of Australia suggesting that the bulk of the population does not migrate beyond Papua New Guinea and Australia. (See Figure 5.34.)

Small numbers of non-breeding birds spend the breeding months on the Pacific Islands (216, 220), in Java (205, 206) and in Borneo (440).

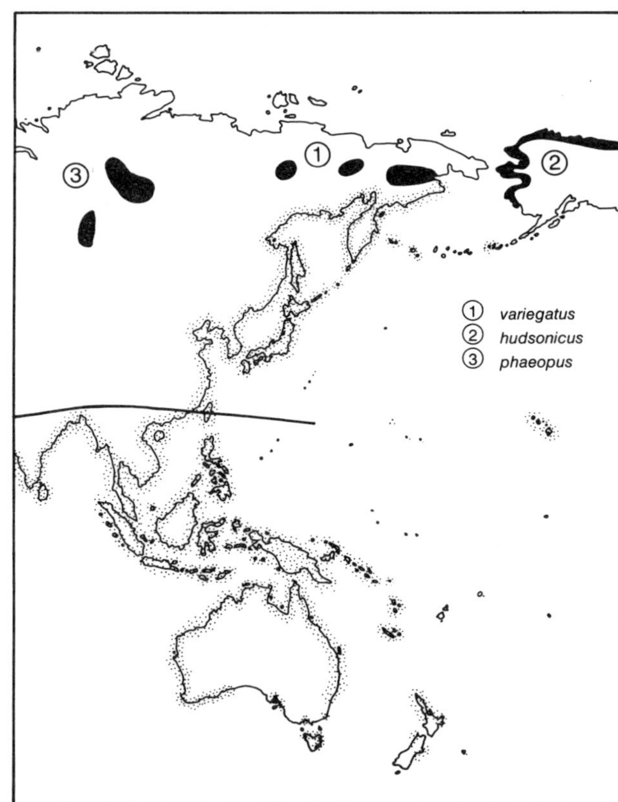


FIG. 5.34.   Breeding and non-breeding range of Whimbrel.

## STATUS AND DISTRIBUTION IN AUSTRALIA

The Whimbrel occurs in greatest numbers on the northern coast of the continent. Occurrences farther south, although regular, are of very small numbers, the maximum being 90 in St Vincent Gulf, South Australia. (See Figure 5.35, Table 5.19.)

FIG. 5.35.   Location of important zones for Whimbrel.

TABLE 5.19.   Ranked listing of zones holding 20 or more Whimbrels. (maximum counts marked thus: *; other counts are averages)

| ZONE | NO. OF COUNTS | NO. OF INDIVIDUALS |
|---|---|---|
| * SE corner, Gulf of Carpentaria, Qld | 1 | 1 100 |
| * Roebuck Bay, WA | 6 | 1 020 |
| * Broad Sound & Shoalwater Bay, Qld | + | 570 |
| * N coast, NSW | 5 | 340 |
| Moreton Bay, Qld | 5 | 280 |
| * Eighty Mile Beach, WA | 5 | 180 |
| Mackay area, Qld | 4 | 140 |
| * Pilbara coast, WA | 3 | 120 |
| * St Vincent Gulf, SA | 4 | 90 |
| * Hervey Bay – Great Sandy Strait, Qld | 5 | 80 |
| Townsville area, Qld | 5 | 70 |
| * Port Hedland Saltworks, WA | 12 | 50 |
| Darwin area, NT | 4 | 40 |
| * NE Arnhem Land coast, NT | + | 40 |
| Hunter estuary, NSW | 4 | 30 |
| Port Stephens, NSW | 4 | 30 |
| Cairns area, Qld | 4 | 30 |
| * S coast, Gulf of Carpentaria, NT | + | 20 |

(+ = aerial count)

The Whimbrel forages at low tide on open tidal mudflats and along the banks of tidal creeks and rivers. On coral islands, it forages on coral rubble and reef flats exposed at low tide (78, 115). On rising tides it roosts either on sandy or rocky beaches, or in mangroves and trees on the coast.

## MOVEMENTS IN AUSTRALIA

The Whimbrel migrates southwards through Roebuck Bay, in north-western Australia, in August and September. There are influxes at sites right along the east coast during southward migration; in August north of 20°S and in September and October south of 20°S, suggesting that it moves along the east coast.

No sites show a peak in numbers between December and February, suggesting that the Whimbrel spreads out along the coast in low densities, flocking only when migrating. It does not use Roebuck Bay on northward migration. There are influxes at sites along the east coast at this time; in February and March south of 20°S and in March and April north of 20°S, suggesting that it moves up the east coast.

## FEEDING

The Whimbrel forages mostly on intertidal mud, probing with its bill to obtain food. At One Tree Reef on the Barrier Reef it feeds on reef flats, where it probes into percolation holes in the hard reef substrate (115). There are no detailed observations of the diet of the Whimbrel in Australia. In the Kermadec Islands, the stomach of a specimen contained two small crabs (312).

## CONSERVATION

The Whimbrel occurs predominantly in northern Australia, where its habitat is relatively free from immediate threats. Recreational and real-estate development along the southern Queensland and northern New South Wales coasts could potentially reduce its feeding and roosting habitats.

# LITTLE CURLEW
## *Numenius minutus*
(Plate 19)

**Other names** Little Whimbrel; Pigmy Curlew; Dough-bird.

No races recognised, although early authors considered the Little Curlew and Eskimo Curlew (*N. borealis*) to be conspecific (7).

## DESCRIPTION

A large to medium-sized, mottled, brown-and-buff shorebird of open grass plains in northern Australia.

**Adult non-breeding plumage** *Crown and nape, dark brown*; hindneck and sides of neck, cream-buff with dark brown streaks; mantle, scapulars and wing-coverts, dark brown, speckled cream-buff; lores and ear-coverts, white, finely streaked dark brown. Chin and throat, white; breast, cream-buff, streaked dark brown; belly and vent, white; flanks, cream-buff, streaked dark brown. Iris, black; bill, long, down-curved with pink base; legs, grey. Sexes similar.

**In flight** Back and rump, dark brown, spotted cream-buff; tail, light brown, barred dark brown; lesser coverts, dark brown; median and greater coverts, dark brown, spotted and barred cream-buff; primaries and secondaries, dark brown with varying amounts of paler brown barring; *shows no distinctive markings when in flight*.

**Voice** Utters loud 'chew chew chew' or 'tlee tlee tlee' when disturbed, various other hoarse calls including one similar to that of the Lesser Golden Plover: 'cheedlechee'.

**Adult breeding plumage** Similar to adult non-breeding plumage, but more buff on underparts; spotting on upperparts more buff.

**Juvenile plumage** Similar to adult non-breeding plumage, but upperparts generally darker and less distinctly spotted.

**Field notes** Feeds in loose flocks in short grassland in northern Australia; sometimes seen in very large numbers; large flocks of panting birds can be seen drinking at waterholes during the hot part of the day; often seen in northern towns on playing fields, airfields and garden lawns.

**Similar species** Whimbrel; Lesser Golden Plover. The Little Curlew is about 30 per cent smaller than the Whimbrel and has a shorter bill. It resembles the Lesser Golden Plover in its behaviour and in flight, but has a longer, down-curved bill with a pink base.

## BREEDING RANGE

The Little Curlew breeds in a zone either side of the Arctic Circle in Siberia from the Indigirka River in the east, westwards to the Kochechumo River. (See Figure 5.36.) Its distribution is thought to be patchy within this region due to its preference for secondary vegetation, which regenerates after fires or storms (239).

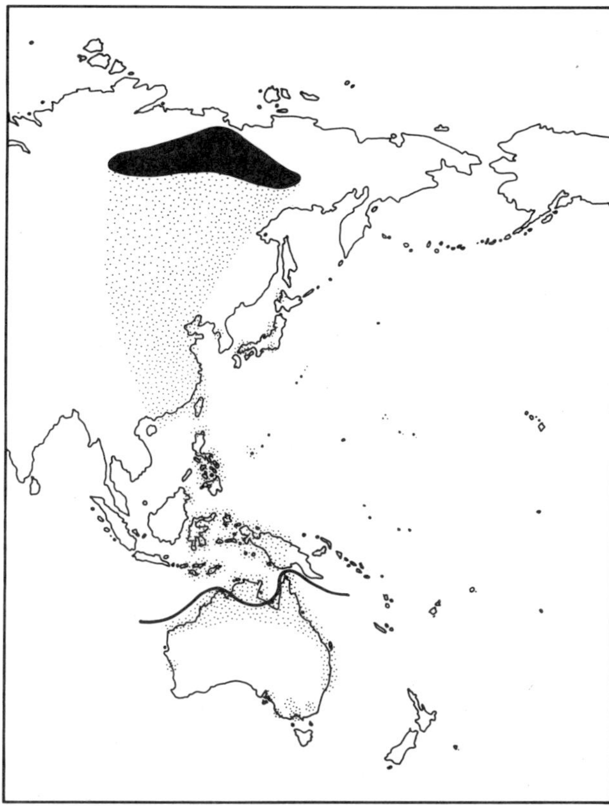

FIG. 5.36.   Breeding and non-breeding range of Little Curlew.

## ASIAN RANGE

The Little Curlew occurs only in Asia and Australia. It is abundant in southern Siberia, Mongolia and China between mid-August and late September on southward migration, arriving in southern China by mid-September, where small numbers stay until early November (179, 237, 239, 255). It is a vagrant in Korea (162), but occurs more regularly in Japan: max. 71, April 1978 (488). In the Philippines, it is a rare straggler (110). It has been recorded in Belau, in western Micronesia (345). South of China, its migration route may lie east of Borneo, where there are a few records between mid-September and late November (440), but only a handful of specimens have been collected in the islands immediately north-west of Australia (487). In New Guinea, however, it is fairly common in the east (184), but on the grassy plains of southern New Guinea it numbers in the thousands from mid-September onwards (36, 139). The finding of a dead Little Curlew at 4450 m on Mount Carstenz (391) suggests that it may migrate across the New Guinea cordillera.

Most of the population of the Little Curlew occurs in northern Australia and possibly southern New Guinea between November and March. (See Figure 5.36.) The extent of movement between northern Australia and southern New Guinea remains unclear.

Little is known of its northward migration, but it occurs again in south-eastern China between late April and mid-May (54, 255) and arrives on the breeding grounds in late May (239).

The lack of reports of large numbers between China and southern New Guinea and Australia and the timing of movements between the two areas suggests that the Little Curlew undertakes a non-stop migration of about 5000 km over New Guinea, eastern Indonesia and the Philippines.

## STATUS AND DISTRIBUTION IN AUSTRALIA

Little Curlews are abundant throughout northern Australia, especially in the top end of the Northern Territory, the Gulf Country of the Northern Territory and Queensland, and in north-western Australia. (See Figure 5.37, Table 5.20.) Small numbers reach southern and eastern Australia (272).

The Little Curlew occurs most frequently on short or sparse grassland, such as that on black soil flood plains during the dry season, as well as on the short grass of airfields, playing fields and garden lawns (261, 416). It rarely uses intertidal mudflats (126). On Eighty Mile Beach, Western Australia, it roosts among other shorebirds.

Once dry grassland becomes wet, it leaves it and moves inland (90). Its preference for dry habitat may explain observations of large flocks drinking at small pools in the heat of the day (29, 153).

Flocks of Little Curlews totalling 250 000 were observed migrating over Fogg Dam near Darwin in late October 1966 (425). There are other reports of flocks numbering in the thousands from other parts of northern Australia (153, 472). They are widespread on the grassy plains of northern Australia and counts did not cover this habitat. Consequently, counts underestimate their abundance.

## MOVEMENTS IN AUSTRALIA

The Little Curlew arrives in northern Australia in mid-September, but the largest influxes have been recorded in October (90, 261, 425). On Eighty Mile Beach, Western Australia, many thousands are present by late October. The subcoastal plains round Darwin

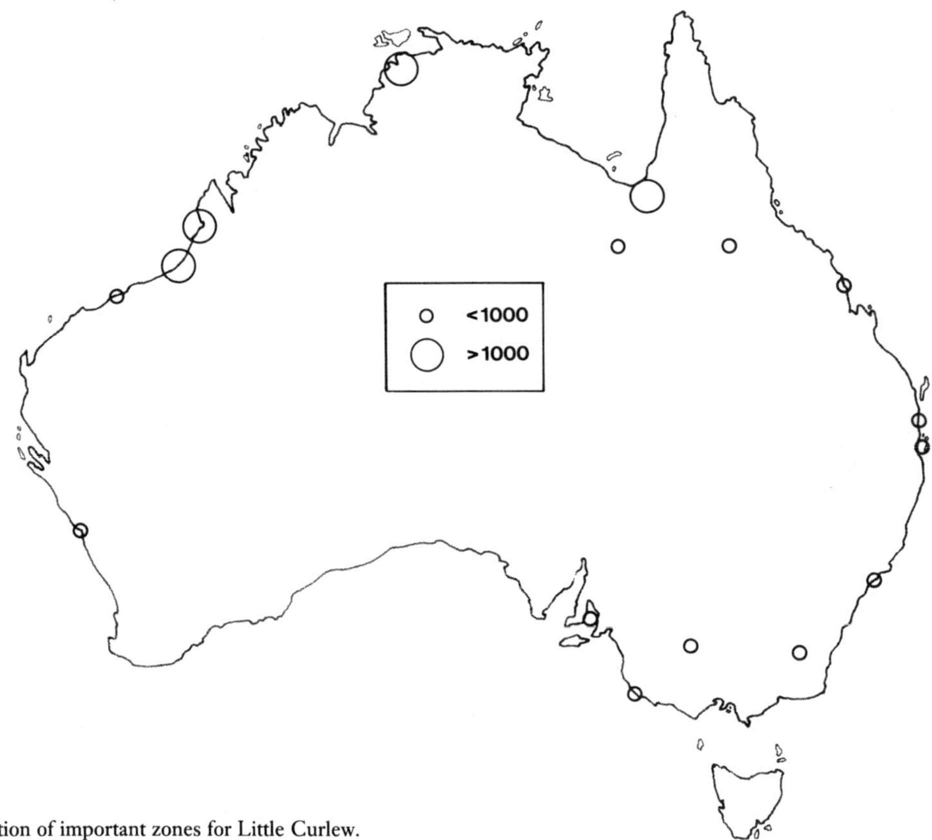



Fig. 5.37.   Location of important zones for Little Curlew.

TABLE 5.20.  Ranked list of zones that held more than 1000 Little Curlews. (all counts are maxima)

| ZONE | NO. OF COUNTS | NO. OF INDIVIDUALS |
|---|---|---|
| Roebuck Plains, WA | 6 | 30 000 |
| Anna Plains, WA | 5 | 12 000 |
| Karumba Plains, Qld | 1 | 6 390 |

NB.  An estimated 250 000 were observed round Fogg Dam near Darwin, NT, in 1966 (425) and this is shown on Figure 5.37.

are considered to be a staging area and, with the onset of the wet season in late November, they disperse inland (90).

Between December and February the Little Curlew has been recorded in largest numbers at Innisfail, north Queensland (416), on the flood plains of the Karumba area of Queensland (153), in the Richmond district, Queensland (29), and on the drier grasslands of the interior of the Northern Territory and Queensland (90).

These observations suggest that the Little Curlew probably moves into northern and north-western Australia prior to the wet season, then spreads southwards and eastwards to as far as the north-eastern coast of Queensland. The movements of the Little Curlew during the non-breeding months and its distribution from year to year may be greatly influenced by rainfall patterns over northern Australia (272).

Large numbers have been observed during northward migration in April and May on the plains between Darwin and Oenpelli (272). More than 30 000 were present on Roebuck Plains, Western Australia, in early April 1985; many were fattening for migration and by the third week of April most had departed. These observations suggest that they move north-westwards again by April.

## FEEDING

The Little Curlew feeds in dry grasslands. At Broome, it uses these during the day, returning to beaches at night (261). Conversely, at Eighty Mile Beach it roosts on beaches during the day, presumably feeding on nearby grassland in largest numbers at night, when insects are most active.

Grasshoppers, beetles, caterpillars and vegetable matter, notably rice husks, are eaten by the Little Curlew (90, 272, 472).

## CONSERVATION

The Little Curlew is very closely related to the possibly extinct Eskimo Curlew (*N. borealis*) and the two have been considered conspecific by some people (7). The two differ slightly in size and plumage (239). The Eskimo Curlew was abundant in the Americas until the 1870s, but has declined dramatically since (7, 28). The last record was in 1980 in Canada (7). In North America it was very fat prior to what was probably a non-stop flight southwards across the Atlantic to South America (28), and it was particularly favoured by hunters. Its tameness, a trait of the Little Curlew as well, made it easy game.

Fortunately, a similar fate has not befallen the Little Curlew, and it is an abundant wader. At present there are no apparent threats to its population here, as favoured habitat in northern Australia is very extensive. Any plans for intensive agricultural development or extensive use of pesticides in northern Australia should be properly investigated as they could affect its favoured habitat.

# WOOD SANDPIPER
## *Tringa glareola*
(Plate 8)

**Other names**
No races recognised.

## DESCRIPTION

A small to medium-sized sandpiper with grey upperparts, spotted white; rarely occurs on the coast.

**Adult non-breeding plumage**  Crown, grey-brown, spotted white; nape, hindneck and sides of neck, grey, streaked white; mantle, scapulars and wing-coverts, dark grey, notched white, *appearing generally spotted*; supercilium, white; lores and ear-coverts, streaked grey. Chin and throat, white; breast, washed grey with darker streaking at sides; belly, flanks and undertail-coverts, clean white. Iris, black; bill, short, straight, black with olive base; legs, pale olive-yellow.

**In flight**  Back, grey; *rump and uppertail-coverts, white; tail, white with fine black barring*; no conspicuous wing-bar; underwing-coverts, white.

**Voice**  Loud metallic whistle of four quick notes, often uttered when disturbed: 'chink chink chin chin'.

**Adult breeding plumage**  Similar to non-breeding plumage, but upperparts more heavily spotted; head and breast show heavier streaking.

**Juvenile plumage**  Like adult non-breeding plumage, but upperparts generally browner and spotted buff.

**Field notes**  Generally solitary, but loose flocks can be seen in the west; occurs almost exclusively away from the coast.

**Similar species**  Sharp-tailed Sandpiper; Common Sandpiper. The Wood Sandpiper is distinguished from the Sharp-tailed Sandpiper by its greyer plumage, cleaner white underparts and, in flight, white, not generally dark, rump. It is distinguished from the Common Sandpiper by being slightly larger, longer-legged and by having more spotted upperparts.

## BREEDING RANGE

The Wood Sandpiper breeds in coniferous forests throughout Eurasia from Britain and Scandinavia to the Koryakski Range and the Kamchatka Peninsula in the USSR, Mongolia and north-eastern China (85, 129).

## Asian Range

The Wood Sandpiper migrates to Africa and southern Asia as far east as Sumatra, Java, Borneo and the Philippines.

The Wood Sandpiper is the most abundant of migratory shorebirds of non-coastal areas in Asia. Between 2000 and 5000 occur on migration in Japan (483). It arrives in Taiwan in August, and passes through Hong Kong, mostly in September and October, and again in April (306). It arrives in Borneo in August and September, most migration occurring in September and October and again in March (440, 477). On the Serangoon Estuary in Singapore, 2000 are estimated to occur (174). It appears to migrate down major river valleys in southern Asia as well as through near-coastal parts of eastern Asia (267). The Wood Sandpiper is not common in New Guinea (184) and is not recorded from New Zealand (134). Only a small proportion of the Asian population reaches Australia. (See Figure 5.38.)

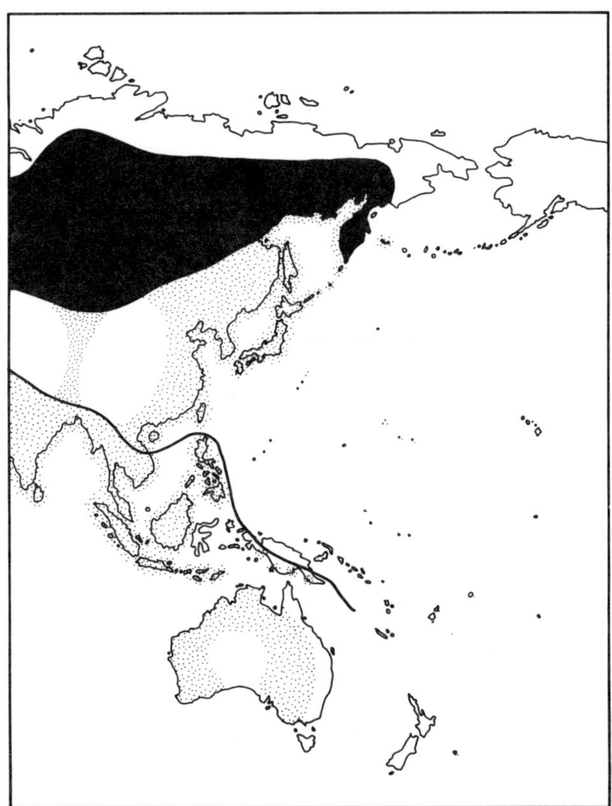

FIG. 5.38. Breeding and non-breeding range of Wood Sandpiper.

## Status and Distribution in Australia

The Wood Sandpiper is recorded regularly in most of the better-watered parts of inland Australia as well as on non-tidal wetlands near the coast. Largest congregations are observed in north-western Australia (immediately south of Java and Borneo, where they are abundant). Recent observations from Western Australia include: 300 in the eastern Kimberleys in December 1985 (214); 30 on the Fitzroy River, western Kimberleys, in October 1983 (50); and 60 on the Vasse estuary, south of Perth, in January 1986 (214). It inhabits shallow freshwater swamps, the margins of lakes and the edges of sewage ponds. The Wood Sandpiper does not occur on coastal mudflats.

## Movements in Australia

Little is known of its movements within Australia. It probably wanders between wetlands, depending on how full these are. At Werribee, near Melbourne, small numbers generally arrive in November and December, and depart in March and April, though single birds have been recorded there as late as May.

## Feeding

It forages in shallow water and on wet mud. There have been no studies of its feeding behaviour and diet, but it has been recorded eating small shellfish and grasshoppers (300).

## Conservation

Wood Sandpipers do not often concentrate, so their survival in Australia depends on an adequate network of shallow non-coastal swamps.

# GREY-TAILED TATTLER

## *Tringa brevipes*

(Plate 6)

**Other names**  Siberian, Polynesian, Asiatic or Asian (Wandering) Tattler; Grey-rumped, Eastern Grey or Short-legged Sandpiper.

It is sometimes considered to be a race of the Wandering Tattler (*T. incana*) and named *T. incana brevipes* (255).

## DESCRIPTION

A small to medium-sized, slender, uniform grey sandpiper with yellow legs.

**Adult non-breeding plumage**  Crown, nape, hindneck, sides of neck, mantle, scapulars and wing-coverts, *uniform grey*; supercilium, white, extending behind eye; lores, dark grey; ear-coverts, grey. Chin and throat, white; breast, washed grey; belly, vent and flanks, white. Iris, black; bill, long, straight, black with olive base; *legs, dull yellow*. Sexes similar.

**In flight**  Back and rump, uniform grey; distinctive in having *uniform grey upperparts and wings*.

**Voice**  Whistle: 'too-eet, too-eet', often uttered in flight; short trilling scratchy whistle, usually less than five notes: 'peepeepeepee . . .'

**Adult breeding plumage**  Similar to adult non-breeding plumage, but *breast, upper belly and flanks, finely barred grey*.

**Juvenile plumage**  Like adult non-breeding plumage, but coverts have buffish edges and dark subterminal bars; difficult to distinguish from adult non-breeding plumage.

**Field notes**  Mostly coastal, foraging on mudflats and often roosting at high tide in exposed sites, such as on rocks, mangroves or man-made structures; more abundant in northern Australia.

**Similar species**  Wandering Tattler; Terek Sandpiper. The Grey-tailed Tattler has slightly paler grey upperparts than the Wandering Tattler, the supercilium extends behind the eye and, in breeding plumage, it lacks barring on most of the belly and the vent; its call is different. It superficially resembles the Terek Sandpiper, but is larger, has duller, more olive-yellow legs and a shorter, straight bill.

## BREEDING RANGE

The Grey-tailed Tattler breeds in alpine tundra in the mountains south of the Taimyr Peninsula, in the Verkhoyanskii highlands and near Anadyr (7, 344). It may breed in the mountains of the Kamchatka Peninsula and on the Kurile Islands (7), but the limits of its breeding range are not completely known.

## ASIAN RANGE

The Grey-tailed Tattler occurs on migration in the western Aleutian Islands (7), Korea (162) and Japan (max. 2760 in September 1978 (488)). In north-eastern China it passes southwards between late July and September–October, but is generally rare in China (179, 255). In Hong Kong it occurs on southward migration between August and October (54). It is rare in Thailand (257) and scarce in Malaysia (305), but common during the non-breeding months in the Philippines (110, 171). It occurs regularly but in small numbers in Vietnam (489), Borneo (440) and western Micronesia (18). In Bali it is common (14) and in the islands north-west of Australia, 45 specimens have been collected, many between December and February (487). They are not abundant in most of New Guinea (36, 140, 184), but in the Solomon Islands they are annual visitors, arriving in large numbers in late August (25, 176). Small numbers have been recorded on other Pacific Islands as far east as Fiji. It occurs annually, but in small numbers, in New Zealand (387, 388).

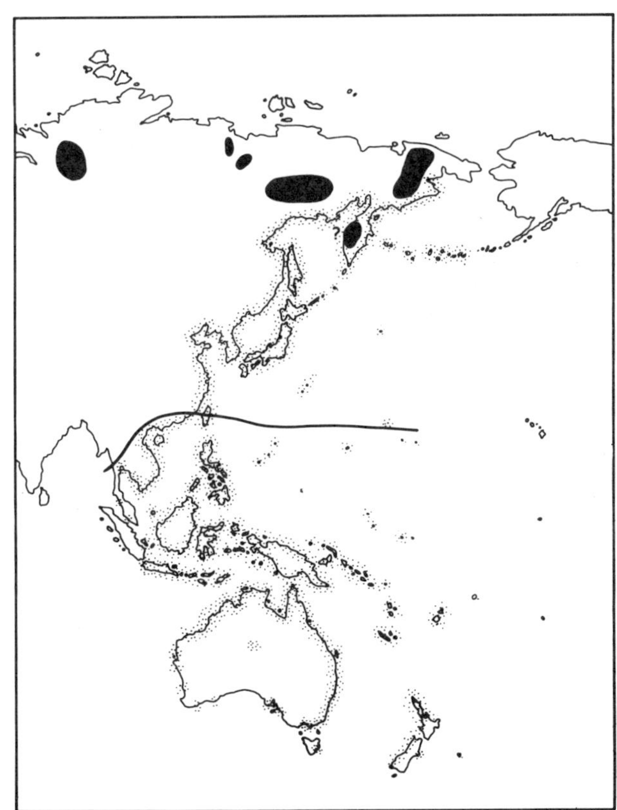

FIG. 5.39.  Breeding and non-breeding range of Grey-tailed Tattler.

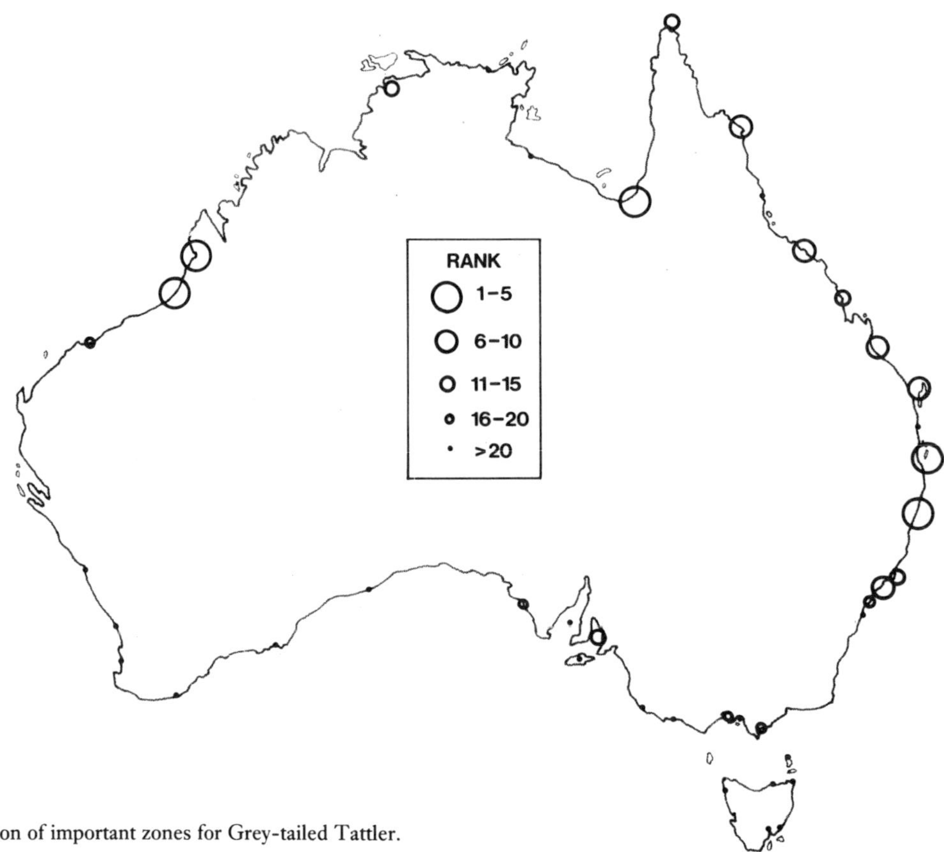


FIG. 5.40.   Location of important zones for Grey-tailed Tattler.

TABLE 5.21.   Ranked listing of top twenty zones for Grey-tailed Tattler. (maximum counts marked thus: *; other counts are averages) (20 or more birds)

| ZONE | NO. OF COUNTS | NO. OF INDIVIDUALS |
|---|---|---|
| * Eighty Mile Beach, WA | 5 | 8 500 |
| * Moreton Bay, Qld | 5 | 3 190 |
| * Roebuck Bay, WA | 6 | 3 180 |
| * SE corner, Gulf of Carpentaria, Qld | 1 | 985 |
| * N coast, NSW | 5 | 380 |
| * Hervey Bay – Great Sandy Strait, Qld | 5 | 180 |
| * Hunter estuary, NSW | 4 | 100 |
| * Capricorn coast, Qld | 2 | 90 |
| * Daintree – Melville, Qld | 2 | 90 |
| Townsville area, Qld | 5 | 80 |
| * Mackay area, Qld | 4 | 80 |
| * Torres Strait, Qld | 3 | 70 |
| * Port Stephens, NSW | 4 | 70 |
| Darwin area, NT | 4 | 60 |
| * St Vincent Gulf, SA | 4 | 60 |
| Botany Bay, NSW | 5 | 60 |
| * Pilbara coast, WA | 3 | 50 |
| Port Phillip Bay, Vic. | 5 | 30 |
| * W coast, Eyre Peninsula, SA | 5 | 30 |
| * Corner Inlet, Vic. | 5 | 20 |

The Grey-tailed Tattler departs Australia and the south-western Pacific Islands by early to mid-April (25) and migrates northwards, passing through Hong Kong in April and May (54). It arrives on Sakhalin Island and the Kamchatka Peninsula in mid to late May (111).

The Grey-tailed Tattler therefore migrates mainly through Japan, the Philippines and the islands north-west of Australia, and across the south-western Pacific Ocean as far south as Fiji and New Zealand: a more westerly range than that of the Wandering Tattler. (See Figure 5.39.) The lower numbers in New Guinea and the Gulf of Carpentaria than on the Solomons and the east coast of Australia or north-western Australia suggest that there may be two migration routes into Australia, either side of New Guinea.

## STATUS AND DISTRIBUTION IN AUSTRALIA

The Grey-tailed Tattler occurs continuously along the northern coast of Australia, from Botany Bay on the east coast to about Karratha in Western Australia (38). The largest concentrations are in the Broome – Port Hedland area, Western Australia, and at Moreton Bay, Queensland. (See Figure 5.40, Table 5.21.) It is possible that significant concentrations occur elsewhere in northern Australia, but coverage there is incomplete. It is found annually in small numbers on the south coast, e.g. Corner Inlet and Westernport Bay, Victoria; the south-east coast, St Vincent Gulf and western Eyre Peninsula, South Australia; and near Perth, Western Australia.

The Grey-tailed Tattler occurs mostly in areas with extensive mangroves and intertidal mudflats. It often roosts at high tide on mangroves or other objects (145, 264). Where there are no mangroves, it roosts on sandy beaches, but prefers rocks if these are available, e.g. Roebuck Bay, Western Australia; Mackay area, Queensland; Swan Bay, Victoria. The small number in south-eastern South Australia forage almost entirely on rocky intertidal reefs. It is one of the few migratory shorebirds living throughout the Great Barrier Reef (228), where it feeds on soft reef flats as well as on the harder substrate of coral rubble areas (115).

## MOVEMENTS IN AUSTRALIA

The Grey-tailed Tattler arrives in northern Australia in late August and early September. Between September and November there are influxes on the east coast, suggesting that it moves along the coast. Inland records (378, 458) suggest that the small numbers that move along the south coast at this time may have crossed the continent from the north coast.

In March and April, temporary increases in numbers occur along the east coast, in parts of South Australia and in Darwin. This suggests that it may migrate northwards across the continent as well as up the east coast.

## FEEDING

The Grey-tailed Tattler forages mostly over intertidal mudflats. There have been no detailed studies of its feeding behaviour and diet in Australia. On the Hunter estuary, New South Wales, it has been observed eating crabs (226).

## CONSERVATION

Most Grey-tailed Tattlers live in northern Australia, where there are few serious threats to its habitat. Its preference for roosting on mangroves requires that these be preserved wherever tattlers occur.

# WANDERING TATTLER
## *Tringa incana*
(Plate 6)

**Other names**   American Wandering Tattler.

This species is sometimes considered conspecific with the Grey-tailed Tattler *T. brevipes* (255).

## DESCRIPTION

The Wandering Tattler is almost identical to and difficult to distinguish from the Grey-tailed Tattler.

**Adult non-breeding plumage**   Like the Grey-tailed Tattler, but slightly darker grey on upperparts; white supercilium less extensive than in the Grey-tailed, *rarely extending far behind the eye.*

**In flight**   Uniform grey upperparts and wings; no distinctive markings.

**Voice**   Descending trill of eight or more notes, similar in quality to the Grey-tailed Tattler.

**Adult breeding plumage**   Like the Grey-tailed Tattler, but the barring on the underparts is heavier and *extends from the belly and flanks onto the undertail-coverts.*

**Field notes**   Often occurs in different habitats to the Grey-tailed Tattler, avoiding mudflats and preferring rocky coasts and coral reefs. It is much less abundant than the Grey-tailed Tattler.

**Similar species**   Grey-tailed Tattler. The Wandering Tattler is difficult to distinguish from the Grey-tailed Tattler. It has a longer nasal groove, slightly darker upperparts (varies with light conditions though), and a different call.

## BREEDING RANGE

The Wandering Tattler breeds in eastern, central and southern coastal Alaska east to the Yukon River and in north-western British Columbia. In eastern Siberia it breeds in Anadyrland and the Chukotski Peninsula (7).

## ASIAN RANGE

The Wandering Tattler spends the non-breeding months on the coasts of south-western North America and Central and South America as far as Peru. It also migrates to the Pacific Islands and the east coast of Australia.

The only reliable records of Wandering Tattlers in Asia come from Japan where they are stragglers (344), and Korea while on migration (129). There is one record from Taiwan (172). It is found throughout much of the central Pacific, including Hawaii, French Polynesia (7), eastern Micronesia (18), in small numbers on the Solomon Islands (25, 173), in Fiji, where they are abundant (406), the Tokelau Islands (467), Samoa (13) and the islands of the south-western Pacific Ocean (120). (See Figure 5.41.) More work is needed in Asia to clarify its distribution. There are no published records of the species in western Micronesia or the Philippines. Non-breeding birds regularly spend the breeding months in the Pacific Islands (218, 219, 220).

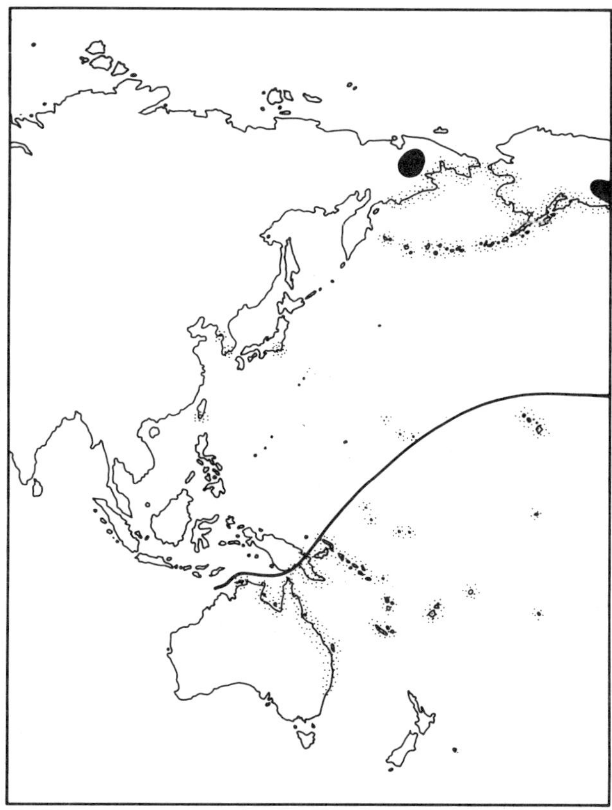

Fig. 5.41. Breeding and non-breeding range of Wandering Tattler.

## Status and Distribution in Australia

The Wandering Tattler occurs singly or in small parties in coastal northern New South Wales, Moreton Bay, in the Rockhampton, Whitsunday and Cairns areas of Queensland, in Torres Strait and at Darwin (38). Its status on the Great Barrier Reef is uncertain.

Difficulty in distinguishing it from the more abundant Grey-tailed Tattler and the lack of coverage during counts of its preferred habitat (rocky ocean shores and coral reefs) may account for the small number of records during the study.

## Movements in Australia

Little is known of the timing and route of its migration in Australia, but its distribution suggests arrival on the north-east coast of Australia from directly across the Pacific Ocean.

## Feeding

Little is known of its feeding behaviour and diet. It has been observed foraging among beach rock and rock platforms and catching a crab and bashing it against a rock to remove the legs (190).

## Conservation

The Wandering Tattler is found along the north-eastern coast of Australia, an area with relatively low human population density, and is probably not subject to any immediate threat. Its status is still to be fully investigated.

100

# COMMON SANDPIPER
## *Tringa hypoleucos*
(Plate 9)

**Other name**   Summer Snipe.
No races recognised.

## DESCRIPTION

A small sandpiper with rich olive-brown upperparts, and clean white belly and flanks; teeters when agitated; weak, low, flutter−glide flight.

**Adult non-breeding plumage**   Crown, nape, hind-neck, mantle, scapulars and wing-coverts, *deep olive-brown* with fine black-and-white edges to feathers; supercilium, white; lores, ear-coverts and sides of neck, paler olive-brown. Chin and throat, white; *broad olive-brown* sides to breast; lower breast, belly, flanks and vent, clean white, contrasting sharply with dark breast markings and upperparts. Iris, black; bill, medium length, slender, grey with pale yellow base; legs, pale yellow. Sexes similar.

**In flight**   Back, olive-brown; rump, clean white with olive-brown central line; tail, olive-brown with white notched edges and paler tips to feathers; bases of inner primaries and tips of secondary coverts, white, appearing as *broad, conspicuous white wing-bar; fluttering flight* interspersed with short glides on stiff, slightly down-swept wings, usually low over water.

**Voice**   Repeated 'chee-wee-wee', often uttered when flushed; softer whistle; 'peee−ooo' and other 'chipping' notes.

**Adult breeding plumage**   Like non-breeding plumage, but upperparts show heavier fine black tips to feathers.

**Juvenile plumage**   Similar to adult non-breeding plumage, but feathers of upperparts have buff tips.

**Field notes**   Usually found singly or in twos or threes; inhabits rocky shores, concrete and gravel edges of artificial ponds and farm dams; frequently encountered along tidal creeks in mangroves in northern Australia.

**Similar species**   Wood Sandpiper; Red-necked Stint; Sharp-tailed Sandpiper. The Common Sandpiper has more uniform upperparts and is smaller than the Wood Sandpiper. It is darker in coloration, has paler legs and a different flight pattern compared with the Red-necked Stint. It is smaller than the Sharp-tailed Sandpiper, and is neither as brown nor as heavily marked on the upperparts.

## BREEDING RANGE

The Common Sandpiper breeds throughout northern Asia and Europe in a wide variety of habitats, except the high-arctic. In eastern Asia it breeds throughout eastern Siberia, and in scattered localities in Korea, Japan and central China (85).

## ASIAN RANGE

The Common Sandpiper migrates to southern Europe, Africa, Asia and most parts of Australia.

The Common Sandpiper occurs commonly throughout eastern and southern Asia on migration. It is more numerous in Japan during southward migration (488). At Hong Kong, numbers peak during both migrations (306), and collections in the islands north-west of Australia suggest a similar pattern (487). At Penang, numbers are greater during northward migration, whereas at Jurong (in Singapore) and at Brunei, numbers are higher during southward migration (305, 477).

The Common Sandpiper spends the non-breeding months (December−February) from the Yangtze River in China (58, 255) and Taiwan (56, 401), southwards to New Guinea, where it is very common (278), and Australia. (See Figure 5.42.) It occurs regularly in

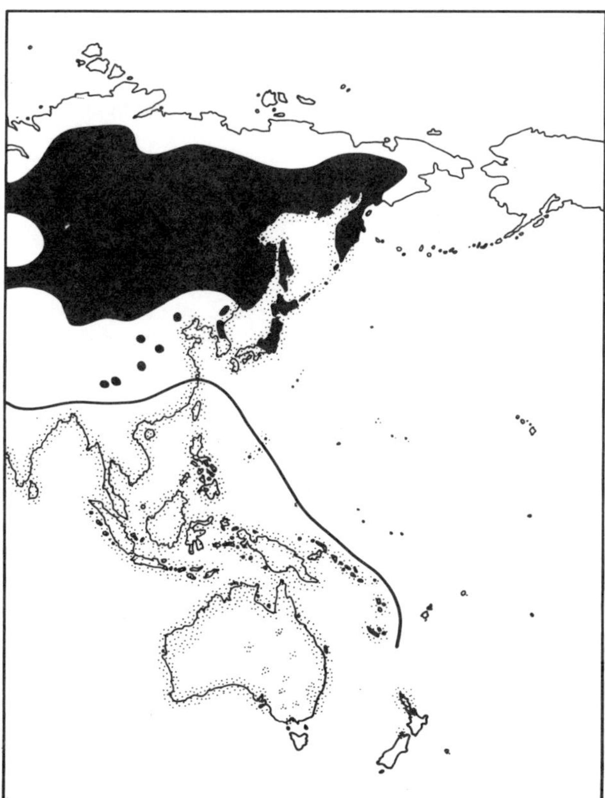

FIG. 5.42.   Breeding and non-breeding range of Common Sandpiper.

101

western Micronesia (18) and is widely distributed in small numbers throughout the south-western Pacific Islands (25, 43, 173). Farther east it is less common (18), and it is not listed for Fiji (406) or Nuie (231). It is a vagrant in New Zealand (134). Considerable numbers of non-breeding birds stay in south-east Asia during the breeding months (267).

A Common Sandpiper banded in the Philippines has been recovered on the Amur River in south-eastern Siberia (267).

## STATUS AND DISTRIBUTION IN AUSTRALIA

The Common Sandpiper is spread in small numbers along the coasts of Queensland, Northern Territory and north-western Australia (38). Fewer occur in South Australia, Victoria and southern New South Wales. It often occurs on inland wetlands in most parts of the continent.

In northern Australia, the common Sandpiper feeds most often among mangroves, on both the open shore and along the margins of tidal creeks and rivers. It also forages on rocky coast, on the margins of lakes and dams, and on the rocky edges of sewage ponds and reservoirs in the inland. At high tide it roosts almost exclusively on rocks or on the roots and branches of mangroves and other coastal trees (67).

The counts did not give a realistic idea of the numbers of Common Sandpipers visiting Australia, as they do not concentrate, being widespread across the north coast of the continent, where there have been fewer counts.

## MOVEMENTS IN AUSTRALIA

Little is known of the movements of the Common Sandpiper in Australia. There are many records of them in inland Australia (159, 234, 340, 458, 486) suggesting that some migrate across the continent.

## FEEDING

There have been no studies of the feeding behaviour and diet of the Common Sandpiper in Australia. It is an active feeder, running and stopping, chasing and catching insects on the surface of mud and probing amongst mangrove roots, rocks and in shallow water for invertebrate prey (38).

## CONSERVATION

The Common Sandpiper is widespread in low densities throughout northern Australia, where there are no immediate or serious threats to its survival.

# GREENSHANK
## *Tringa nebularia*
(Plate 9)

**Other name**  Common Greenshank.
No races recognised.

## DESCRIPTION

A large to medium-sized, clean, grey and white, long-legged and long-necked sandpiper.

**Adult non-breeding plumage**  Crown, nape, hind-neck and ear-coverts, white with dark grey streaking; mantle, scapulars and wing-coverts, grey with white speckling; supercilium, indistinct, white; lores, dark grey. *Underparts, clean white.* Iris, black; bill, long and *slightly upcurved*, grey with olive-green base; legs, long, olive-green. Sexes similar.

**In flight**  *Rump and back, conspicuous clean white*, contrasting with uniform grey wings; primaries and outer wing-coverts slightly darker than secondaries and inner wing-coverts; *feet trail only a short distance beyond tail.*

**Voice**  Vocal: loud 'tew-tew-tew', uttered mostly in flight, but occasionally when agitated; single 'tew' sometimes uttered while feeding.

**Adult breeding plumage**  Darker feathers on back and wing-coverts and heavier streaking on the head, throat and, to a lesser degree, the breast.

**Juvenile plumage**  Similar to adult non-breeding plumage, but feathers of upperparts darker with buff edges.

**Field notes**  A nervous bird, quick to fly when disturbed; often feeds by running and jabbing in an urgent manner; spreads over mudflats in dispersed flocks; occurs singly or in small parties on inland and coastal wetlands.

**Similar species**  Marsh Sandpiper; Bar-tailed Godwit; Black-tailed Godwit. The Greenshank is about 30 per cent larger than the Marsh Sandpiper, has a heavier, two-toned bill and, in flight, legs that trail only a short distance beyond the tail. It is slightly smaller, greyer and more slender than the godwits.

## BREEDING RANGE

The Greenshank breeds over a large part of the taiga forest and northern steppe zones of the Palaearctic between 50°N and 70°N, from northern Scotland and Scandinavia eastwards to the Kamchatka Peninsula, but excluding extreme north-eastern Siberia (85).

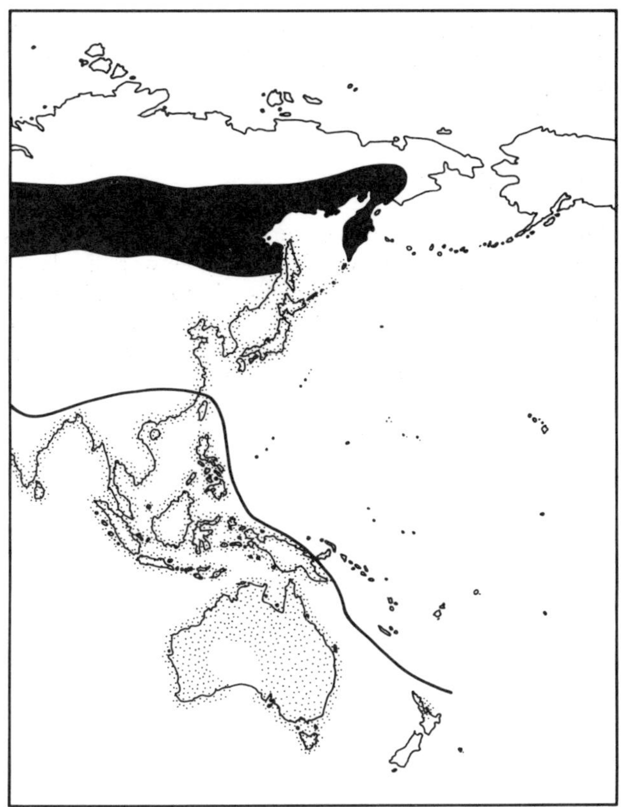

Fig. 5.43.   Breeding and non-breeding range of Greenshank.

TABLE 5.22.   Ranked listing of top twenty zones for Greenshank. (maximum counts marked thus: *; other counts are averages) (130 or more birds)

| ZONE | NO. OF COUNTS | NO. OF INDIVIDUALS |
|---|---|---|
| * Eighty Mile Beach, WA | 5 | 2 440 |
| * SE corner, Gulf of Carpentaria, Qld | 1 | 1 240 |
| * St Vincent Gulf, SA | 4 | 1 130 |
| * The Coorong, SA | 1 | 720 |
| * W coast, Eyre Peninsula, SA | 5 | 580 |
| * Hunter estuary, NSW | 4 | 560 |
| * Roebuck Bay, WA | 6 | 560 |
| Port Phillip Bay, Vic. | 5 | 460 |
| * Spencer Gulf, SA | 5 | 320 |
| * Kangaroo Island, SA | 5 | 260 |
| Westernport Bay, Vic. | 5 | 260 |
| * Anderson's Inlet, Vic. | 4 | 240 |
| * Pilbara coast, WA | 3 | 230 |
| Corner & Shallow inlets, Vic. | 5 | 220 |
| SE coast, SA | 5 | 220 |
| * Peel Inlet, WA | 5 | 160 |
| * Riverina, NSW | 4 | 150 |
| * Port Hedland Saltworks, WA | 12 | 150 |
| * Albany coast, WA | 4 | 130 |
| * Murray mouth lakes, SA | 3 | 130 |

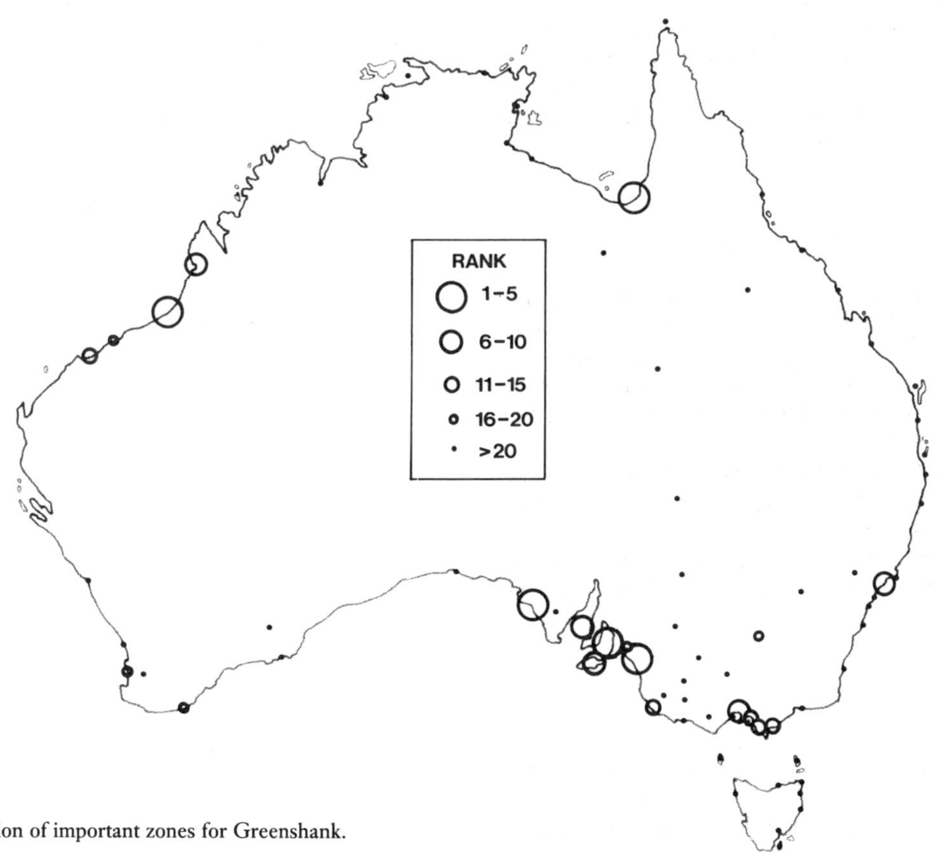


Fig. 5.44.   Location of important zones for Greenshank.

## ASIAN RANGE

The Greenshank migrates to Europe, Africa, the Middle East, the Indian subcontinent, south-east Asia, Australia and New Zealand, and is widespread on both inland and coastal wetlands.

It is common in most of eastern Asia except the Philippines (110). (See Figure 5.43.) It occurs on migration in Korea (162), Japan (344) and, in August and April, north-eastern China (255, 266, 404). Throughout the non-breeding months it is found in southern China (255), Taiwan (37) and southwards to Australia. It is irregular in New Zealand (134) and on the Pacific Islands (recorded from Micronesia (216, 302, 345)). In southern New Guinea, it is a common migrant (36, 138). Non-breeding birds often remain in Indonesia (205), New Guinea and Australia during the breeding months.

## STATUS AND DISTRIBUTION IN AUSTRALIA

Largest numbers of Greenshanks occur in coastal areas of northern Australia. (See Figure 5.44, Table 5.22.) In the south, greatest numbers are recorded in coastal areas of Victoria and South Australia. It is frequently seen inland in northern and western Victoria, western New South Wales and Queensland, and in south-western Australia.

It occurs on near-coastal saline wetlands, including saltmarshes, and intertidal mudflats. On the northern coast of Australia, it lives on wide intertidal mudflats, roosting in pools and puddles in saltflats behind mangroves or on sandy beaches.

## MOVEMENTS IN AUSTRALIA

The Greenshank arrives in Australia in August. Influxes occur in South Australia and at Darwin in August and September, whereas it reaches the east coast mainly in November. This suggests that it arrives in the western half of Australia before moving to other areas by November.

Numbers fluctuate in South Australia and Victoria between December and February, indicating that they move about in this period.

Northward migration occurs in March and particularly April, when influxes are recorded in Victoria, South Australia and up the east coast.

## FEEDING

The Greenshank usually feeds in shallow water, often running rapidly to peck at food items. There have been no detailed studies of its feeding behaviour and diet in Australia, but its choice of a wide range of wetlands suggests that it is able to exploit a variety of food sources.

## CONSERVATION

The Greenshank is more wary than other shorebirds and may be adversely affected by lower levels of disturbance than most species. Large numbers of Greenshank use commercial saltworks in Australia, showing that it is able to adapt to some artificial changes in wetlands.

# MARSH SANDPIPER
*Tringa stagnatilis*
(Plate 8)

**Other names**   Little Greenshank; Marshy.
   No races recognised.

## DESCRIPTION

A medium-sized to small, slender, grey and white sandpiper, with a black, needle-like bill and long legs.

**Adult non-breeding plumage**   Crown, grey-brown, finely streaked black; nape, hindneck and sides of neck, white, streaked grey; mantle, scapulars and wing-coverts, grey with white edges to feathers; supercilium, white; lores and ear-coverts, finely streaked grey. Chin and throat, white; breast, white, finely streaked grey at sides; belly, flanks and vent, white. Iris, black; *bill, needle-like and black*, with hint of green at base; legs, grey-green. Sexes similar.

**In flight**   *Back and rump, white*; tail, white, finely barred black towards tip; outer wing-coverts and primaries, dark grey, contrasting with paler grey inner wings; *legs and feet trail well beyond tail*.

**Voice**   High-pitched piping calls including 'cheek cheek cheek . . .' uttered when in flight.

**Adult breeding plumage**   Heavier dark grey streaking on head, neck and breast; upperparts more heavily marked with dark grey bars on feathers.

**Juvenile plumage**   Similar to adult non-breeding plumage, but feathers of upperparts more heavily marked: dark grey with buff edges.

**Field notes**   Feeds over mudflats of coast and inland lakes with *rapid, jerky movements*; in southern Australia more often on inland wetlands; when flushed, it takes off with distinctive high-pitched call.

**Similar species**   Greenshank; Curlew Sandpiper; Wood Sandpiper. The Marsh Sandpiper is smaller than the Greenshank and has a more slender bill, higher-pitched call and, in flight, legs that trail farther beyond the end of the tail. It is similar in size to the Curlew Sandpiper, but has a straight bill, longer legs and paler upperparts. It is paler, and has longer legs and bill than the Wood Sandpiper.

## BREEDING RANGE

The Marsh Sandpiper breeds in eastern Europe, southern Siberia as far east as Lake Baikal, Mongolia and northern China (58, 144, 232, 237). The precise limits of its breeding range are poorly known (85). It no longer breeds at the eastern extremity of its breeding range in Hungary (85).

## ASIAN RANGE

During the non-breeding months it occurs south of the Sahara Desert in Africa, along the Nile Valley and round the Red Sea. It also visits the Persian Gulf and the Indian subcontinent, and up to 100 000 have been seen in Sri Lanka (181). It also visits the Bay of Bengal, coastal south-east Asia, Indonesia, New Guinea and Australia. It has not been recorded in large numbers in eastern China or Taiwan.

On southward migration the Marsh Sandpiper occurs in central Mongolia (237) and central and north-eastern China, where it is rare (58, 129, 255). In Korea, it is rare and occurs only during southward migration (162). In Japan, it occurs only on southward migration (488). It occurs on migration in Hong Kong between late August and October, and again in late March and April (54). It occurs regularly in Burma, but is not common (439). It migrates through Malaysia, and small numbers occur there throughout the non-breeding months (305). In Thailand, it is a common visitor (257). In the Philippines, it is rare (110), but on the west coast of southern Sulawesi it is very common (128). Elsewhere in Indonesia and in New Guinea, it is not common (14, 184, 440, 487), although 400 have been observed on Bali (235) and up to 200 in the Port Moresby area of New Guinea, where it arrives in September (140). Smaller groups have been seen between September and December in south-eastern Irian Jaya (36), and small numbers regularly visit Bougainville Island (173). It is a vagrant in New Zealand (134), and is not regularly recorded in the Pacific Islands (18), although there are records from western Micronesia (127, 345). (See Figure 5.45.)

## STATUS AND DISTRIBUTION IN AUSTRALIA

The extensive intertidal mudflats of northern Australia hold the largest numbers of Marsh Sandpipers. Its status on the extensive seasonal wetlands of inland northern Australia remains to be investigated, but it occurs on the wetlands of the Barkly Tableland (38). In southern Australia, it prefers inland saline wetlands, sewage treatment works and, on the coast, the saltworks round Port Phillip Bay and St Vincent Gulf. It also occurs on the wetlands along the rivers flowing west from the Great Dividing Range, e.g. Bulloo and Paroo, Macquarie, Murray (38). (See Figure 5.46, Table 5.23.)

Counts are likely to underestimate the numbers of Marsh Sandpipers visiting Australia, as not all their

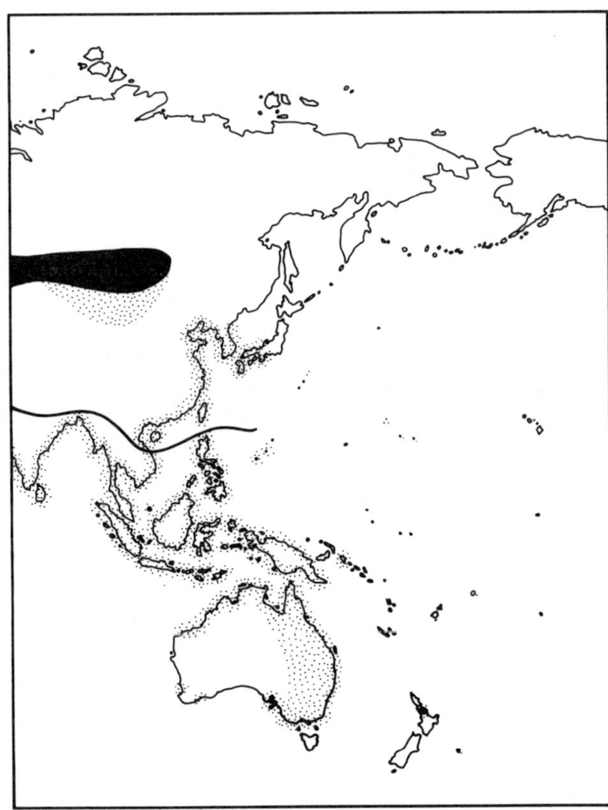

FIG. 5.45.   Breeding and non-breeding range of Marsh Sandpiper.

TABLE 5.23.   Ranked listing of top twenty zones for Marsh Sandpiper. (maximum counts marked thus: *; other counts are averages) (30 or more birds)

| ZONE | NO. OF COUNTS | NO. OF INDIVIDUALS |
|---|---|---|
| * SE corner, Gulf of Carpentaria, Qld | 1 | 1 150 |
| * Port Hedland Saltworks, WA | 12 | 500 |
| * Far SW lakes, Qld | 1 | 500 |
| * Hunter estuary, NSW | 4 | 280 |
| * Eighty Mile Beach, WA | 5 | 140 |
| * Parkes region, NSW | 4 | 130 |
| Port Phillip Bay, Vic. | 5 | 80 |
| * Charters Towers area, Qld | 1 | 70 |
| * Far NW lakes, NSW | 1 | 70 |
| Kerang–Swan Hill, Vic. | 5 | 70 |
| * W coast, Eyre Peninsula, SA | 5 | 60 |
| * N coast, NSW | 5 | 60 |
| Moreton Bay, Qld | 5 | 60 |
| * Murray mouth lakes, SA | 3 | 60 |
| Riverina, NSW | 4 | 50 |
| * Lower Darling River lakes, NSW | 4 | 40 |
| St Vincent Gulf, SA | 4 | 40 |
| * Southern tablelands, NSW | 4 | 40 |
| * Pilbara coast, WA | 3 | 40 |
| Darwin area, NT | 4 | 30 |

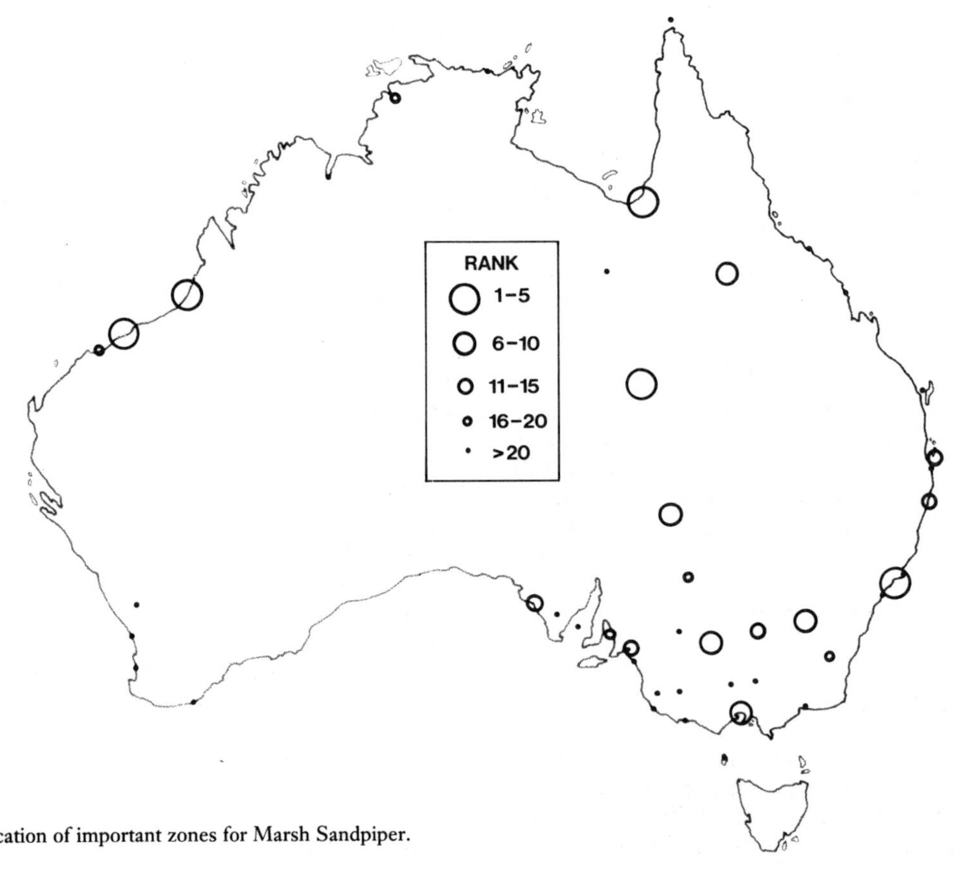



FIG. 5.46.   Location of important zones for Marsh Sandpiper.

106

inland habitats (38) have been covered and many coastal areas of northern Australia have not been surveyed on the ground.

## MOVEMENTS IN AUSTRALIA

The Marsh Sandpiper arrives in Australia in September. Between then and November, temporary influxes occur at some sites on the east coast and near Mount Isa. Maximum numbers are attained in December at Parkes in New South Wales, and in St Vincent Gulf in South Australia. This suggests that it migrates southwards between September and November. Numbers vary on the Hunter estuary between December and February.

It departs northwards in March and April, when temporary influxes occur at some east coast sites. It moves to the northern end of St Vincent Gulf before departure. It leaves the Port Hedland Saltworks, in north-western Australia, in mid-April.

## FEEDING AND FOOD

The Marsh Sandpiper feeds while running through shallow water, pecking rapidly or probing in wet mud. Very little is known of its diet.

## CONSERVATION

Its occurrence on most commercial saltworks west of the Great Dividing Range suggests that the Marsh Sandpiper can adapt to changes in wetlands.

# TEREK SANDPIPER
## *Tringa terek*
(Plate 6)

**Other names** Eastern Terek Sandpiper; Avocet-Sandpiper.
No races recognised.

## DESCRIPTION

A small to medium-sized grey sandpiper with conspicuous orange legs and a long, upturned bill; feeds very actively.

**Adult non-breeding plumage** Crown and nape, grey; hindneck, mantle and scapulars, grey, finely streaked black, *the streaks are often thicker on the scapulars*; supercilium, white; lores and ear-coverts, finely streaked grey. Chin and throat, white; breast, white, finely streaked grey; belly, flanks and vent, white. Iris, black; *bill, long, upturned,* black with orange base; *legs, orange.* Sexes similar.

**In flight** *Back and rump, grey, finely streaked black;* narrow white sides to rump; tail, grey; lesser coverts, dark grey; median coverts, grey, finely streaked black; greater coverts, dark grey; inner primaries and secondaries, grey, broadly tipped white, showing as *white trailing edge to wing;* outer primaries, black.

**Voice** Loud, high-pitched whistling: 'terr-errr-eeet, terr-errr-eeet'; trilling and flute-like calls.

**Adult breeding plumage** Similar to non-breeding plumage, but with heavier black streaking on upperparts; legs, duller orange to red.

**Juvenile plumage** Similar to adult non-breeding plumage, but with slightly heavier streaking on upperparts; feathers of upperparts have buff fringes, mostly lost by the time it reaches Australia.

**Field notes** Exclusively coastal; feeds on soft mudflats, especially near mangroves, with *frantic running and chasing behaviour;* roosts on beaches, but often prefers the branches of mangrove trees; calls when disturbed.

**Similar species** Curlew Sandpiper; Red-necked Stint; Grey-tailed Tattler. The Terek Sandpiper differs from the Curlew Sandpiper by having an upturned bill. It is larger than the Red-necked Stint, but is smaller, shorter-legged and longer-billed than the Grey-tailed Tattler. The Terek Sandpiper is unmistakable with its bright orange legs and upturned bill.

## BREEDING RANGE

The Terek Sandpiper breeds in the taiga zone of the Palaearctic, from the Baltic coast of Finland, eastwards through Siberia, north of Lake Baikal to the Kolyma River and possibly as far as the Chukotski Peninsula, extending south into the steppe zone and north into sub-arctic tundra regions (85). It nests inland in lowland valleys and avoids mountainous areas. Nests are located in damp marshy sites as well as on open ground, generally away from forest (85).

## ASIAN RANGE

The Terek Sandpiper spends the non-breeding months on coasts from southern Africa, round east Africa, the Arabian Gulf, India and south-east Asia to New Guinea and Australia.

In Asia, it migrates southwards through Korea (162), Japan (max. September 2700 (488)), Mongolia (237) and north-eastern China between mid-July and late September. It is fairly common along the east coast of China and in Taiwan between mid-August and October (56, 255) and passes southwards through Hong Kong in September and October (54). It migrates regularly through western Micronesia (127, 345) and through Borneo (440). Its status in Indonesia is not known.

After southward migration it occurs commonly in Burma (439) and Thailand (257) and in small flocks throughout Vietnam (489), Malaysia (157, 348) and the Philippines (110). In Borneo at this time it is scattered generally around the coast (440). It is absent from most of the Pacific Islands (one record for Fiji (406)), but in New Guinea and nearby islands it is widespread in small numbers (173, 184). A few reach New Zealand each year (230). (See Figure 5.47.)

Measurements of birds that visit north-western Australia and presumably the rest of the continent are the same as those of birds that visit eastern Indonesia. They probably come from a more northerly breeding area than the slightly smaller birds that visit western Indonesia and south-east Asia (85).

On northward migration it passes through Hong Kong, China, Taiwan and Japan in April and May (54, 56, 255, 488). In May, it moves through Korea and Mongolia in small numbers (162, 237) and through north-eastern China as late as June (179).

## STATUS AND DISTRIBUTION IN AUSTRALIA

The Terek Sandpiper occurs regularly on the northern Australian coast from Shark Bay in Western Australia round to the estuaries of northern New South Wales, south to Botany Bay in Sydney. Elsewhere a few occur

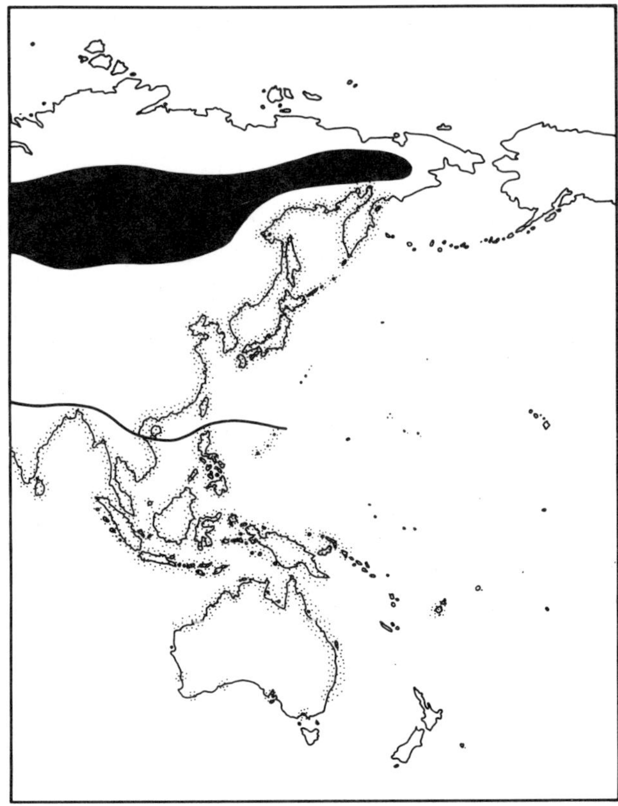

FIG. 5.47.   Breeding and non-breeding range of Terek Sandpiper.

at Corner Inlet and Westernport Bay, in Victoria; St Vincent Gulf in South Australia; and in the Albany and Geraldton areas of Western Australia. It is exclusively coastal.

It is most abundant on the wide intertidal mudflats of northern Australia, especially in the north-west. (See Figure 5.48, Table 5.24.) The Terek Sandpiper forages on soft mud, often near mangroves or in tidal creeks. It roosts on or among mangroves and sometimes on sandy beaches.

The abundance of these birds in many parts of northern Australia may have been underestimated due to the lack of ground-based coverage.

## MOVEMENTS IN AUSTRALIA

The Terek Sandpiper arrives at Darwin and Cairns in August and in north-western Australia in the first week of September. There appear to be two waves of migration down the east coast, one in September and one in November.

Temporary influxes in March and April at sites on the east coast suggest that it migrates northwards via the east coast. It does not migrate northwards through Darwin, but there is an influx in north-western Australia at this time (247). It does not depart from the north-west until late April.

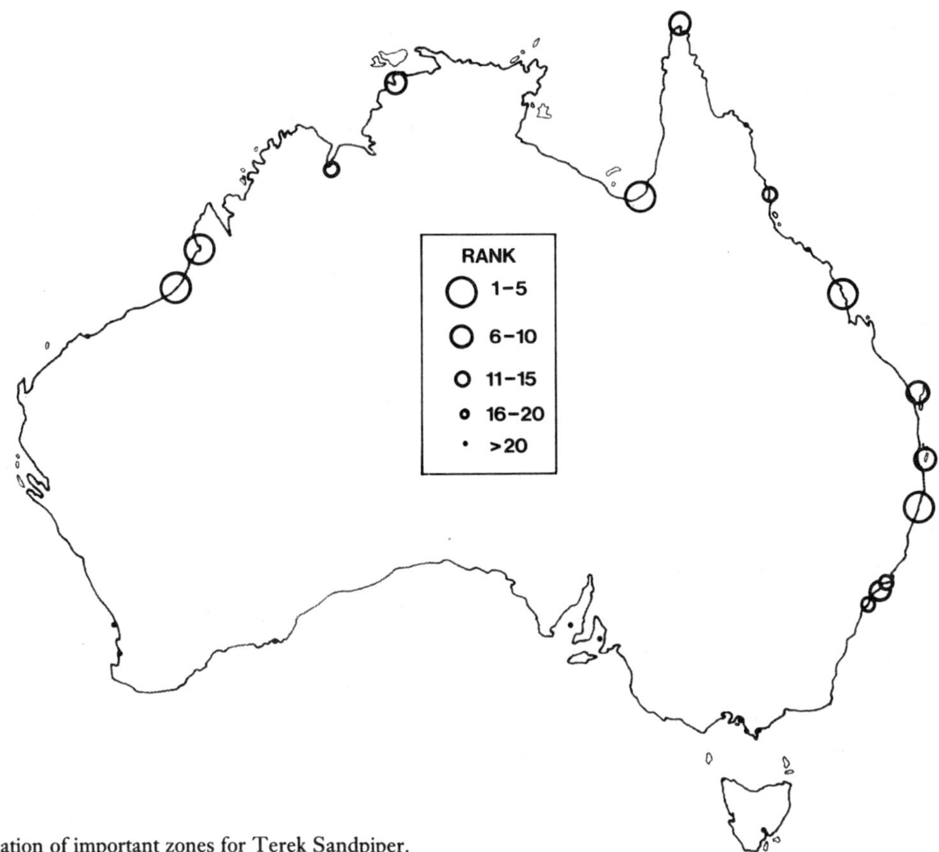



FIG. 5.48.   Location of important zones for Terek Sandpiper.

TABLE 5.24.   Ranked listing of zones that held 20 or more Terek Sandpipers. (maximum counts marked thus: *; other counts are averages)

| ZONE | NO. OF COUNTS | NO. OF INDIVIDUALS |
|---|---|---|
| * Eighty Mile Beach, WA | 5 | 3 000 |
| * SE corner, Gulf of Carpentaria, Qld | 1 | 900 |
| * Roebuck Bay, WA | 6 | 820 |
| * N coast, NSW | 5 | 150 |
| Mackay area, Qld | 4 | 120 |
| Moreton Bay, Qld | 5 | 90 |
| * Torres Strait, Qld | 3 | 80 |
| * Hervey Bay–Great Sandy Strait, Qld | 5 | 60 |
| * Central coast, NSW | 5 | 40 |
| Darwin area, NT | 4 | 40 |
| * Kimberley coast, WA | 1 | 30 |
| Hunter estuary, NSW | 4 | 30 |
| * Botany Bay, NSW | 5 | 20 |
| Cairns area, Qld | 4 | 20 |

# FEEDING AND FOOD

The Terek Sandpiper forages by running rapidly over wet mud, seldom near the edge of the water, and pecking at items of food. There have been no detailed studies of its feeding behaviour and diet in Australia.

# CONSERVATION

Most of Australia's Terek Sandpipers live on the mudflats and in the mangrove woodlands of the north coast where there are no immediate threats to their habitat.

# LATHAM'S SNIPE

## *Gallinago hardwickii*
(Plate 10)

**Other names** Japanese, Australian, Common or Jack Snipe; Bleater; Long Bill.
No races recognised.

## DESCRIPTION

A medium-sized, long-billed shorebird found in freshwater wetlands with rank vegetation; cryptic, retiring and *mostly seen when flushed; flies rapidly in an irregular path uttering harsh calls.*

**Adult non-breeding plumage** Crown and nape, dark brown with buff central stripes; supercilium, pale buff; lores and ear-coverts, dark brown; hindneck, sides of neck, mantle, scapulars, back and rump, dark brown, heavily streaked and scalloped buff. Chin, throat and breast, buff, heavily marked dark brown; belly and vent, pale buff; flanks, buff, finely barred dark brown. Iris, dark brown; *bill, long, straight* and brown, darker near tip, legs, olive-green. Sexes similar.

**In flight** Tail, pale chestnut with black subterminal bar and white tip; shows no contrasting areas of colour on wings and upperparts; underwings white, finely barred dark brown.

**Voice** A harsh 'chuck . . .' sometimes repeated; most often uttered when disturbed and in flight.

**Adult breeding plumage** Similar to adult non-breeding plumage.

**Juvenile plumage** Similar to adult non-breeding plumage; difficult to distinguish in the field.

**Field notes** *Seldom seen away from the shelter of rank vegetation*; usually roosts during the day and feeds at night; although many can occur on one wetland, they behave individually rather than as a flock; occasionally perch on posts and logs in wetlands.

**Similar species** Two other *Gallinago* snipe visit Australia regularly: Pintail Snipe (*G. stenura*) and Swinhoe's Snipe (*G. megala*). Neither occurs regularly where Latham's Snipe occurs. Differences are described in the rare species accounts.

## BREEDING RANGE

Latham's Snipe breeds predominantly on Hokkaido, but smaller numbers breed on Honshu (146, 327). Near Hokkaido, there are records of Latham's Snipe breeding in the southern Kurile Islands (Kunashiri Island) and on the island of Sakhalin (327). It has the most restricted breeding range of all the world's *Gallinago* snipe (146).

In Hokkaido, it breeds in natural grasslands, re-afforestation areas, in both lowlands and mountains, and cultivated areas. In contrast to its habits in Australia it avoids wetlands and rice paddies (327). It arrives on the breeding grounds in mid-April, nesting probably starts in early to mid-May, and by the end of August it has usually departed (327).

## ASIAN RANGE

Latham's Snipe occurs on migration in Honshu and occasionally Taiwan (56). Apart from these places, there are no other areas between Japan and New Guinea with confirmed records. A number were banded in the Philippines in the 1960s (268), but the identification was not supported by adequate descriptions, which are neccessary because a number of similar species are abundant there. Latham's Snipe departing northwards from northern New South Wales carries sufficient fat to fly non-stop to the Philippines, but not to Japan (245). More investigation is needed of its status in the Philippines.

There is a possible record of one in the Marshall Islands, Micronesia (8). In New Guinea, specimens have been collected from widely separate locations from the lowlands up to 3550 metres in the highlands in August–October and April–May (146, 184). This suggests that it flies across the New Guinea mountains and occurs there mostly on migration. There have been a few records of them in New Zealand (230). (See Figure 5.49.)

## STATUS AND DISTRIBUTION IN AUSTRALIA

Latham's Snipe is distributed widely throughout coastal and nearby inland parts of Queensland, New South Wales, Victoria, Tasmania and south-eastern South Australia. Wetlands associated with coastal plains on the east coast and along major rivers on the western side of the Great Dividing Range appear to hold the largest numbers. Examples include the region around Port Phillip Bay, the La Trobe and Goulburn rivers and the south-western coastal plain in Victoria, the Lachlan, Nepean, Hawkesbury, Richmond and Clarence rivers in New South Wales and the Brisbane River in Queensland (38, 146, 326). As well, many occur in the alpine plains of south-eastern Australia and Tasmania (146, 326).

Latham's Snipe is occasionally found on beaches and other saline and brackish habitats (146), but it occurs mostly on freshwater wetlands, particularly those with rank rushes, sedges and reeds.

Latham's Snipe is cryptic and retiring. Furthermore, it occurs on numerous wetlands not inhabited by other

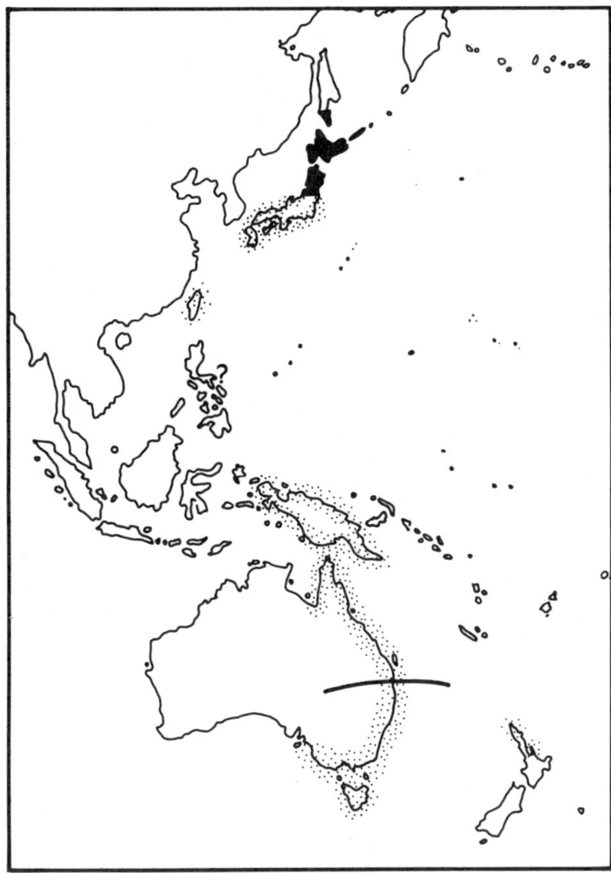

FIG. 5.49.   Breeding and non-breeding range of Latham's Snipe.

shorebirds. For these reasons, very few were recorded during national counts: rarely were more than 100 recorded on a single wetland. Naarding (326) estimates that there are approximately 20 000 in Australia, but draws attention to the problems of accurately determining numbers.

## MOVEMENTS IN AUSTRALIA

Latham's Snipe arrives in north Queensland as early as late July, most passing through in September (146). In October and November it moves through inland New South Wales. It arrives in Victoria and New South Wales as early as late August, but numbers do not peak until December—January, by which time few are present in Queensland.

Latham's Snipe concentrates on permanent wetlands towards the end of January, and most have departed from south-eastern Australia by the end of February (245, 326). In March and April it migrates northwards through Queensland, and the latest record is 25 May, although most have gone by mid-April (146). Studies show that Latham's Snipe does not fatten much before departing from south-eastern

Australia (146, 245). Its earlier departure from south-eastern Australia and the presence of good numbers farther north in later months, together with the results of these studies, indicates that it undergoes most pre-migratory fattening in northern New South Wales and Queensland (245). It has been suggested that this is an adaptation to the drying of southern wetlands in summer and the filling of northern wetlands with the onset of the wet season (245).

## FEEDING

Latham's Snipe feeds by probing its bill into soft mud and soil such as occurs in freshwater meadows and marshes, alpine plains and, in Tasmania, heathy button-grass plains. Richly organic subtrates in which there is an abundance of invertebrates are preferred (146).

The diet of Latham's Snipe has been reviewed (146). It feeds on seeds of sedges and rushes, beetles (adults and larvae), crane fly larvae and pupae (Tipulidae), spiders and probably most often on earthworms.

## CONSERVATION

Until recently Latham's Snipe was shot as game in most Australian States where it occurs. Most States have placed a moratorium on snipe shooting in recent years, which will be reviewed in the light of results from current research. In southern Queensland it can still be hunted at some times of year.

Habitat destruction is a major cause for concern. In the region west of Port Phillip Bay and in South Gippsland, two regions of Victoria, 34 per cent and 95 per cent respectively of the area of freshwater wetland has been destroyed (79, 80). The species occurs in the better watered parts of eastern Australia, but unfortunately this is where agricultural development has led to and will continue to cause destruction of freshwater wetlands. To survive, Latham's Snipe requires a network of freshwater wetlands in eastern Australia.

# ASIAN DOWITCHER
*Limnodromus semipalmatus*
(Plates 10 and 11)

**Other names** Asiatic Dowitcher; Snipe-billed Godwit.
No races recognised.

## DESCRIPTION

A large, long-billed shorebird, similar to the Bar-tailed Godwit.

**Adult non-breeding plumage** Crown, nape, hindneck and sides of neck, white, streaked grey-brown, *more heavily on crown*; mantle, scapulars and wing-coverts, grey-brown with white edges; supercilium, white; line through eye, dark grey. Chin and throat white; breast and flanks, white, sparsely streaked dark grey; belly and vent, white. Iris, black; bill, long, straight, *slightly swollen towards tip, black with brown base*; legs, black.

**In flight** Back, grey-brown; rump, white, finely barred grey-brown; tail, white with brown barring; white tips to greater coverts appear as narrow wing-bar; secondaries, grey-brown, tipped white; primaries, dark grey.

**Voice** A single, plaintive note: 'yow'.

**Adult breeding plumage** Crown, nape, hindneck and sides of neck, deep chestnut, slightly darker than in Bar-tailed Godwit; mantle and scapulars, black with deep chestnut edges; supercilium, deep chestnut. Chin, throat and breast, deep chestnut; sides of breast and flanks, finely barred black and chestnut; belly and vent, white with sparse chestnut barring.

**Juvenile plumage** Similar to adult non-breeding plumage, except wing-coverts have buff edges; sides of breast and flanks washed pale buff, spotted and barred grey.

**Field notes** Exclusively coastal; usually occurs in small flocks or alone, nowhere abundant; feeds by walking in shallow water, repeatedly probing deeply into underlying mud.

**Similar species** Bar-tailed Godwit; Black-tailed Godwit. The Asian Dowitcher is distinguished from the godwits by having greyer, heavier markings on upperparts, more conspicuous supercilium, slightly shorter legs and an all-dark bill, slightly swollen towards tip rather than tapering.

## BREEDING RANGE

The Asian Dowitcher breeds in small colonies at a handful of sites in the steppe regions of central and eastern Siberia, Mongolia and north-eastern China (111, 144). It nests in flooded meadows, often near the edges of lakes, and colonies move depending on lake levels (160). Drainage of wetlands for agriculture may be reducing the area of suitable breeding habitat (338).

## ASIAN RANGE

The Asian Dowitcher spends the non-breeding months on the coast of Asia from the Persian Gulf eastwards to Malaysia, Sumatra and Thailand, and on the north-western coast of Australia.

It is rare thoughout its range and little is known of its migration. It occurs in small numbers during migration in Japan and China (260). Recently, sizeable flocks have been reported: 400 in Thailand, 1200 in Sumatra and, during northward migration in April, 400 at Hong Kong (347). There are records of smaller flocks from Malaysia and the Philippines (311). (See Figure 5.50.)

There may be two migration routes in Asia; one via the Chinese coast and the other across the interior of Asia (260).

## STATUS AND DISTRIBUTION IN AUSTRALIA

Australia holds very few Asian Dowitchers; the first record was from the Darwin area in October 1971 (87). Single birds have been observed at Cairns and Moreton Bay, Queensland, and at Shoalhaven Heads south of Sydney (253), as well as in Port Phillip Bay (426).

In north-western Australia, flocks of up to 130 occur annually at Port Hedland Saltworks. Up to 60 have been seen elsewhere in the region (212). (See Figure 5.51, Table 5.25.)

It occurs only on the coast, feeding on intertidal mudflats and roosting, often with godwits, on sandy beaches or in shallow lagoons, e.g. Port Hedland Saltworks.

## MOVEMENTS IN AUSTRALIA

Little is known of the movements of the Asian Dowitcher in Australia. It arrives in the north-west in early September and departs in the third week of April.

## FEEDING

The Asian Dowitcher forages in shallow water over intertidal mudflats, repeatedly probing the underlying mud. Its diet has not been described.

## CONSERVATION

Other than the sites in Hong Kong, Thailand and Sumatra, the north-west coast of Australia is the only place where the Asian Dowitcher is known to occur regularly in some numbers. The protection and appropriate management of these sites is therefore desirable.

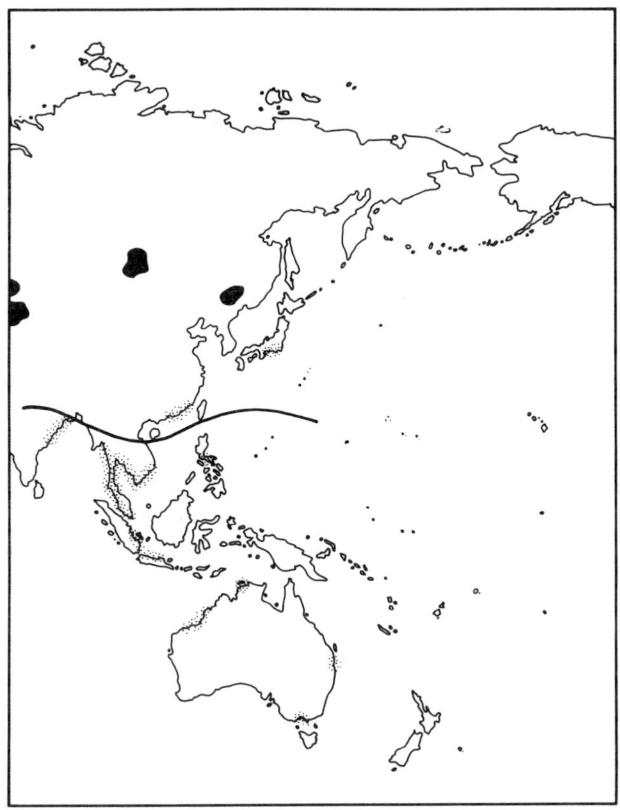

FIG. 5.50. Breeding and non-breeding range of Asian Dowitcher.

TABLE 5.25. Ranked listing of zones that held more than 20 Asian Dowitchers. (all counts are maxima)

| ZONE | NO. OF COUNTS | NO. OF INDIVIDUALS |
|---|---|---|
| Port Hedland Saltworks, WA | 12 | 130 |
| Roebuck Bay, WA | 6 | 57 |
| Eighty Mile Beach, WA | 5 | 35 |

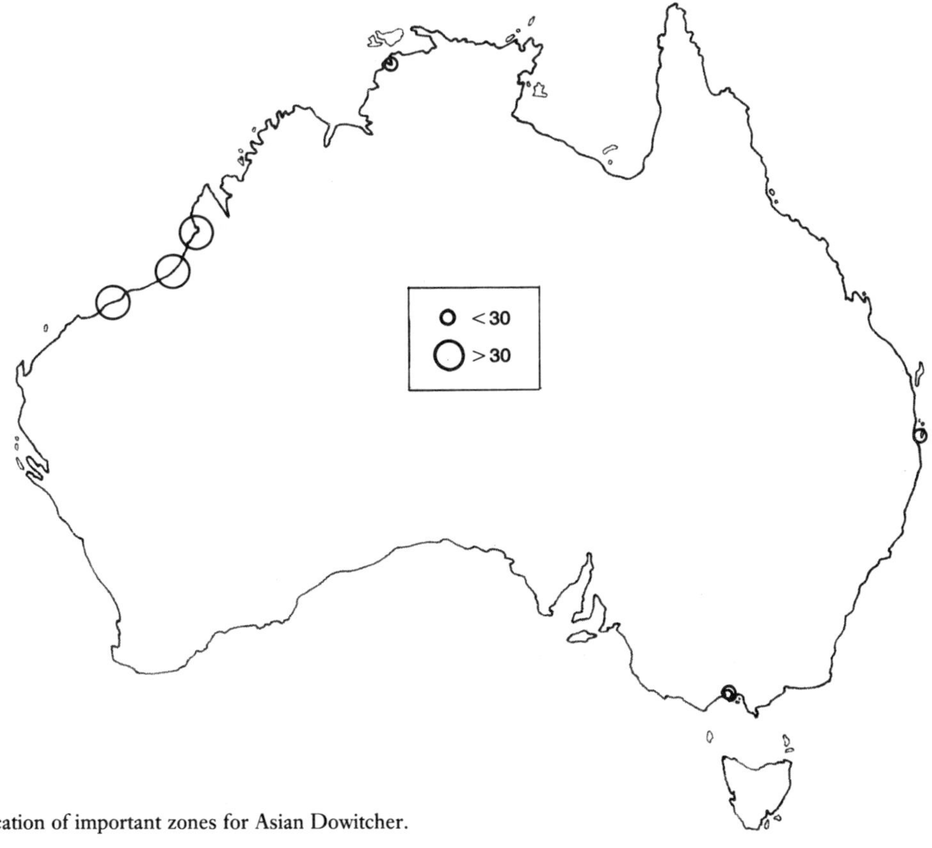


FIG. 5.51. Location of important zones for Asian Dowitcher.

# BLACK-TAILED GODWIT

## *Limosa limosa*
(Plates 10 and 11)

**Other names** Eastern Black-tailed Godwit; Blackwit.

There are three races of Black-tailed Godwits: *limosa*, *islandica* and *melanuroides*; only the smaller *melanuroides* visits Australia.

## DESCRIPTION

A large, long-legged, long-billed shorebird with uniform grey-brown upperparts and, in flight, conspicuous black-and-white markings.

**Adult non-breeding plumage** Crown, nape, hindneck, sides of neck, mantle, scapulars and inner wing-coverts *uniform grey-brown*; supercilium, buffish white; lores and ear-coverts, streaky grey-brown. Chin, white; throat, breast and flanks, washed uniformly grey-brown; belly and vent, off white. Iris, dark brown; bill, long, straight, tapering and *black with pink base*; legs, black. Sexes similar in non-breeding plumage, but the female is slightly larger than the male.

**In flight** Back, grey-brown; *rump, white; tail, black*; outer wing-coverts, dark grey; secondaries and inner primaries, black with broad white bases showing as *conspicuous white wing-bar in flight*; outer primaries, black.

**Voice** Quiet, hoarse 'ritta-rit'.

**Adult breeding plumage** Head, neck and upper breast, chestnut; chin, white; mantle and scapulars, chestnut, spotted black; wing-coverts, grey-brown; *lower breast and flanks, off-white, barred chestnut and black*; belly and vent, white; females duller with chestnut not extending as far down on breast.

**Juvenile plumage** Like adult non-breeding plumage, but more buff round head and neck; feathers of mantle and scapulars have buffish chestnut notched edges.

**Field notes** Occurs in flocks on mudflats and sandy beaches, often with Bar-tailed Godwit; found on inland wetlands.

**Similar species** Bar-tailed Godwit; Whimbrel; Greenshank. The Black-tailed Godwit is distinguished from the Bar-tailed Godwit by being slightly smaller, having more uniform upperparts, a grey-brown wash on underparts and a more striking flight pattern, including a black, not barred, tail. It is smaller than a Whimbrel, has more uniform upperparts and a straight, not down-curved, bill. It is browner than the Greenshank, which is smaller and lacks the white wing-bar and black tail.

## BREEDING RANGE

Those of the nominate race (*limosa*) breed between 40°N and 60°N in Europe and western Siberia, and those of the race *islandica* in Iceland and parts of Britain and Norway. The race that visits Australia (*melanuroides*) breeds in the taiga zone of eastern Siberia between 45°N and 65°N from the Yenesey River eastwards and in northern Mongolia (85). The precise boundaries of its breeding range are unknown.

## ASIAN RANGE

The Black-tailed Godwit occurs on the wetlands of Europe, Africa, Asia and Australia in the non-breeding months.

Those of the race *melanuroides* are common migrants in Korea (162), Japan, where they are more numerous on southward migration (max. 917 (488)), and in eastern China, where 'large flocks' are seen from July to September and from March to May (255). In northeastern China, it is outnumbered by the Bar-tailed Godwit (404). Small numbers regularly occur during

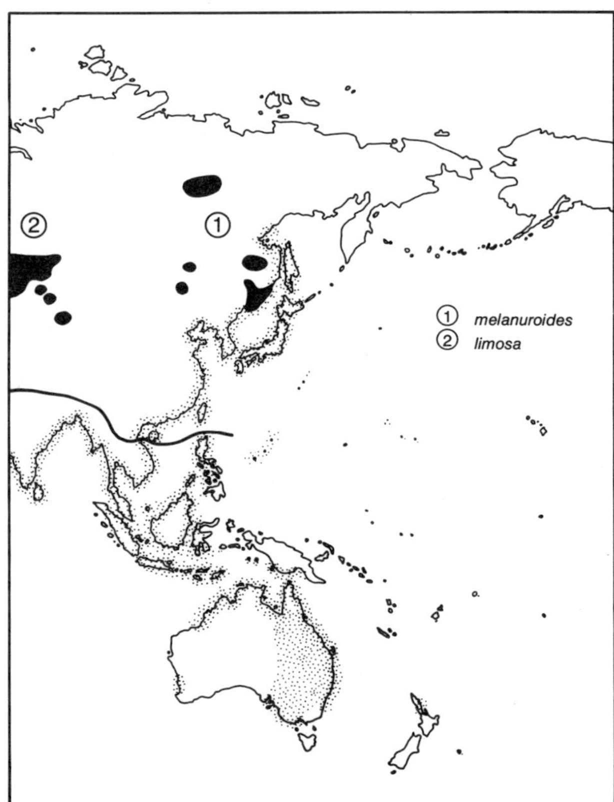


FIG. 5.52.   Breeding and non-breeding range of Black-tailed Godwit.

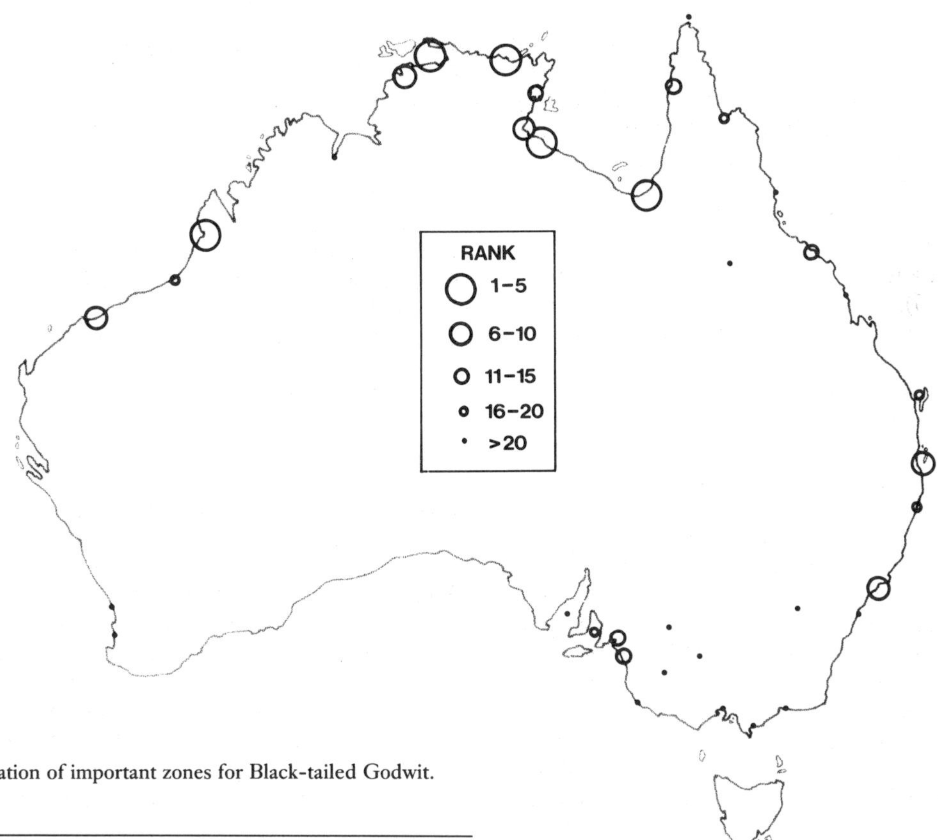



FIG. 5.53.   Location of important zones for Black-tailed Godwit.

TABLE 5.26.  Ranked listing of top twenty zones for Black-tailed Godwit. (maximum counts marked thus: *; other counts are averages) (40 or more birds)

| ZONE | NO. OF COUNTS | NO. OF INDIVIDUALS |
|---|---|---|
| * SE corner, Gulf of Carpentaria, Qld | 1 | 49 400 |
| * NE Arnhem Land, NT | + | 11 400 |
| * S coast, Gulf of Carpentaria, NT | + | 1 360 |
| * Van Diemen Gulf, NT | + | 870 |
| * Roebuck Bay, WA | 6 | 500 |
| Hunter estuary, NSW | 4 | 470 |
| * SW corner, Gulf of Carpentaria, NT | + | 450 |
| Darwin area, NT | 4 | 360 |
| * Moreton Bay, Qld | 5 | 260 |
| * Pilbara coast, WA | 3 | 200 |
| * The Coorong, SA | 1 | 180 |
| * Townsville area, Qld | 5 | 180 |
| * NW Cape York, Qld | + | 160 |
| * Murray mouth lakes, SA | 3 | 120 |
| * W coast, Gulf of Carpentaria, NT | + | 120 |
| * Eighty Mile Beach, WA | 5 | 110 |
| St Vincent Gulf, SA | 4 | 90 |
| * N coast, NSW | 5 | 90 |
| * Princess Charlotte Bay, Qld | + | 50 |
| * Hervey Bay–Great Sandy Strait, Qld | 5 | 40 |

(+ = aerial count)

migration in Hong Kong, where they are more abundant during northward migration (54). It is an uncommon visitor to Taiwan between October and May (401), to the Philippines (110, 171) and Borneo (440); only three were recorded at Brunei in 1980−81 (477). There is one record from western Micronesia (215). In Malaysia they are numerous, and up to 2800 have been seen on the west coast (348).

In Thailand it is common (257), but in southern Vietnam it is 'very rare' (489). It is regularly observed in Papua New Guinea, and flocks of more than 400 have been seen on the south-east coast of Irian Jaya (36), just north of its main Australian non-breeding area, the Gulf of Carpentaria and Arnhem Land. Small numbers occasionally visit New Zealand (134). (See Figure 5.52.)

## STATUS AND DISTRIBUTION IN AUSTRALIA

Despite being one of the most abundant waders in Australia, the Black-tailed Godwit has a restricted range, most occurring on the coast of Arnhem Land and the Gulf of Carpentaria. (See Figure 5.53, Table 5.26.) Only small numbers visit eastern and south-eastern Australia. Single birds or small flocks occur regularly at inland sites.

On the coast, it feeds at low tide on wide intertidal mudflats, mostly along the water's edge, and roosts at high tide on occasional sandy beaches, on mudbanks in shallow water, and on areas of saltflat behind mangroves. It also frequently visits inland wetlands, including sewage treatment works.

Estimates of the sizes of three other populations of Black-tailed Godwits are as follows: Atlantic European, 44 000; Atlantic African, 30 000 and Mediterranean, 12 000 (85). The sizes of the East African and southern and eastern Asian populations are not known, but Australia is unquestionably of major importance for the Black-tailed Godwit.

## MOVEMENTS IN AUSTRALIA

The Black-tailed Godwit arrives in Australia in late August, reaching the north-west coast first. In November, it passes through Darwin and does not reach maximum numbers in the Gulf of Carpentaria until December. Observations show that it is still arriving in the Gulf of Carpentaria in early December (49). Between September and November small numbers pass through the Mount Isa region and down the east coast as far as the Hunter estuary in New South Wales. Numbers remain relatively stable in the Gulf of Carpentaria and at the Hunter estuary between December and February.

Northward migration occurs up the east coast of Queensland in March and April, but it does not pass through Darwin or north-western Australia at this time. This suggests that the Black-tailed Godwit departs northwards directly into Asia from its main non-breeding area, the Gulf of Carpentaria and the north coast of Arnhem Land.

## FEEDING

The Black-tailed Godwit feeds by probing its bill into soft mud near the edge of water. No detailed study has been made of its feeding behaviour or diet in Australia. The gut of a bird collected in north-western Australia in 1908 had two small round seeds in it (300).

## CONSERVATION

Australia holds a substantial proportion of the world population of the Black-tailed Godwit, and probably most of those of the race *melanuroides*. Therefore the preservation of the important coastal habitat in Arnhem Land and the Gulf of Carpentaria is most important for its future survival. Fortunately, there are no immediate threats to these remote areas.

# BAR-TAILED GODWIT
## *Limosa lapponica*
(Plates 10 and 11)

**Other names**   Eastern Bar-tailed Godwit; Barwit.

There are two races of Bar-tailed Godwits: *lapponica*, breeding in northern Europe and western Siberia as far east as the Taimyr Peninsula, and *baueri*, which breeds in north-eastern Siberia, from the Taimyr Peninsula eastwards to the Chukotski Peninsula and Alaska (85). Only the latter comes to Australia.

## DESCRIPTION

A large, long-necked, long-legged shorebird with long, straight bill; generally mottled grey-brown.

**Adult non-breeding plumage**   Feathers of crown, nape, hindneck, sides of neck, sides of breast, mantle, scapulars, tertials and wing-coverts, mid grey-brown with white edges, giving the upperparts a *mottled appearance*; pale supercilium. Rest of underparts, off-white. Iris, black; bill, long, straight *and tapering, grey with pink base*; legs, grey. Sexes similar in non-breeding plumage, but female slightly larger than male.

**In flight**   *Rump and tail, off-white, barred dark brown*; primaries and secondaries, dark grey to brown; median and greater coverts have white tips appearing as a diffuse pale wing-bar.

**Voice**   Rather quiet, hoarse croaking calls are uttered by individuals; flocks can be noisy at times.

**Adult breeding plumage**   In male, feathers of crown, nape, hindneck, sides of neck, sides of breast and flanks, chestnut-red with fine black streaking; supercilium, white. Throat, foreneck, breast, belly and vent, *rich chestnut-red*; feathers of mantle, scapulars and tertials, black with chestnut edges. In female, underparts are streaked black on a pale chestnut background, except for the belly and vent, which are white.

**Juvenile plumage**   Similar to adult non-breeding plumage, but upperparts browner with buff edges to feathers.

**Field notes**   Occurs in flocks on mudflats and sandy beaches, often with knots; plumage transitional between breeding and non-breeding is often seen when birds first arrive (September–October) and just before departure (February–March); some young birds that do not return to the breeding grounds can show partial breeding plumage, especially on the underparts and scapulars.

**Similar species**   Black-tailed Godwit; Whimbrel; Greenshank. The Bar-tailed Godwit has more mottled markings than the Black-tailed Godwit, has a barred, not black, tail, and lacks the conspicuous wing-bar. It has a straight bill, not a down-curved one like that of the Whimbrel, and is browner than the Greenshank, which has a white, not barred, rump and tail.

## BREEDING RANGE

Of the two races, only the eastern race, *baueri*, visits Australia and New Zealand. It breeds from the eastern Taimyr Peninsula to the Chukotski Peninsula in northern Siberia and in western Alaska from Wales to Barrow (7), preferring the tundra zone north of the Arctic Circle.

## ASIAN RANGE

Those of the race *lapponica* spend the non-breeding months on the Atlantic coast of Europe and Africa (85) and in lesser numbers on the coast of the Arabian Sea to Bombay (5). The race *baueri* occurs in eastern Asia, the Pacific Islands, Australia and New Zealand.

The Bar-tailed Godwit occurs in large numbers on southward migration in Korea (162) and Japan, where on average three times the number occur compared with northward migration (488). In Taiwan, mainland China, Hong Kong, Vietnam and Thailand, it is uncommon to scarce (54, 255, 257, 489), and it is a vagrant in Malaysia (157). It is common in the Philippines (110), but not so in Borneo (440) or Bali (14). In the south-western Pacific Islands, it occurs regularly on migration and during the non-breeding months; up to 250 have been seen during northward migration in Fiji (406). At sites in south-eastern Irian Jaya, up to 700 have been reported (36), but elsewhere in New Guinea it is an uncommon annual visitor (184). The small number in New Guinea contrasts with their abundance in Australia (139). In New Zealand, they are the most abundant migratory shorebird, and about 70 000 visit annually (387, 388). (See Figure 5.54.) On northward migration it occurs in smaller numbers in far eastern Asia than on southward migration, suggesting a more direct return route to the breeding grounds.

The absence of large numbers in south-east Asia and its frequent occurrence on Pacific Islands during migration suggests that the Bar-tailed Godwit migrates long distances across the Pacific Ocean between north-eastern Asia and Australia and New Zealand, a distance of over 6000 kilometres.

## STATUS AND DISTRIBUTION IN AUSTRALIA

The Bar-tailed Godwit occurs in largest numbers in two discrete areas of Australia: along the east and south coasts to St Vincent Gulf in South Australia, and on the north-west coast from Eighty Mile Beach east to at least Darwin. On the Queensland coast, they were the most abundant shorebird. Most were confined to a small number of key areas: Hervey Bay,

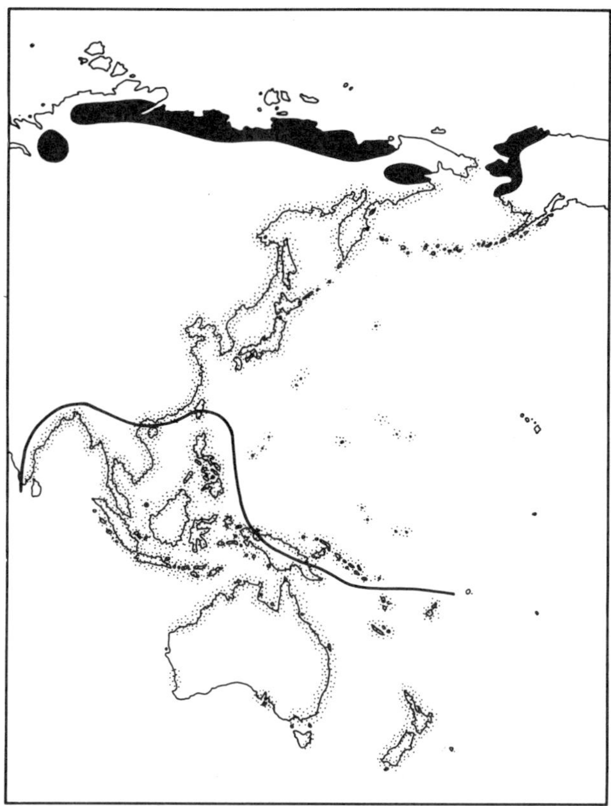

FIG. 5.54.   Breeding and non-breeding range of Bar-tailed Godwit.

Great Sandy Strait and Moreton Bay, south-eastern Queensland; Corner Inlet, Victoria; and Eighty Mile Beach and Roebuck Bay, Western Australia. (See Figure 5.55, Table 5.27.) In the south-eastern corner of the Gulf of Carpentaria it is greatly outnumbered by the Black-tailed Godwit.

The Bar-tailed Godwit is exclusively coastal. It forages on extensive intertidal flats, including those with seagrass meadows (Zosteraceae). It roosts at high tide on sandy beaches and spits.

At least 100 000 Bar-tailed Godwits occur in north-western Australia, 34 000 on the east coast and 70 000 in New Zealand, giving a total for Australasia of over 200 000. This compares with a population of the race *lapponica* in Europe and Africa of about 650 000 (85).

## MOVEMENTS IN AUSTRALIA

The Bar-tailed Godwit arrives in Australia, in the north-west and along the east coast, in August. Few appear to move on from the north-west, but on the east coast temporary influxes occur as far south as the Hunter estuary, New South Wales, until November, suggesting southward movement. Similar movements occur in coastal Victoria and South Australia.

From December to February, numbers are stable on most bays and estuaries along the east coast, e.g. Moreton Bay, Queensland; Botany Bay, New South Wales.

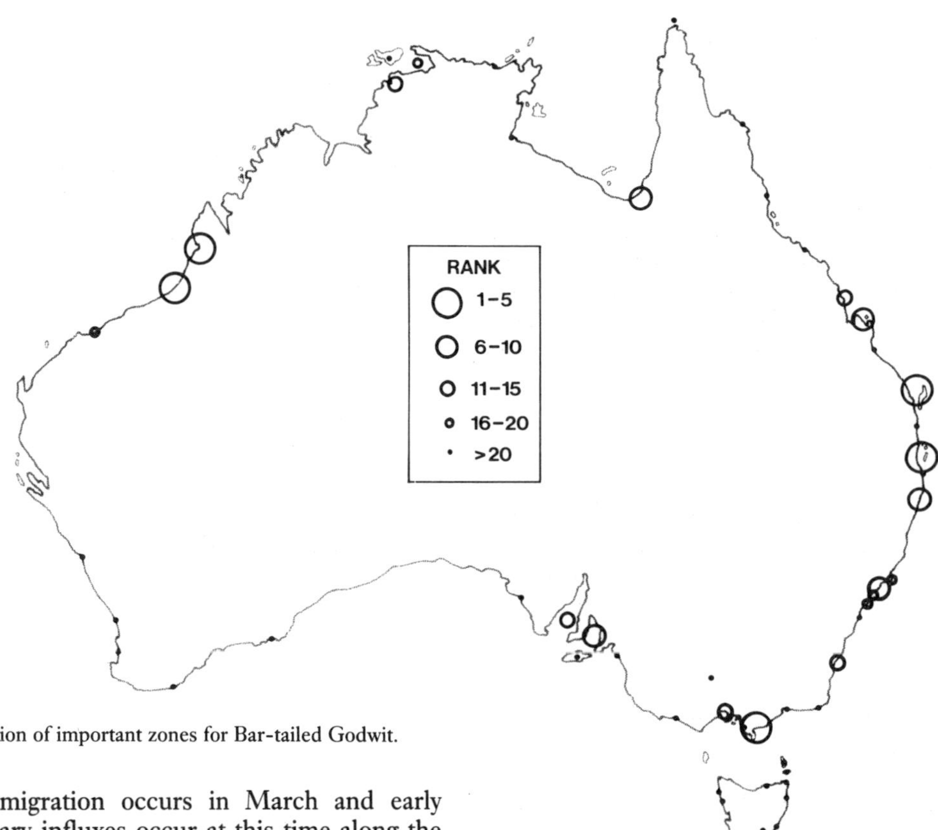



Fig. 5.55.  Location of important zones for Bar-tailed Godwit.

Northward migration occurs in March and early April. Temporary influxes occur at this time along the east coast as far north as Townsville. It appears not to pass through either Darwin or the north-west coast. In the latter area, the Bar-tailed Godwit departs in the first weeks of April.

After departure, influxes of non-breeding young birds occur in June and July on the east coast of Queensland and at Darwin. Furthermore, July numbers in northern New South Wales, Queensland and round Darwin average about 55 per cent of February numbers, whereas in southern New South Wales, Victoria and Tasmania, July numbers are less than 10 per cent of February numbers. These observations suggest that young Bar-tailed Godwits move out of south-eastern Australia after the adults have departed.

## FEEDING

The Bar-tailed Godwit feeds by repeatedly probing its bill into soft mud and sand near the edge of water. At Westernport Bay, Victoria, it has been observed eating polychaetes and sentinel crabs (*Macrophthalmus latifrons*). Here, food-robbing by gulls reduced foraging success (number of food items ingested per unit time) by 36 per cent (95).

## CONSERVATION

Most eastern race Bar-tailed Godwits in the world probably come to Australia and New Zealand, so the adequate reservation and management of the small number of sites that hold most of them is essential to the continued existence of this unique form.

TABLE 5.27.  Ranked listing of top twenty zones for Bar-tailed Godwit. (maximum counts marked thus: *; other counts are averages) (370 or more birds)

| ZONE | NO. OF COUNTS | NO. OF INDIVIDUALS |
|---|---|---|
| * Roebuck Bay, WA | 6 | 65 000 |
| * Eighty Mile Beach, WA | 5 | 34 300 |
| * Hervey Bay – Great Sandy Strait, Qld | 5 | 7 700 |
| Corner Inlet, Vic. | 5 | 7 300 |
| * Moreton Bay, Qld | 5 | 6 300 |
| * SE corner, Gulf of Carpentaria, Qld | 1 | 2 210 |
| * N coast, NSW | 5 | 1 830 |
| Hunter estuary, NSW | 4 | 1 300 |
| St Vincent Gulf, SA | 4 | 1 240 |
| * Broad Sound & Shoalwater Bay, Qld | + | (1 070) |
| Mackay area, Qld | 4 | 890 |
| Darwin area, NT | 4 | 750 |
| * S coast, NSW | 5 | 620 |
| * Spencer Gulf, SA | 5 | 600 |
| Port Phillip Bay, Vic. | 5 | 570 |
| Botany Bay, NSW | 5 | 470 |
| * Central coast, NSW | 5 | 450 |
| * Pilbara coast, WA | 3 | 390 |
| * Van Diemen Gulf, NT | 2 | 380 |
| Port Stephens, NSW | 4 | 370 |

(+ = aerial count)

# RED KNOT

## *Calidris canutus*
(Plate 15)

**Other names** Lesser, Common or Grey-crowned Knot; Knot; Knot-Snipe.

Four races are recognised (85): *islandica* and *rufa*, which breed in arctic Canada and Greenland; *canutus*, breeding in central Siberia; and *rogersi*, which breeds in eastern Siberia and western Alaska. Those of the race *rogersi* visit Australia and New Zealand.

## DESCRIPTION

A medium-sized dumpy grey sandpiper with a straight, black bill.

**Adult non-breeding plumage** Crown, nape, hind-neck and sides of neck, grey, finely streaked black; mantle, scapulars and wing-coverts, grey, *feathers thinly edged white*; supercilium, white, streaked grey; lores and ear-coverts, grey, streaked black. Chin, white; throat and breast, white, finely streaked and spotted grey; belly and vent, white; *flanks, white with dark grey chevrons on feathers*; Iris, black; *bill, medium length, straight and black*; legs, greenish-grey. Sexes similar.

**In flight** *Back and rump, white, diffusely barred dark grey*; tail, grey; primaries and secondaries, dark grey with white bases to inner webs, which show as *diffuse wing-bar* together with white-tipped greater and secondary coverts.

**Voice** Clucking calls when feeding and roosting; a quiet 'nyoot-nyoot' in flight.

**Adult breeding plumage** Crown, nape and hind-neck, streaked black, white and chestnut; mantle, scapulars and wing-coverts, chestnut with black spots and white edges to feathers; wing-coverts, grey with white edges; supercilium, chestnut; lores and line behind eye, streaked black and chestnut. Ear-coverts, chin, throat, sides of neck and rest of *underparts, rich chestnut-red*, banded black on flanks; vent, white.

**Juvenile plumage** Like adult non-breeding plumage, but upperparts browner and more heavily marked; scapulars and coverts have buff edges and black subterminal bars; birds in fresh plumage show buff wash on flanks.

**Field notes** Flies, roosts and feeds in tight flocks; feeds with head down, rapidly and repeatedly jabbing mud with bill. Upon arrival (August—October) and before departure (February—April) birds moulting out of and into breeding plumage are seen.

**Similar species** Great Knot; Curlew Sandpiper. The Red Knot is slightly smaller than the Great Knot, has more uniform upperparts and lacks a distinct white rump. It has a shorter neck, legs and bill, giving it a generally dumpier appearance than either the Great Knot or the Curlew Sandpiper. It is slightly larger than the Curlew Sandpiper.

## BREEDING RANGE

The Red Knot breeds across the arctic in localised populations (85). Those of the race *rogersi*, which visit Australia, have been recorded breeding from the Taimyr Peninsula in north-central Siberia eastwards to western Alaska. The limits of their breeding range are not completely known. Dorogoi (117) documents their breeding on Wrangel Island in far-eastern Siberia. Once breeding is completed, they move to coastal intertidal habitats (85).

## ASIAN RANGE

The Red Knot is rare in Asia and the Pacific despite being abundant in Australia and New Zealand. No more than 124 have been seen in Japan on migration (488), and it is an uncommon passage migrant in Korea (162) and China (255). In Hong Kong, it is scarce during migration (54), and only small numbers have been observed in Taiwan (401), Thailand (257), the Philippines (110), Sulawesi (128), Bali (14) and New Guinea (184). Small groups occur during southward migration in September in Borneo (440). (See Figure 5.56.)

Elsewhere in the world, other races use only a few stopover areas during migration (85). Its general rarity

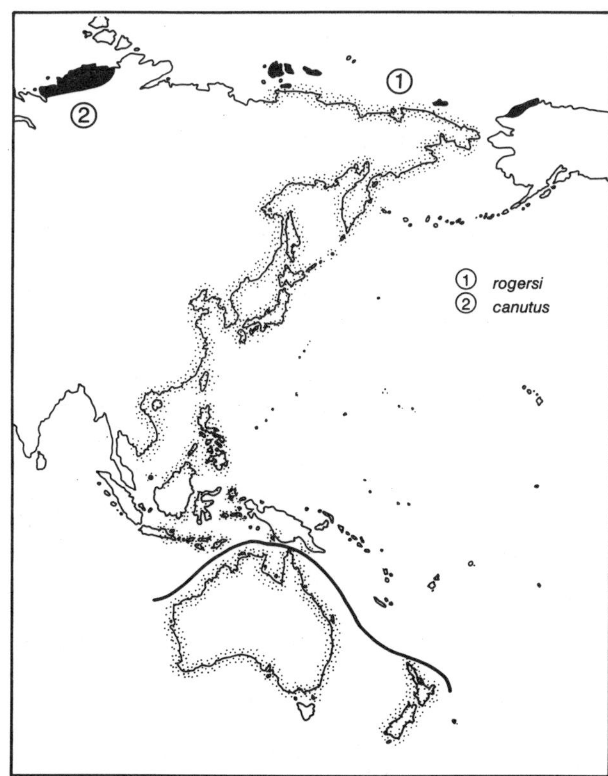


FIG. 5.56.   Breeding and non-breeding range of Red Knot.

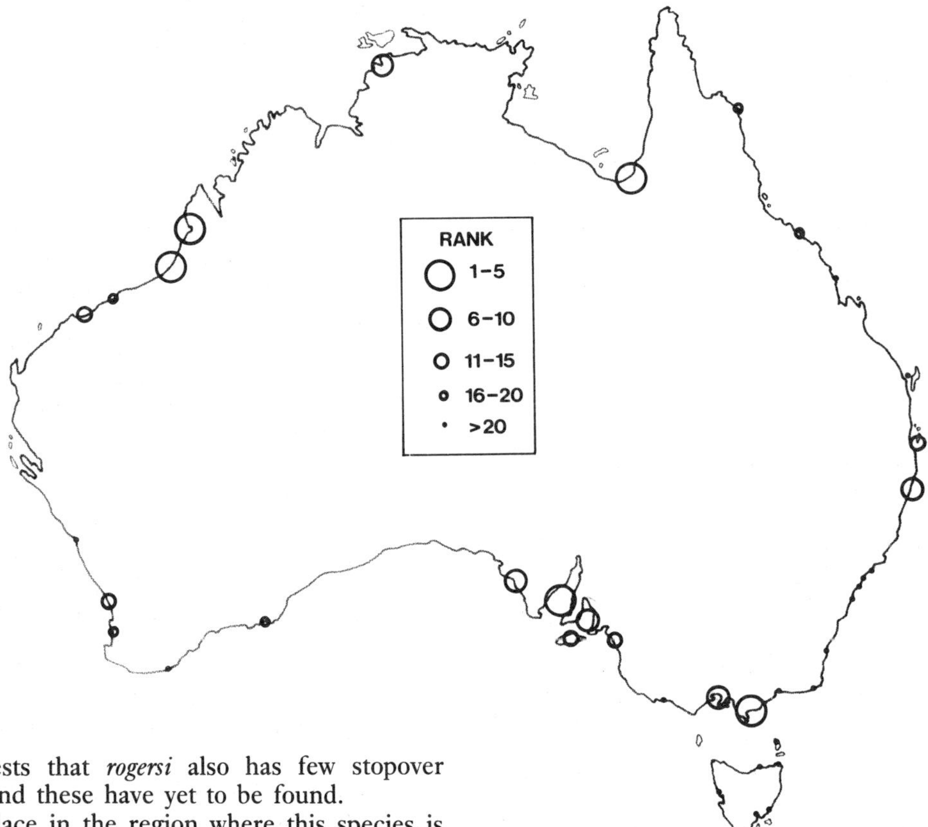


FIG. 5.57.   Location of important zones for Red Knot.

in Asia suggests that *rogersi* also has few stopover sites in Asia and these have yet to be found.

The only place in the region where this species is common outside Australia is in New Zealand, where over 45 000 spend the non-breeding months (387, 388). Five banding recoveries confirm movement between New Zealand and Australia. This, combined with its general absence in the Pacific, suggests that New Zealand's Red Knots migrate through Australia (see 'Movements in Australia').

## STATUS AND DISTRIBUTION IN AUSTRALIA

Red Knots are very abundant during southward migration in the Broome—Port Hedland area of Western Australia and in the south-eastern corner of the Gulf of Carpentaria in Queensland. Later in the year, numbers are much lower in these places. Less than 400 occur on the New South Wales coast and less than 20 in Tasmania. In eastern Queensland, it is recorded at only a few sites, mostly on migration. Corner Inlet and Port Phillip Bay, in Victoria, and the Gulfs in South Australia each hold a thousand or more, and a few hundred occur regularly between Esperance and Perth in Western Australia. (See Figure 5.57, Table 5.28.) It is almost exclusively coastal.

The Red Knot feeds on intertidal mudflats in large flocks. The only record of substantial numbers away from intertidal habitat is at Lake Reeve, Victoria, a near-coastal saline wetland, where over 1000 have been seen (82). At high tide it roosts in tight flocks on sandy beaches and spits. In St Vincent Gulf, South Australia, there is little sandy shore and it roosts in shallow saline ponds in the Price and St Kilda Salt-

TABLE 5.28.   Ranked listing of top twenty zones for Red Knot. (maximum counts marked thus: *; other counts are averages) (50 or more birds)

| ZONE | NO. OF COUNTS | NO. OF INDIVIDUALS |
|---|---|---|
| * Eighty Mile Beach, WA | 5 | 80 700 |
| * SE Corner, Gulf of Carpentaria, Qld | 1 | 79 100 |
| * Roebuck Bay, WA | 6 | 11 200 |
| * Spencer Gulf, SA | 5 | 4 800 |
| Corner Inlet, Vic. | 5 | 2 720 |
| * St Vincent Gulf, SA | 4 | 2 000 |
| Port Phillip Bay, Vic. | 5 | 730 |
| * W coast, Eyre Peninsula, SA | 5 | 710 |
| Darwin area, NT | 4 | 340 |
| * N coast, NSW | 5 | 310 |
| * Kangaroo Island, SA | 5 | 300 |
| * Pilbara coast, WA | 3 | 250 |
| * The Coorong, SA | 1 | 170 |
| Moreton Bay, Qld | 5 | 100 |
| Swan coastal plain, WA | 5 | 80 |
| Peel Inlet, WA | 5 | 80 |
| * Port Hedland Saltworks, WA | 12 | 80 |
| * Daintree—Melville, Qld | 2 | 80 |
| Townsville area, Qld | 5 | 60 |
| Esperance coast, WA | 4 | 50 |

works. At Whyalla Saltfields in Spencer Gulf, South Australia, it roosts on a sandy beach rather than with other waders in the shallow ponds.

It appears that few Red Knots spend the non-breeding months in Asia. The Australian and New Zealand counts compare with over 800 000 in Europe–West Africa (85) and at least 160 000 in North and South America (178).

## MOVEMENTS IN AUSTRALIA

The Red Knot arrives in Australia in late August. In the Broome–Port Hedland area, about 90 000 arrive in late August and early September. Banding studies show that it is fattening rapidly at this time, getting ready for onward migration, and by November more than half have moved on. Maximum numbers occur in Darwin in September and October. At the same time, in the Gulf of Carpentaria, sample counts (150) suggest that tens of thousands pass through. National counts show that less than 10 000 Red Knots occur in southern Australia. There have been five recoveries of banded birds between Australia and New Zealand and they occur on the eastern coast of Queensland in small flocks between September and November. These observations suggest that many migrate along the north coast of Australia, probably on their way to New Zealand, where they are very abundant (387, 388).

Small flocks of Red Knots migrating northwards appear on the eastern coast of Queensland in mid-March. Sample counts in the Gulf of Carpentaria in April (150) suggest that they move through that area again, during northward migration. However, they do not occur at Darwin at this time. In mid-April there is a small influx in the Broome–Port Hedland area, probably of birds from south-eastern Australia (247). This suggests that Red Knots from New Zealand might return via the Gulf, while those from south-eastern Australia may fly northwards via the north-west coast.

## FEEDING

A characteristic behaviour of the Red Knot is its habit of feeding by rapidly drilling soft sand and mud at the edge of water. It is a tactile rather than a visual feeder. At Lake Reeve, Victoria, it feeds in this manner over wet algal mat in shallow water, taking gastropod molluscs and amphipod crustacea (100).

## CONSERVATION

Most Red Knots in Asia probably come to Australia and New Zealand, so the survival of the race *rogersi*, which occurs nowhere else, depends upon the preservation of adequate habitat, particularly in northern Australia, where most use the coast either during migration or throughout the non-breeding months.

There are no immediate serious threats to their habitat in northern Australia, and most of their important habitats in southern Australia have been reserved.

# GREAT KNOT
## *Calidris tenuirostris*
(Plate 15)

**Other names** Eastern, Japanese, Slender-billed or Stripe-crowned Knot; Great Sandpiper.
No races recognised.

## DESCRIPTION

A stocky, medium-sized to large sandpiper with a longer bill, legs and neck than the Red Knot; scalloped grey and white upperparts and grey blotches on breast which are black in breeding plumage.

**Adult non-breeding plumage** *Crown, heavily streaked* black on grey; nape, hindneck, mantle, scapulars, wing-coverts and tertials, grey-brown, *feathers edged white giving scaly appearance*; supercilium, pale but indistinct; ear-coverts, grey with white edges. Chin and throat, white; sides of neck, *breast and flanks, blotched grey*; belly and vent, white. Iris, black; *bill, medium length, slightly down-curved* and black; legs, grey. Sexes similar.

**In flight** Back, grey-brown, feathers having white edges; *rump, white*; tail, grey; median and greater wing-coverts have narrow white borders that in flight show as a *narrow white wing-bar*; primaries and secondaries, dark grey.

**Voice** Generally silent; occasionally utters calls similar to the Red Knot: double-noted whistle, 'nyut-nyut'.

**Adult breeding plumage** Crown, nape and hindneck, streaked black and white; scapulars and tertials, mostly black with rich chestnut-and-white edges; wing-coverts remain grey; supercilium, less noticeable. Chin and throat, white; *sides of neck, lower throat, breast and flanks*, spotted black, often so heavy as to give the appearance of a completely black breast; belly and vent, white.

**Juvenile plumage** Difficult to distinguish: slightly darker upperparts than adult non-breeding plumage.

**Field notes** Usually in tight flocks, often in the company of godwits and Red Knots, most abundant in northern Australia; often seen in plumage transitional between breeding and non-breeding, black breast blotches being especially obvious in August–November and February–April. Some spend the breeding season (May–July) in Australia, a few showing partial breeding plumage.

**Similar species** Red Knot. The Great Knot is slightly larger than the Red Knot, has longer legs, neck and bill, which is slightly down-curved, and more scalloped grey upperparts. In breeding plumage it is conspicuously different.

## BREEDING RANGE

The Great Knot breeds in north-eastern Siberia. The location of the few known nests suggests a breeding range extending from the Verkhoyanskii Range in the west to the lower Kolyma and Anadyr Rivers and the Koryaksky Range in the east (5, 85). It breeds in dry alpine tundra and feeds alone or in small flocks in bogs up to 3 kilometres from the nest site (325).

## ASIAN RANGE

In the non-breeding months it occurs occasionally from the Persian Gulf eastwards and more regularly from Malaysia and China, south to Australia.

The Great Knot is an uncommon migrant in Korea (162), although 1200 have been counted on the Nakdong estuary (364). Up to 1900 have been counted during southward and 3000 during northward migrations in Japan (488). It occurs in north-eastern China, mostly during southward migration (54, 179, 255), and south-eastern China, mostly during northward migration. It is uncommon in Vietnam (58), Thailand (257), Malaysia (157) and the Philippines (110). Before about 1940, they were abundant in Borneo, but have declined since (440). Very few have been seen in Bali (14). Up to 180 have been seen on the coast of south-eastern Irian Jaya (36), but it is rare in the Port Moresby district of Papua New Guinea (138, 278). Up to 20 000 have been seen on northward migration on the Kamchatka Peninsula, eastern Siberia (325). This is the only report of large numbers away from Australia.

It spends the non-breeding months from southern China and Taiwan (58), south to Australia. (See Figure 5.58.) It is very rare in New Zealand (134).

The absence of reports of large numbers in Asia compared to its great abundance in northern Australia suggests that it undertakes long migration flights between a limited number of stopover places, and that most of the world population spends the non-breeding months in Australia.

## STATUS AND DISTRIBUTION IN AUSTRALIA

Tens of thousands of Great Knots occur on Eighty Mile Beach and Roebuck Bay in Western Australia and in the south-eastern corner of the Gulf of Carpentaria in Queensland. Hundreds also inhabit the east coast of Queensland as far south as Brisbane. They occur in smaller numbers (up to 2000) in the South Australian gulfs and in Port Phillip Bay, Victoria, and 200−300 spend the non-breeding months on the south coast of Western Australia. (See Figure 5.59, Table 5.29.) The count of 6500 at Darwin in July 1982 shows that good numbers of young Great Knots stay in northern Australia during the breeding months. They are exclusively coastal.

FIG. 5.58.   Breeding and non-breeding range of Great Knot.

The Great Knot forages at low tide on intertidal mudflats. At high tide it roosts on sandy spits and beaches (e.g. north-western Australia, Victoria), on saltflats among scattered mangroves (e.g. south-eastern corner of the Gulf of Carpentaria) or in shallow non-coastal wetlands (e.g. in saltworks on St Vincent Gulf, South Australia).

The Great Knot was considered rare up until the 1970s, but this study discovered very large numbers on the north coast of Australia. Many of the shorebirds observed during aerial surveys of more remote parts of the north coast were Great Knots. It was one of Australia's most abundant shorebirds; indeed it could be described as the Red-necked Stint of the north.

## MOVEMENTS IN AUSTRALIA

Large numbers of Great Knots arrive in the Broome−Port Hedland area in late August and early September. Some of these move on by November, when there is a temporary influx at Darwin. There are many Great Knots in the Gulf of Carpentaria, but maximum numbers are not reached there until December. Sample counts in the Gulf (150) show that few pass through. Great Knots are not abundant in New Guinea, small numbers occur there in October and November. These observations suggest that they arrive in the north-west first and spread eastwards

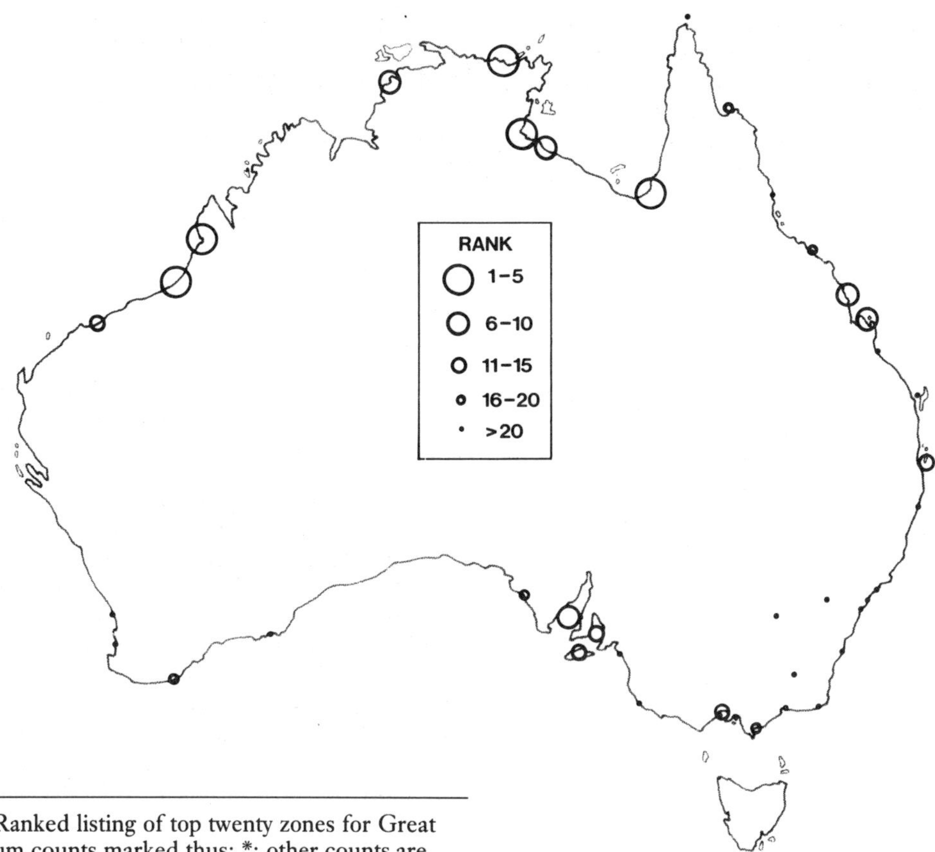



FIG. 5.59.   Location of important zones for Great Knot.

TABLE 5.29.  Ranked listing of top twenty zones for Great Knot. (maximum counts marked thus: *; other counts are averages) (80 or more birds)

| ZONE | NO. OF COUNTS | NO. OF INDIVIDUALS |
|---|---|---|
| * Eighty Mile Beach, WA | 5 | 160 000 |
| * SE corner, Gulf of Carpentaria, Qld | 1 | 68 000 |
| * NE Arnhem Land, NT | + | 65 200 |
| * Roebuck Bay, WA | 6 | 22 600 |
| * SW corner, Gulf of Carpentaria, NT | + | 21 400 |
| * Darwin region, NT | + | 14 800 |
| * S coast, Gulf of Carpentaria, NT | + | 11 200 |
| * Broad Sound & Shoalwater Bay, Qld | + | 4 200 |
| * Mackay area, Qld | + | 4 000 |
| * Spencer Gulf, SA | 5 | 1 300 |
| Moreton Bay, Qld | 5 | 910 |
| * Kangaroo Island, SA | 5 | 800 |
| St Vincent Gulf, SA | 4 | 520 |
| Port Phillip Bay, Vic. | 5 | 480 |
| * Pilbara coast, WA | 3 | 430 |
| * W coast, Eyre Peninsula, SA | 5 | 230 |
| Townsville area, Qld | 5 | 230 |
| Corner Inlet, Vic. | 5 | 210 |
| Albany coast, WA | 5 | 150 |
| * Princess Charlotte Bay, Qld | + | 80 |

(+ = aerial count; figure refers only to unidentified medium-sized shorebirds and may not be as precise an estimate of Great Knot numbers as in other areas)

along the north coast of Australia. Smaller numbers (hundreds) also appear on the east coast of Queensland, between September and November, but only south of Townsville, suggesting that some migrate south-eastwards from the Gulf of Carpentaria to the east coast. No more that 2000 regularly reach South Australia and Victoria, usually in October and November. Tens of thousands occur in the north-west and in the Gulf of Carpentaria between December and February.

Northward migration occurs in March and April. There are temporary influxes at this time at most places on the east coast of Queensland, but not at Darwin. The Great Knot does not appear to migrate northwards through the Broome–Port Hedland area from whence it departs between mid-March and early April (247). These observations suggest that there is some northward migration along the eastern coast of Queensland, but that most depart the northern coast directly rather than returning via the north-west.

## FEEDING

The Great Knot feeds mainly on intertidal mudflats by rapidly and repeatedly jabbing its bill into mud, often at the edge of water on the falling tide. It also feeds in shallow water. In the Gulf of Carpentaria, the Great Knot feeds in similar densities on the sandy muds at river mouths, on the softer muds between sand bars and along intervening stretches of mangrove-fringed shore. Total numbers there were highest near river-mouths, and it avoided flats covered with seagrass (Zosteraceae). Nothing is known of its diet in Australia.

## CONSERVATION

As Australia appears to hold most of the world population of the Great Knot, the preservation and correct management of its principal migratory stopovers in the north-west and of other sites along the north coast are important to its survival. Fortunately, there are no immediate threats to its habitat in Australia as its favoured haunts are remote and largely wilderness areas.

# SHARP-TAILED SANDPIPER
## *Calidris acuminata*
(Plate 12)

**Other names**  Asiatic or Siberian Pectoral Sandpiper; Sharp-tailed Stint; Snipe; Sharpie.

## DESCRIPTION

A small sandpiper with scaly brown upperparts and olive-green legs; often found on inland wetlands as well as the coast.

**Adult non-breeding plumage**  *Crown, chestnut with black streaks*; nape and hindneck, pale grey-brown streaked black; supercilium, white; ear-coverts, sides of neck and sides of breast, grey-brown streaked black; feathers of mantle, scapulars and tertials, dark brown, broadly edged buff; wing-coverts, dark brown edged paler buff. This gives upperparts a *generally heavy scalloped pattern. Breast, washed pale buff with* brown speckling; belly, flanks and vent, off-white; undertail-coverts, off-white, sparsely streaked brown. Iris, black; bill, short, straight, blackish with olive base; legs and feet, olive-green. Sexes similar in plumage, but male slightly larger than female.

**In flight**  Back and rump, dark brown with thin buffish edges to feathers; *narrow white area on sides of rump*; tail, brown with pointed central feathers that *protrude slightly beyond end of tail*; wings generally mottled brown above, with little suggestion of a wing-bar; primaries and secondaries, dark brown.

**Voice**  Soft, metallic double whistle: 'pleep-pleep' uttered when disturbed, at other times a variety of quiet whistles and peeps.

**Adult breeding plumage**  Like non-breeding plumage, except that throat, breast and flanks are washed buff, heavily marked with dark brown chevrons; margins of feathers on upperparts are brighter buff.

**Juvenile plumage**  Like adult non-breeding plumage, only more strikingly marked: crown, redder; margins of feathers on upperparts, orange; breast, washed pale orange, spotted brown on sides. Difficult to distinguish from adult after December.

**Field notes**  Equally likely to occur on coastal habitats as on inland swamps, lakes and sewage treatment works. Often seen in partial breeding plumage when it first arrives (August–October) and before departure (January–April).

**Similar species**  Red-necked Stint; Curlew Sandpiper; Pectoral Sandpiper; Long-toed Stint. Distinguished from the stints by its larger size and from the Curlew Sandpiper, which is the same size, by more scalloped, browner upperparts, shorter, straight bill and shorter, olive (cf. black) legs. See also, Pectoral Sandpiper and Long-toed Stint in rare species accounts, Chapter 6.

## Breeding Range

The nest of the Sharp-tailed Sandpiper was first discovered in 1957 (238). Since then its breeding range has been shown to extend through the high-latitude arctic lowlands of north-eastern Siberia from the delta of the Lena River in the west to just east of the Kolyma River (325, 468). It nests in damp sedgeland, selecting the wettest habitat of all sandpipers that breed in the high arctic (325).

## Asian Range

Most come to Australia, but small numbers also occur in Borneo (440), the Philippines (171), New Guinea (184), Micronesia (18) and New Zealand (387, 388) during the non-breeding months. (See Figure 5.60.)

The adult Sharp-tailed Sandpiper migrates directly southwards to China, while young birds probably take a more easterly route down the Pacific coast (468). Only during southward migration are they abundant in northern China, whereas during northward migration they are abundant only in southern China (255). In Korea, they occur in thousands on the west coast during southward (August–October) and northward (April–May) migration (162). Only small numbers occur in Japan, mostly during northward migration (344, 488). They are scarce in Burma and Malaysia (229), and have not been recorded from Thailand (257). They are uncommon in the Philippines (110, 171). On the islands of central and western Micronesia, and in Bali, they are common during migration (14, 18, 59, 60). They are the most abundant shorebirds in New Guinea during southward migration, and up to 3000 have been seen near Port Moresby (138, 184). Up to 670 have been counted in south-eastern Irian Jaya (36). In New Zealand they are uncommon, but annual, visitors (387, 388).

To summarise: the Sharp-tailed Sandpiper probably migrates southwards through the island chains of Micronesia and possibly the Philippines to reach New Guinea and Australia directly from Korea and nearby northern China. On northward migration it probably takes the same route. The lack of records in New Guinea at this time contrasts with their abundance in the Gulf of Carpentaria in Australia, and they may overfly New Guinea and Micronesia to southern and eastern China, then move to Siberia, some going via Korea.

The absence of records of large numbers in Asia and northern Australia between December and February suggests that most of the world population of Sharp-tailed Sandpipers comes to southern Australia.

## Status and Distribution in Australia

Most Sharp-tailed Sandpipers are found in the wetlands of Victoria, south-western New South Wales and eastern South Australia. (See Figure 5.61, Table 5.30.) Only small numbers occur in Tasmania, despite their abundance in nearby Victoria. South-western Australia annually holds up to 2000. A substantial proportion of those in Victoria and New South Wales occur on inland wetlands. They use ephemeral inland salt lakes, such as Lake Eyre, when these hold water. The lakes of the lower Murray and Darling drainage systems hold thousands of them at times. During southward migration they are abundant in the north-west, and on northward migration large numbers occur in the Gulf Country of Queensland.

The Sharp-tailed Sandpiper prefers non-tidal wetlands, especially freshly exposed mudflats in drying lakes, saltworks and sewage treatment works, and it ventures onto intertidal mudflats between December and March, probably because inland wetlands in southern Australia dry out at this time.

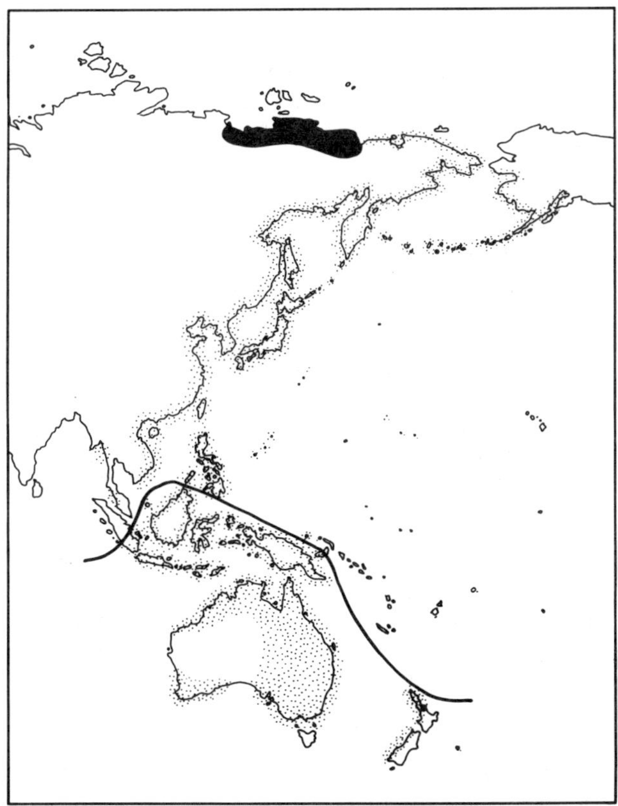

Fig. 5.60. Breeding and non-breeding range of Sharp-tailed Sandpiper.

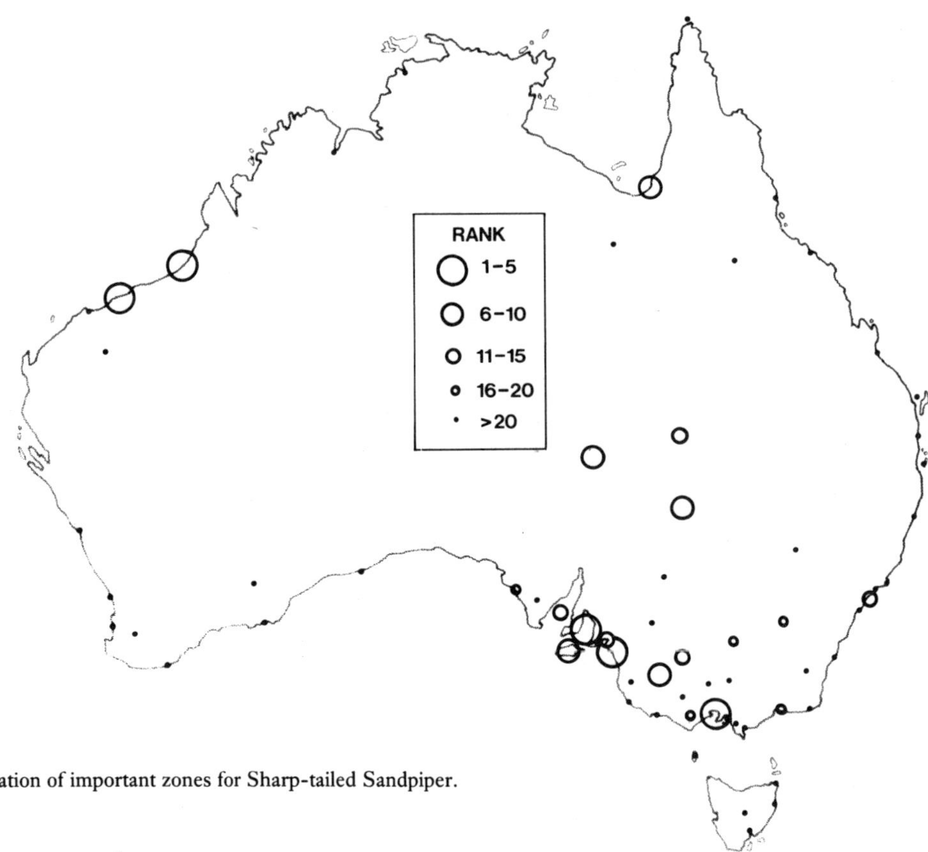



FIG. 5.61.  Location of important zones for Sharp-tailed Sandpiper.

## MOVEMENTS IN AUSTRALIA

Small numbers of Sharp-tailed Sandpipers arrive in the Broome–Port Hedland area, Western Australia, and in southern Australia in mid-August. They first arrive in their thousands in the north-west in early September. Temporary influxes occur between August and December at Darwin (88), near Mount Isa and in most inland wetlands in eastern Australia. Maximum numbers occur on the coast of Victoria and South Australia in January and early February. (See Chapter 4, Figure 4.7.) These observations suggest that they take their time migrating southwards, using inland wetlands as they cross the continent from north to south. They move about once they reach southern Australia, and their numbers are generally not stable anywhere in Australia.

Northward migration starts in the south in mid-February. Few pass through inland parts of eastern Australia or through Darwin or the Broome–Port Hedland area at this time. Before departing from Victoria it does not fatten sufficiently to fly directly out of Australia (476). In April, after most have left southern Australia, thousands have been observed in the wet-season swamps near the Gulf of Carpentaria (150). These observations show that it departs from southern Australia in February and March and flies to the wetlands and the Gulf of Carpentaria and fattens there before departing directly from Australia some time in April.

TABLE 5.30.  Ranked listing of top twenty zones for Sharp-tailed Sandpiper. (maximum counts marked thus: *; other counts are averages) (more than 1000 birds)

| ZONE | NO. OF COUNTS | NO. OF INDIVIDUALS |
|---|---|---|
| * The Coorong, SA | 1 | 55 700 |
| * Eighty Mile Beach, WA | 5 | 25 000 |
| * Port Hedland Saltworks, WA | 12 | 20 000 |
| * St Vincent Gulf, SA | 4 | 17 000 |
| Port Phillip Bay, Vic. | 5 | 8 800 |
| * Far NW lakes, NSW | 1 | 7 400 |
| * SE corner, Gulf of Carpentaria, Qld | 1 | 7 300 |
| * Lake Eyre North, SA | 1 | 4 000 |
| * Kangaroo Island, SA | 5 | 3 150 |
| Horsham area lakes, Vic. | 5 | 3 000 |
| * Spencer Gulf, SA | 5 | 2 730 |
| * Murray mouth lakes, SA | 3 | 2 630 |
| Kerang–Swan Hill, Vic. | 5 | 2 260 |
| * Far SW lakes, Qld | 1 | 2 010 |
| * Central coast, NSW | 5 | 1 830 |
| * W coast, Eyre Peninsula, SA | 5 | 1 740 |
| * Parkes region, NSW | 4 | 1 590 |
| Western District lakes, Vic. | 5 | 1 510 |
| Riverina, NSW | 4 | 1 180 |
| Gippsland lakes, Vic. | 5 | 1 090 |

## FEEDING

The Sharp-tailed Sandpiper feeds in mud or shallow water in a wide variety of wetlands, from hypersaline lakes (e.g. the salinity of Lake Eyre in September 1984, when 6000 were present, was four times that of sea-water) to freshwater swamps and intertidal mudflats. It probably eats a great variety of food. It has been recorded eating aquatic and terrestrial insects, small shellfish and grass seeds (256, 300, 472), as well as polychaete worms (97).

## CONSERVATION

Its choice of a large variety of wetlands and its extensive use of artificial habitats, such as sewage farms and saltworks, indicates that it is able to adapt to changes to wetlands wrought by human activities. However, in midsummer, most occurred in south-eastern Australia, where the most people live. The preservation and correct management of an adequate area of coastal and inland wetlands is essential for its future survival. This is especially important given that most of the world population comes to this part of Australia.

# RED-NECKED STINT

## *Calidris ruficollis*
(Plate 14)

**Other names**  Rufous-necked Stint or Sandpiper; Little Stint.
No races recognised.

## DESCRIPTION

The smallest of Australia's migratory shorebirds; plump, short-billed, short-legged; often seen in large tight flocks.

**Adult non-breeding plumage**  Crown, nape, hind-neck, mantle, scapulars and wing-coverts, grey-brown, feathers edged white; supercilium, white; lores and ear-coverts, streaked grey. Chin and throat, white; breast, white with variable amounts of grey flecking; belly and vent, white. Iris, black; *bill, short and black*; *legs, short and black*. Sexes similar.

**In flight**  Back like rest of upperparts; rump, white with central black line; tail, grey; tips of greater coverts, white, appearing as a noticeable wing-bar; primaries and secondaries, dark grey.

**Voice**  High-pitched 'weet' or in flight a high-pitched rolling 'chirrip'.

**Adult breeding plumage**  *Head, neck and breast, rufous-orange*, streaked black; white at base of bill; lower breast, belly and vent, white; mantle, scapulars, coverts and tertials, black with rufous-orange edges, fringed white when new.

**Juvenile plumage**  Like adult non-breeding plumage, but with more olivey-brown appearance; crown, markedly browner; tertials and wing-coverts, fringed buffish orange.

**Field notes**  The most abundant shorebird in southern Australia; feeds, roosts and flies generally in large tight flocks on coastal and inland wetlands; common in the north during migration.

**Similar species**  Curlew Sandpiper; Sharp-tailed Sandpiper; Broad-billed Sandpiper; Sanderling; Long-toed Stint. The Red-necked Stint is smaller, shorter-legged and shorter-billed than the Curlew Sandpiper and Sharp-tailed Sandpiper with which it often associates. It is slightly smaller than the Broad-billed Sandpiper, which has a longer down-curved bill and a double supercilium. It is browner and slightly smaller than the Sanderling. For distinctive features of the Long-toed Stint, see the rare species accounts, Chapter 6.

## BREEDING RANGE

The Red-necked Stint breeds in far north-eastern Siberia, from the eastern Taimyr Peninsula to the Bering Sea, especially on the Lena River and Chukotski Peninsula (325). It breeds sporadically at Point Barrow and the Seward Peninsula in Alaska (7). Scrubby and mossy tundra areas on foothills and mountains near barer stony areas are its favoured nesting habitats. It often flies long distances to feed in wet areas.

## ASIAN RANGE

Red-necked Stints are common migrants in Japan, where up to 8200 have been counted on southward and 4400 on northward migrations (488). They are also abundant during migration in Korea (116, 162, 364), China (255) and common in southern Taiwan (401). In southern Taiwan, they migrate southwards between early August and mid-October and northwards from late April to mid-May, when they are more common (37). In Hong Kong, they are the most abundant stints, being much more numerous during northward migration in April and May (54). Banding studies show that they fatten before northward departure from Hong Kong (307). In Vietnam, they occur in great numbers on migration (142, 489), and they are abundant in the Philippines (171), maximum southward migration occurring in September (2), and in Borneo (440). They are common in Malaysia (157) and Thailand (257). At Brunei, they occur on migration: in 1980–81 numbers peaked at 500 on southward (late August) and 120 on northward (early May) migration (477). In New Guinea, up to 1400 have been seen on the coast of south-eastern Irian Jaya (36) and up to 2500 in the Port Moresby area in October and November (138). They are rare on the islands of the south-western Pacific (120), but occur regularly in far-western Micronesia during migration (18). Less than 300 occur in New Zealand each year (387, 388). (See Figure 5.62.)

## STATUS AND DISTRIBUTION IN AUSTRALIA

Red-necked Stint are the most abundant of the migratory shorebirds in Australia and they occur in large numbers in all coastal areas except the east coast of Queensland. Although large numbers occur in coastal areas, they are frequently encountered inland. (See Figure 5.63, Table 5.31.)

They occur on a wide variety of wetlands, from inland salt lakes and freshwater swamps to intertidal mudflats and sandy ocean beaches. They are most abundant on the coast, where they forage mostly on intertidal mudflats and roost on sandy beaches and spits, often in large tight flocks.

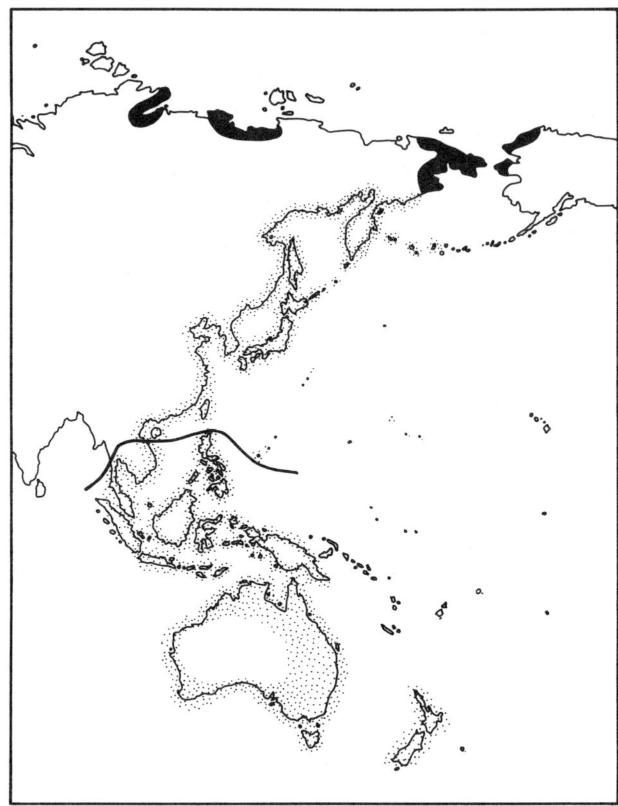

FIG. 5.62.   Breeding and non-breeding range of Red-necked Stint.

## MOVEMENTS IN AUSTRALIA

Red-necked Stints first arrive in Australia in late August, but large numbers first appear in early September. They pass through the north-west in September, but fewer are present in October. A second influx occurs there in November. Colour-dyeing shows that many more migrate southwards across the continent in September than in November. Between September and November they occur in flocks on most of the southern and eastern coasts of Australia. Two influxes occur at some places, one in September and another in November. Many are observed on inland wetlands in the south-east in October and November. These observations suggest that Red-necked Stints arrive on the north-west coast and cross the continent to the eastern, south-eastern and southern coasts. There are probably two waves of migration.

They move to non-breeding sites in the major marine embayments of southern and south-eastern Australia by December, and numbers are stable at these places until early March or in some cases April. Banding shows that birds return to the same sites each year, and little movement occurs between these at this time (98).

Northward migration starts in early March and continues until mid-April. Influxes occur on the central coast of New South Wales and large numbers pass through South Australia, suggesting some short-

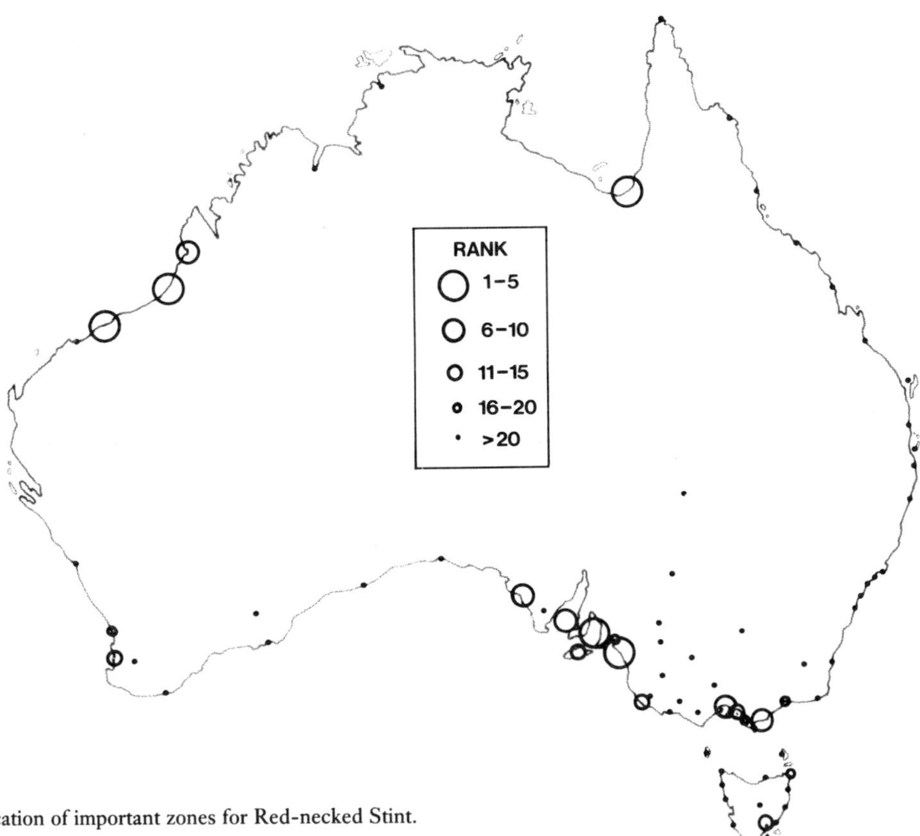


FIG. 5.63.   Location of important zones for Red-necked Stint.

distance movement of birds from Victoria and Tasmania before departure. Small influxes occur in late March in north-eastern Australia, and in the north-west a constant turnover of birds occurs in March and April (247). In the north-west many fewer birds pass through on northward migration compared with southward migration. Birds departing from Victoria fatten sufficiently to fly to northern Australia, but not beyond (476). These observations suggest that Red-necked Stints depart from south-eastern Australia and stop on a wide area of the northern coast before finally leaving Australia.

Winter numbers of Red-necked Stints in inland south-eastern Australia are sometimes higher than summer numbers. Young birds move inland in winter, often in response to flooding, some distance from the coast, e.g. to Lake Eyre in winter 1984. Colour-dyeing of stints in Tasmania shows that many young birds move to the mainland in winter.

# FEEDING

The Red-necked Stint feeds in many different wetland types. On the coast it forages on intertidal mudflats in flocks, rapidly jabbing its bill into mud in search of prey. It usually feeds for the entire period for which mudflats are exposed, and in cold or windy weather continues feeding on near-coastal wetlands over high tide (102, 131). It feeds mostly on wet or drying mud above the edge of the water (351, 461).

TABLE 5.31.   Ranked listing of top twenty zones for Red-necked Stint. (maximum counts marked thus: *; other counts are averages) (1580 or more birds)

| ZONE | NO. OF COUNTS | NO. OF INDIVIDUALS |
|---|---|---|
| * The Coorong, SA | 1 | 63 800 |
| * Eighty Mile Beach, WA | 5 | 60 000 |
| * SE corner, Gulf of Carpentaria, Qld | 1 | 35 200 |
| * St Vincent Gulf, SA | 4 | 24 000 |
| * Port Hedland Saltworks, WA | 12 | 23 000 |
| Port Phillip Bay, Vic. | 5 | 22 600 |
| * Roebuck Bay, WA | 6 | 19 800 |
| Corner & Shallow inlets, Vic. | 5 | 15 600 |
| * W coast, Eyre Peninsula, SA | 5 | 8 500 |
| * Spencer Gulf, SA | 5 | 7 600 |
| SE coast, SA | 5 | 6 800 |
| * Kangaroo Island, SA | 5 | 5 600 |
| Peel Inlet, WA | 5 | 4 100 |
| Westernport Bay, Vic. | 5 | 3 800 |
| Derwent estuary & Pittwater, Tas. | 4 | 3 000 |
| Swan coastal plain, WA | 5 | 2 870 |
| Gippsland lakes, Vic. | 5 | 2 120 |
| * Murray mouth lakes, SA | 3 | 2 010 |
| Anderson's Inlet, Vic. | 4 | 1 940 |
| Cape Portland coast, Tas. | 5 | 1 580 |

At Lake Reeve, a non-tidal site in eastern Victoria, the Red-necked Stint feeds in October and December on amphipod and ostracod crustacea, the gastropod mollusc *Coxiella striatula*, insect larvae and pupae (mostly flies and midges), and the seeds of the aquatic plant *Ruppia maritima* (366). In April the relative abundance of potential prey changes; insect larvae and pupae become dominant and forming the principal food of stints.

On the Coorong, South Australia, the guts of collected birds contained beetle and midge larvae, nematode worms and the seeds and turions (small root stocks) of *Ruppia* (351). At Werribee, in Victoria, the Red-necked Stint was recorded as eating the polychaete worm *Ceratonereis erythraeensis*, preferring smaller ones than were taken by Curlew Sandpiper and Sharp-tailed Sandpiper feeding on the same mudflat (103). A similar range of food was found in the stomachs of birds collected at Hobart, Tasmania (461), and elsewhere in Australia (256, 300). Ants, caterpillars and herbaceous plant seeds have also been recorded in their diet (472).

## CONSERVATION

Australia probably receives most of the world population of Red-necked Stints and most come to south-eastern Australia, where the human population density is highest. Its survival depends on the adequate reservation and management of its important non-breeding sites in the south-east, as well as areas used during migration in northern Australia, particularly in the north-west.

# CURLEW SANDPIPER
*Calidris ferruginea*
(Plate 13)

**Other names**  Curlew Stint; Pigmy Curlew.
No races recognised.

## DESCRIPTION

A small, slender, grey sandpiper, with a long, down-curved bill and, in flight, a white rump.

**Adult non-breeding plumage**  Feathers of crown, nape, hindneck, mantle, scapulars, tertials and wing-coverts, grey-brown, thinly edged white, giving slightly scalloped effect; supercilium, white; lores and ear-coverts, white, sparsely streaked grey. Chin and throat, white; sides of neck and breast, white, finely streaked and spotted pale grey; belly and vent, white. Iris, black; *bill, down-curved, about one-and-a-half times length of head*, black; legs, black.

**In flight**  Back, like rest of upperparts; *rump, white*; tail, grey; greater coverts have white tips, which show as a narrow wing-bar; primaries and secondaries, dark grey.

**Adult breeding plumage**  Strikingly different from non-breeding plumage: crown, nape and hindneck, speckled black, chestnut and white; feathers of mantle and scapulars, rich chestnut with large black central blotches and thin white edges; wing-coverts remain grey-brown with thin white edges; lores and throat, white. Supercilium, ear-coverts, neck and rest of *underparts, except vent, rich chestnut*; sides of breast and flanks, chestnut, feathers edged black-and-white; vent, white.

**Voice**  High-pitched, metallic 'trrrrp . . .' repeated, particularly in flight; quieter 'wheet' or 'wit' at other times.

**Juvenile plumage**  Like adult non-breeding plumage, but upperparts are more olivey-brown, particularly on the crown; wing-coverts have dark subterminal bars and buff margins and appear as a noticeable panel of feathers below the scapulars on the folded wing; very recent arrivals (October) can show a pink-buff wash on the sides of the breast.

**Field notes**  Feeds on mudflats in flocks with other small waders; often seen after arrival and before departure in partial breeding plumage, traces of chestnut feathers showing on the underparts. Often gather in flocks of hundreds, which fly in unison, alternately flashing grey upperparts and white underparts.

**Similar species**  Red-necked Stint; Sharp-tailed Sandpiper; Broad-billed Sandpiper; Sanderling. The

Curlew Sandpiper can be distinguished by its longer, down-curved bill, slightly longer legs and, in flight, white rump.

## BREEDING RANGE

The Curlew Sandpiper breeds in central, high-arctic Siberia, from the Yenesey River east through the Taimyr Peninsula to the Kolyma River, Cape Baranov and the New Siberia Islands (7, 85). A small population (less than 10 pairs) has bred at Barrow, Alaska, over 1300 kilometres east of the normal breeding range (202). It prefers boggy, grassy tundra, nests close to water and avoids sparsely or densely vegetated places.

## ASIAN RANGE

The Curlew Sandpiper is common in southern Africa (288), eastern Africa (287), India and Sri Lanka (5), parts of south-eastern Asia and in Australia. It is uncommon in New Zealand. It is uncommon on migration in Korea (162), Japan (max. 55 (488)) and China (255). There have been twelve movements of Australian-banded Curlew Sandpiper, to Java, Thailand, China and Siberia, mostly during northward migration (15). Hundreds visit Hong Kong during northward migration in April and May (306). In the Philippines, it is a rare visitor (only 25 out of 17 600 waders banded there in the 1960s were Curlew Sandpipers (268)). It is quite common in Burma (439) and Thailand (257), and in Malaysia about 4000 have been counted on the west coast in September (348). A maximum population of 2500 has been estimated for the Serangoon Estuary, Singapore (174). There have been two movements to Australia of Curlew Sandpipers banded in Malaysia during southward migration; one to southern Victoria and one to Port Hedland, Western Australia. Banding studies show that in Malaysia in August and September some fatten enough to fly direct to north-western Australia; others commence moulting and stay in Malaysia for the non-breeding months (348). It is common on the coast of Borneo, especially on migration (440), and in Bali (14, 235). By contrast, no more than a hundred or so have occurred in New Guinea during southward migration (September–November) and Finch (138) states 'in comparison to the numbers found in Australia it is inexplicably scarce, vast numbers must either overfly or by-pass [New Guinea]'. It is not listed for the south-western Pacific (302), and in New Zealand it is a regular summer visitor, but no more than 100 have been counted there (387, 388). (See Figure 5.64.)

The low numbers in north-eastern Australia, New Guinea and far-eastern Asia suggest that it migrates to Australia via a western route through Burma and Malaysia. Recoveries of banded birds and observations in eastern Asia suggest that northward migration is via

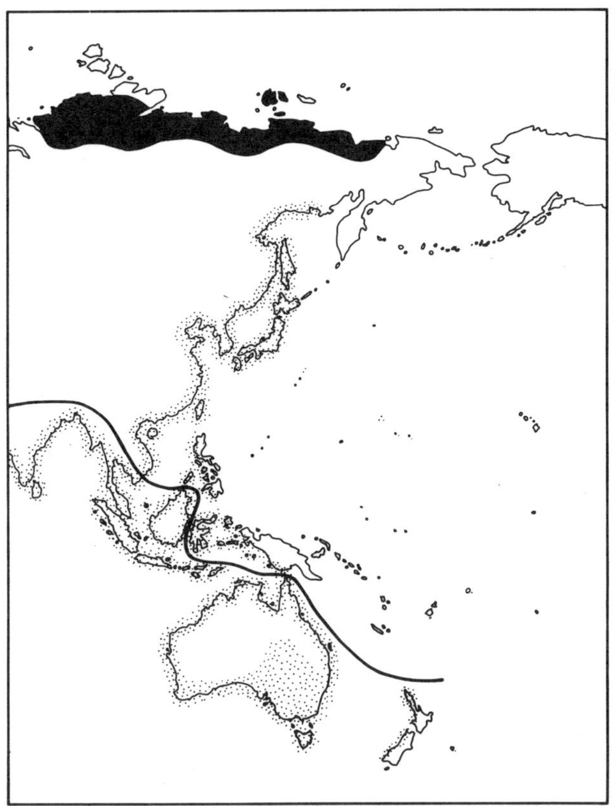

FIG. 5.64.   Breeding and non-breeding range of Curlew Sandpiper.

a more easterly route. More research is needed to confirm this.

## STATUS AND DISTRIBUTION IN AUSTRALIA

The Curlew Sandpiper is not as widespread in Australia as the Red-necked Stint and Sharp-tailed Sandpiper, and it is confined more to coastal areas. It occurs predominantly on the south-eastern coast and during migration on the north-western coast. Numbers are low in the Darwin Area, on the east coast of Queensland and at most inland sites. Largest numbers inland are recorded in the Riverina in southern New South Wales, and in northern and western Victoria. (See Figure 5.65, Table 5.32.)

In Australia, the Curlew Sandpiper lives on intertidal mudflats, roosting on sandy spits and beaches, saline near-coastal wetlands and the ponds of saltworks and sewage treatment works.

An increase in numbers has occurred in the Hobart area between 1965 and 1981 (335).

## MOVEMENTS IN AUSTRALIA

The Curlew Sandpiper arrives in most parts of Australia in late August. It migrates through the

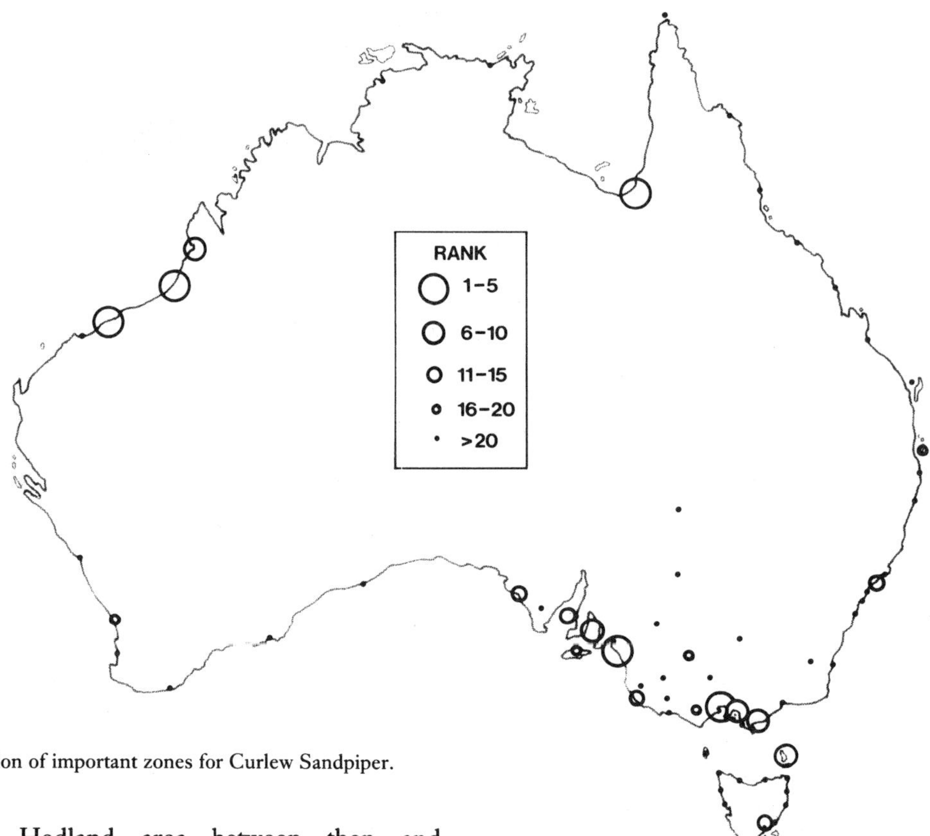



FIG. 5.65.  Location of important zones for Curlew Sandpiper.

Broome–Port Hedland area between then and November. Birds colour-dyed there in this period were sighted in South Australia and Victoria about five weeks later. Flocks appear on the southern and eastern coasts and in most inland parts of eastern Australia between September and November. These observations suggest that large numbers cross the continent from the north-west to the south-east and that southward migration finishes by the end of November.

In south-eastern Australia, most have moved to marine embayments by December, and numbers are stable there until February. Banding in Victoria shows that they return to non-breeding sites each year and do not move out of them until northward departure (98).

Northward migration starts in March and influxes occur in March and April on the east coast north to about 30°S. Similar influxes occur in South Australia and in the Broome–Port Hedland area. The Curlew Sandpiper in Victoria and Tasmania fattens sufficiently before northward departure to fly to the north-west coast and in many cases directly out of Australia (22, 476). These observations suggest that it undergoes short movements before departing from south-eastern Australia, and that it probably flies across the continent to the north-west coast.

## FEEDING

On intertidal flats, the Curlew Sandpiper usually feeds in mixed flocks with the Red-necked Stint. Whereas stints more often feed on bare mud or at the edge of

TABLE 5.32.  Ranked listing of top twenty zones for Curlew Sandpiper. (maximum counts marked thus: *; other counts are averages) (620 or more birds)

| ZONE | NO. OF COUNTS | NO. OF INDIVIDUALS |
|---|---|---|
| * Eighty Mile Beach, WA | 5 | 60 000 |
| * The Coorong, SA | 1 | 40 000 |
| * Port Hedland Saltworks, WA | 12 | 25 000 |
| Port Phillip Bay, Vic. | 5 | 17 700 |
| * SE corner, Gulf of Carpentaria, Qld | 1 | 8 500 |
| * Roebuck Bay, WA | 6 | 6 000 |
| Corner & Shallow inlets, Vic. | 5 | 6 000 |
| St Vincent Gulf, SA | 4 | 5 900 |
| Westernport Bay, Vic. | 5 | 3 100 |
| * Furneaux Islands, Tas. | 3 | 2 640 |
| * Spencer Gulf, SA | 5 | 2 500 |
| SE coast, SA | 5 | 1 880 |
| Hunter estuary, NSW | 4 | 1 570 |
| Derwent estuary & Pittwater, Tas. | 4 | 1 280 |
| * W coast, Eyre Peninsula, SA | 5 | 920 |
| * Kangaroo Island, SA | 5 | 840 |
| Kerang–Swan Hill, Vic. | 5 | 760 |
| Swan coastal plain, WA | 5 | 710 |
| Western District lakes, Vic. | 5 | 670 |
| Moreton Bay, Qld | 5 | 620 |

water, the Curlew Sandpiper feeds in belly-deep shallow water, probing the underlying mud and often completely submerging its head (461). At Hobart, both species took the same prey species of similar size (461). At Werribee, Victoria, it has been recorded as taking slightly larger prey than stints, feeding predominantly on the polychaete worm *Ceratonereis erythraeensis*, which is very abundant in Werribee's enriched mudflats. At non-tidal sites, the Curlew Sandpiper has been observed feeding mainly on wet, bare mud, probing deeply for oligochaete worms and pecking at the surface to take amphipods, gastropods and insect larvae and pupae (100). In the Hobart area, its diet has shifted from insect larvae to molluscs and crustacea by March and April (461). At the Coorong, it feeds on dry and wet muddy areas, and two specimens had been eating beetle larvae and the seeds of the common aquatic plant *Ruppia* (351).

## CONSERVATION

The Curlew Sandpiper occurs in largest numbers in south-eastern Australia, where most people live, so it is more vulnerable to possible habitat alteration than some other shorebirds. The maintenance of undisturbed feeding and roosting habitat along the southeast coast and at sites used during migration on the north-west coast is necessary for it to survive at current population levels.

# SANDERLING
## *Calidris alba*
(Plate 15)

**No races recognised.**

## DESCRIPTION

A small, stocky, pale sandpiper that runs and feeds along the edge of the sea on sandy beaches.

**Adult non-breeding plumage** Crown, nape and upperparts, pale grey with darker grey mottling. Underparts and *forehead, clean white*. Iris, black; bill, short and black; legs, black; *lacks hind toe*.

**In flight** Back, pale grey with darker mottling; rump, white with dark central line; *leading edge of wing, dark grey, contrasting with white wing-bar* formed by white bases to primaries.

**Adult breeding plumage** Upperparts and breast become mottled dark reddish-brown.

**Juvenile plumage** Crown, dark grey with white streaks; rest of upperparts, dark grey with white notches and fringes to feathers.

**Field notes** Occurs in flocks on sandy ocean beaches on the southern and western coasts; feeds by following receding surf and pecking at the sand.

**Similar species** Red-necked Stint; Curlew Sandpiper; Broad-billed Sandpiper. The Sanderling is slightly larger than the Red-necked Stint, has paler plumage, a white forehead and, in flight, a more conspicuous white wing-bar. It is smaller than the Curlew Sandpiper and has shorter legs and a shorter, straight bill. The Sanderling is the same size, but is paler and has a shorter, straight bill compared with the Broad-billed Sandpiper.

## BREEDING RANGE

The Sanderling breeds in high arctic regions in widely scattered locations, including Spitzbergen, Greenland, the northern Canadian islands and near Barrow in Alaska, as well as the Taimyr Peninsula, the delta of the Lena River in Siberia and on the nearby islands in the Arctic Ocean (7).

## ASIAN RANGE

The Sanderling is found most commonly on migration in Korea, eastern China and Japan (116, 162, 255, 364, 488). (See Figure 5.66.) Elsewhere in Asia it is uncommon or rare, although information is available from only a few countries. It occurs regularly, but in

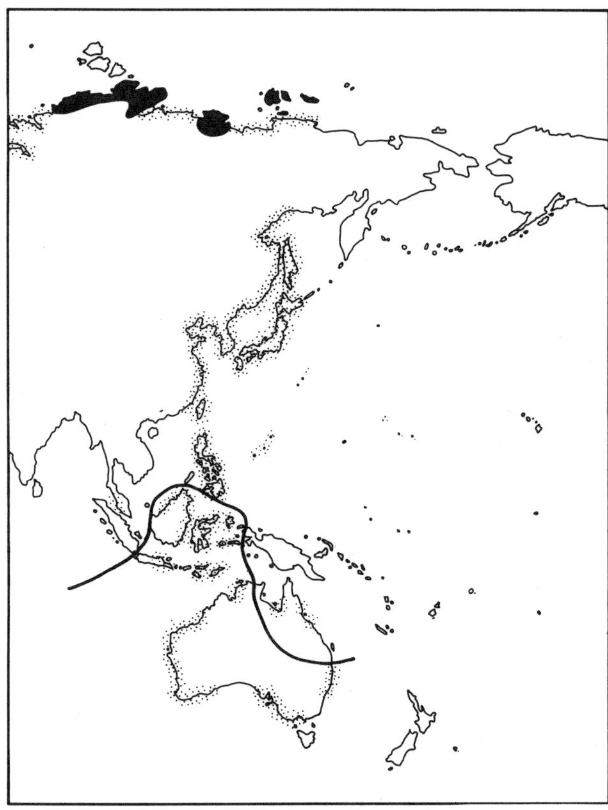

FIG. 5.66.   Breeding and non-breeding range of Sanderling.

TABLE 5.33.  Ranked listing of zones holding 20 or more Sanderlings. (maximum counts marked thus: *; other counts are averages)

| ZONE | NO. OF COUNTS | NO. OF INDIVIDUALS |
|---|---|---|
| * Roebuck Bay, WA | 6 | 1 510 |
| * The Coorong, SA | 1 | 930 |
| * SE coast, SA | 5 | 780 |
| W coast, Eyre Peninsula, SA | 5 | 510 |
| SW coast, Vic. | 5 | 430 |
| Ocean Beach, Strahan, Tas. | 3 | 330 |
| Corner & Shallow inlets, Vic. | 5 | 180 |
| * SE corner, Gulf of Carpentaria, Qld | 1 | 180 |
| * Furneaux Islands, Tas. | 3 | 160 |
| Swan coastal plain, WA | 5 | 110 |
| * Spencer Gulf, SA | 5 | 100 |
| * Eighty Mile Beach, WA | 5 | 100 |
| * Albany coast, WA | 4 | 100 |
| Darwin area, NT | 4 | 80 |
| * Central coast, NSW | 5 | 40 |
| * Esperance coast, WA | 4 | 30 |
| * N coast, NSW | 3 | 30 |
| * Port Hedland Saltworks, WA | 12 | 30 |
| * Central NW coast, Tas. | 1 | 20 |

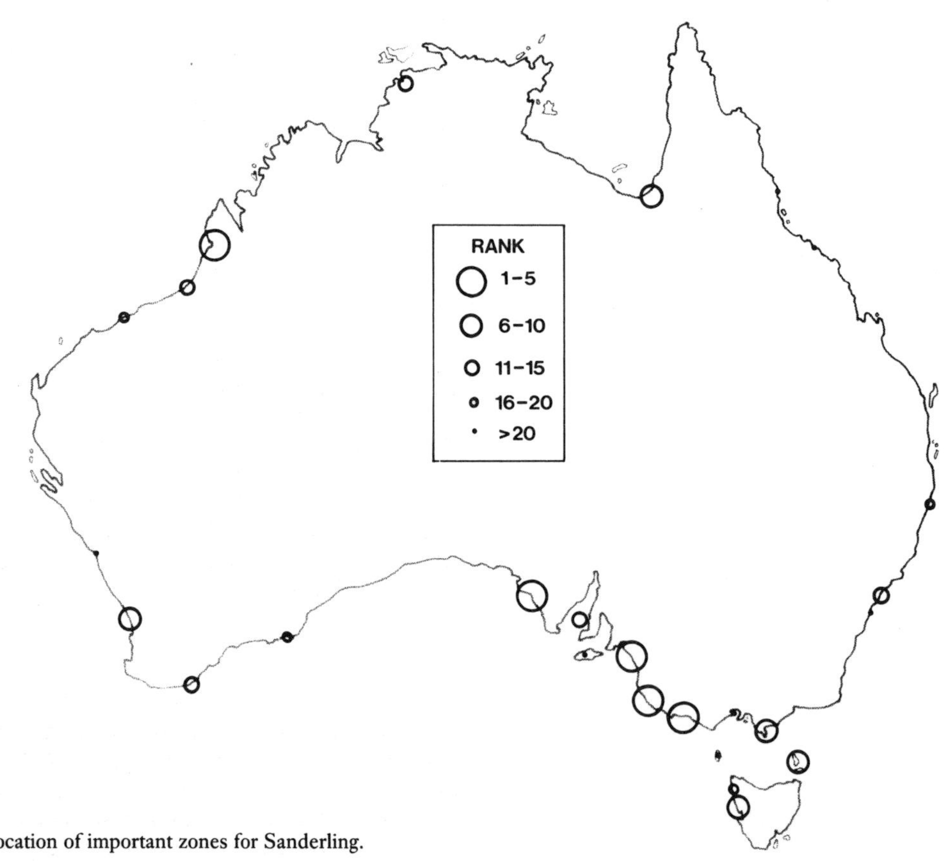


FIG. 5.67.   Location of important zones for Sanderling.

134

small numbers, in most of Micronesia between September and February (10, 216, 345), but mostly during southward migration in the Phoenix and Line islands farther east (59, 60). It is a regular visitor to Borneo in the non-breeding months (440), but is rare in Java, occurring mostly on the south coast (206). No more than three or four occur in places in New Guinea, and few reach New Zealand (134, 184).

## STATUS AND DISTRIBUTION IN AUSTRALIA

The Sanderling occurs in largest numbers on ocean shores in south-eastern Australia and in the Broome—Port Hedland area of Western Australia. (See Figure 5.67, Table 5.33.) It is most abundant in South Australia and occurs in two places in Victoria: between Warrnambool and Portland (51) and round Wilson's Promontory (75, 223). There is a regular flock on the west coast of Tasmania, near Strahan. In Western Australia, there are smaller regular flocks in the Albany and Esperance areas and on the islands off Perth. Smaller numbers visit parts of the New South Wales coast at times, these being the only records of flocks of Sanderlings on the east coast of Australia. There are occasional records of individuals along the east coast of Queensland. It also occurs in the Gulf of Carpentaria and at Darwin. It is almost exclusively coastal.

The Sanderling is found most frequently on sandy ocean beaches, where it feeds in the wave-washed zone at low tide. At high tide it sometimes roosts on rocky reefs (51). In South Australia, substantial numbers of birds roost behind the Younghusband Peninsula on the sheltered shores of the Coorong.

## MOVEMENTS IN AUSTRALIA

The Sanderling arrives in Australia in September. It migrates through Roebuck Bay in Western Australia, Darwin, and Eyre in southern Australia between September and November. Few occur on the east coast or in New Guinea. This suggests that it arrives in the north-west then crosses the continent to the south coast.

Northward migration occurs in March and April, but few pass through Roebuck Bay. It occurs again at Eyre and Darwin at this time. This suggests that it moves westwards along the south coast before departing northwards across the continent, at least some landing on the north coast.

## FEEDING

Sandy ocean beaches are the principal feeding habitat of the Sanderling in Australia. It follows receding waves, pecking in the wet sand for prey. There have been no detailed studies of the feeding behaviour and diet of the Sanderling in Australia. The gut of a bird collected in South Australia contained the remains of caterpillars, beetles, ants, spiders and worms (256).

## CONSERVATION

The Sanderling has been recorded as occurring on less disturbed ocean beaches and moving to undisturbed refuges when disturbance becomes too intense (51, 188). Its habitat in Australia is popular for summer recreation. The most important beaches must be kept free from human disturbance. In south-western Victoria, the wise location of access points away from important beaches has ensured the future availability of relatively undisturbed habitat there. Such enlightened coastal management is needed at the other important Sanderling haunts in southern Australia.

# BROAD-BILLED SANDPIPER
## *Limicola falcinellus*
(Plate 13)

**Other names**   Eastern Broad-billed Sandpiper.

Two races of the Broad-billed Sandpiper are recognised: *falcinellus* and *sibirica*. The latter comes to Australia.

## DESCRIPTION

A small, grey-brown, short-legged sandpiper with a down-curved bill.

**Adult non-breeding plumage**   Crown, nape, hindneck and sides of neck, streaked grey-brown; mantle, scapulars and wing-coverts, grey-brown, broadly edged white; *supercilium, white, with second line extending to sides of crown*; lores and ear-coverts, streaked grey-brown. Chin and throat, white; breast, white with grey streaking and spotting, more dense at sides; belly, flanks and vent, white. Iris, black; *bill, longer than head, broad at base, down-curved at tip* and black; legs, dark grey. Sexes similar.

**In flight**   Back, grey-brown, with white edges to feathers; rump, white, with broad, dark grey central line; greater coverts, tipped white, appearing as a narrow white wing-bar; primaries and secondaries, dark grey.

**Voice**   Short trilling 'trit-trit-trit', uttered repeatedly or singly; 'chetter-chetter'.

**Adult breeding plumage**   Upperparts darker, with feathers showing black centres, chestnut edges and white tips; wing-coverts remain grey-brown; supercilium and lateral crown stripe become more pronounced; breast more heavily streaked.

**Juvenile plumage**   More heavily marked, browner upperparts than adult non-breeding plumage; scapulars and wing-coverts have broad buff edges.

**Field notes**   Found mostly in northern Australia in small numbers on intertidal mudflats, particularly near mangroves.

**Similar species**   Red-necked Stint; Curlew Sandpiper. The Broad-billed Sandpiper is distinguished from the Red-necked Stint by its longer bill, down-curved at its tip. It is smaller than the Curlew Sandpiper, which has longer legs and more uniform grey upperparts.

## BREEDING RANGE

Those of the race *falcinellus* breed in north-eastern Europe and western Siberia, and those of the race *sibirica*, which come to Australia, breed in widely scattered places in eastern Siberia (85). The breeding grounds of *sibirica* are poorly known, and its breeding schedule has not been recorded.

## ASIAN RANGE

Those of the race *falcinellus* migrate through eastern Europe, the Caspian Sea and adjacent areas to northern and eastern Africa and the coast of the Arabian Sea. Those of the race *sibirica* migrate to the Bay of Bengal, south-east Asia, Indonesia, New Guinea, Australia and occasionally New Zealand.

The Broad-billed Sandpiper has probably been overlooked in many parts of Asia because it so resembles the Red-necked Stint (161). It is a regular migrant in Korea (137) and in Japan, where it is more abundant on southward migration (max. 187 (488)). It appears in north-eastern China in the last week of July and moves on by mid-September (179). Elsewhere in China, it is abundant on parts of the east coast (e.g. Fokhien) in September (255) and in south-western Taiwan (56). In Hong Kong, it passes through in September and October (54). It is rare in the Philippines (110), but occurs in large numbers in Vietnam (489). The Broad-billed Sandpiper has been recorded as an uncommon visitor to Thailand (257), but banding studies indicate that it was probably overlooked (309). Small numbers regularly visit Malaysia (157, 348). In Burma and West Bengal it is common (5, 439). In Micronesia, the only record is from Belau (18, 345). In Borneo, small numbers visit regularly (161, 440), and at Brunei it has been recorded in September and October (477). There are two October specimen records from the islands immediately north-west of Australia (487). It is regular but uncommon in New Guinea (184). (See Figure 5.68.)

At Brunei, there are records in May (477), and it passes through Hong Kong, the east Chinese coast, Taiwan and Japan in April and May (37, 54, 56, 255, 488). It is rare on northward migration in Korea (137), and may move inland from the Chinese coast, taking a more westerly route to the breeding grounds.

## STATUS AND DISTRIBUTION IN AUSTRALIA

The Broad-billed Sandpiper occurs annually in small numbers on most parts of the northern coast. Largest numbers are observed in north-western Australia, particularly at Port Hedland Saltworks. (See Table 5.34.) Occasionally, single birds are seen near Sydney, Melbourne and Adelaide (187).

Coastal mudflats, particularly near mangroves, are the preferred foraging habitat of the Broad-billed Sandpiper. It roosts on sandy beaches (e.g. Roebuck Bay, Western Australia) or on saltflats behind mangroves (e.g. Gulf of Carpentaria). At Port Hedland

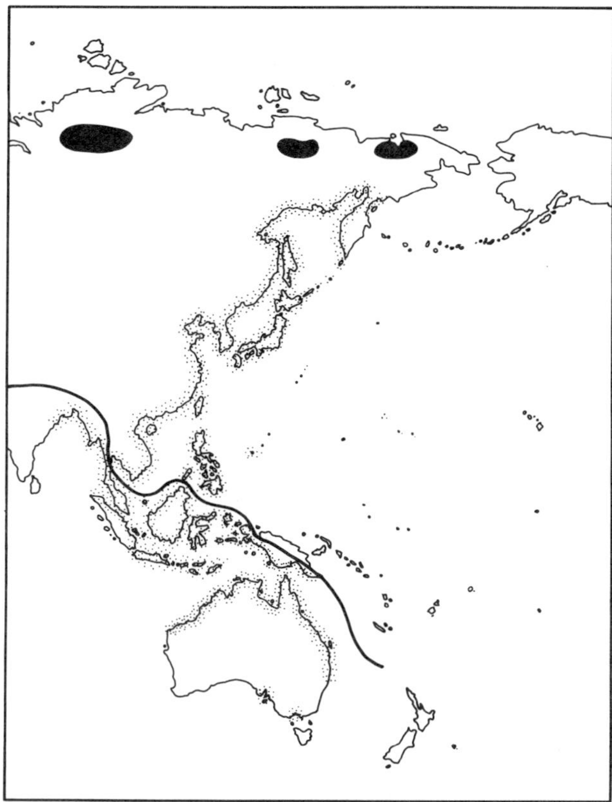

Fig. 5.68.   Breeding and non-breeding range of Broad-billed Sandpiper.

TABLE 5.34.   Ranked listing of zones that held more than 50 Broad-billed Sandpipers. (all counts are maxima)

| ZONE | NO. OF COUNTS | NO. OF INDIVIDUALS |
| --- | --- | --- |
| Port Hedland Saltworks, WA | 12 | 1 800 |
| SE corner, Gulf of Carpentaria, Qld | 1 | 1 740 |
| Roebuck Bay, WA | 6 | 110 |
| Pilbara coast, WA | 3 | 100 |
| Eighty Mile Beach, WA | 5 | 55 |

Saltworks, it inhabits saline ponds, mostly near the intake channel, and avoids ponds of higher salinity, which other small sandpipers use in large numbers. Due to the difficulty of separating Broad-billed Sandpipers in large flocks of Red-necked Stints, counts may have underestimated their abundance in northern Australia.

## Movements in Australia

The Broad-billed Sandpiper arrives at Port Hedland Saltworks by mid-October, and at the south-eastern corner of the Gulf of Carpentaria by late September. It departs Port Hedland Saltworks in the second and third weeks of April (247). Little is known of its migration within Australia.

## Feeding

There have been no detailed studies of the feeding behaviour and diet of the Broad-billed Sandpiper in Australia. At Port Hedland Saltworks, it obtains food by rapidly and repeatedly jabbing its bill into soft mud.

## Conservation

Largest concentrations of Broad-billed Sandpipers occur along the north coast, particularly at Port Hedland Saltworks and in the Gulf of Carpentaria. The preservation and appropriate management of these areas is essential if it is to survive there.

# ORIENTAL PRATINCOLE

## *Glareola maldivarum*
(Plate 3)

**Other names**   Small Indian, Eastern or Eastern Collared Pratincole; Grasshopper Bird; (Little) Stormbird; Swallow-plover; Swarmer.
No races recognised.

## DESCRIPTION

A medium-sized to large, slender shorebird with golden-brown plumage and swallow-like flight.

**Adult non-breeding plumage**   Upperparts, entirely golden-brown. *Chin and throat, cream, edged by a narrow band of black streaks*; breast, golden-brown, grading to *cream on belly and flanks*. Iris, brown; bill, short, broad at base, black with dull red base; legs, black. Sexes similar.

**In flight**   Back and rump, golden-brown; *tail, forked, white with black tip*; outer primaries, dark brown; inner primaries and secondaries, golden-brown; underwing-coverts, chestnut; flight is swallow-like and buoyant.

**Voice**   Harsh, tern-like call: 'chik chik'; softer whistles.

**Adult breeding plumage**   Similar to adult non-breeding plumage, but chin and throat completely edged black; bill, black with bright red base and gape.

**Juvenile plumage**   Similar to adult non-breeding plumage, but feathers of upperparts and wing-coverts have buff tips with dark subterminal bars; most have moulted into adult non-breeding plumage by the time they reach Australia.

**Field notes**   Occurs between December and February on the grassy plains of northern Australia; hawks insects when in flight, but feeds mostly on the ground.

**Similar species**   Australian Pratincole. The Oriental Pratincole is darker brown and shorter-legged than the Australian Pratincole and has a forked, white tail with a black tip rather than a square, dark tail with a white tip. It also lacks the chestnut belly and flanks of the Australian Pratincole and has chestnut, not black, underwing-coverts.

## BREEDING RANGE

The Oriental Pratincole breeds from Pakistan, throughout the Indian subcontinent (5) and east of the Himalayas to northern China (179, 266), southern China (129) and Taiwan (56). It probably breeds in Mongolia (473). There are also breeding records from the Philippines (110), Vietnam, Thailand and Burma (129), but the exact northern boundary of its breeding range is unknown (111). It breeds colonially on the banks of lakes, rivers and rice fields (111).

## ASIAN RANGE

The Oriental Pratincole is common on migration in central and north-eastern China (58, 179), especially during southward migration (July–September). It is a vagrant in Korea (162), uncommon in Japan during migration (344) and is a passage migrant in northern Taiwan (37). It passes through Hong Kong in September and October, and again between March and May (54, 306). The Oriental Pratincole is described as a 'common resident' in Thailand (257), but around Bangkok it arrives in February and departs in August and September (1). Small numbers visit the Philippines (171), and in Malaysia it occurs throughout the non-breeding months (157). It is irregular in Micronesia (345). In Borneo, it occurs in small numbers on southward migration (July–December) (440), and it passes rapidly through the islands north-west of Australia (487). It is a very rare straggler to New Zealand (134). (See Figure 5.69.)

Between December and February it occurs in northern Australia (38). Its absence between December and February immediately north of Australia (487) suggests that the entire migratory population may spend this period on the continent. At the same time it is present in the Indian subcontinent (5), Burma, Thailand, Indochina (129, 257), the Philippines (110) and southern Taiwan (56, 401). More research is needed to determine whether those coming to Australia breed in the northern part of their range, and if the southern Asian birds are entirely sedentary, as their movements in Asia suggest (162).

## STATUS AND DISTRIBUTION IN AUSTRALIA

The north-west is probably the centre of the Australian distribution of the Oriental Pratincole. As well as the records shown in Table 5.35, a large congregation has been observed north-east of Port Hedland (38). (See Figure 5.70.) There is a record of 5000 at Mount Isa in December (48). Small numbers (up to 27) reach the southern and eastern parts of Australia (74, 76, 141, 197, 343, 411).

It is a bird of the open grassy plains and the muddy flood plains and pools of northern Australia. Occasionally it visits beaches and intertidal mudflats (e.g. Eighty Mile Beach), but only in small numbers.

The Oriental Pratincole occurs mainly away from the coast in northern Australia and is rarely counted. Consequently, it is not possible to accurately estimate

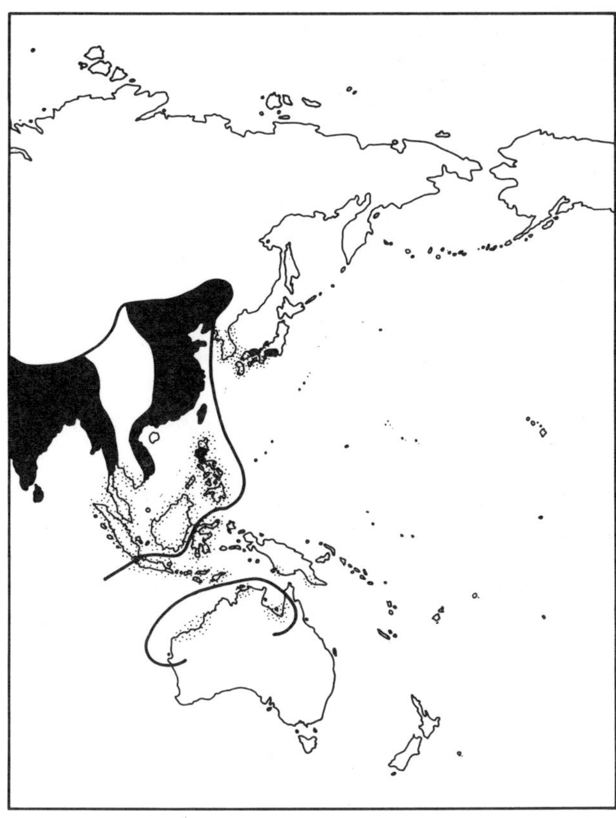

FIG. 5.69. Breeding and non-breeding range of Oriental Pratincole.

the numbers that visit Australia. At their maximum they possibly number tens of thousands in northern Australia.

## MOVEMENTS IN AUSTRALIA

The Oriental Pratincole arrives in northern Australia later than most migratory shorebirds. Its late arrival may be due to habitat being unsuitable until the northern wet season commences in late November and December.

It is observed annually, but at variable times, between November and April in the Darwin area (88), and there are records of large numbers spending short periods in the Gulf Country of north-western Queensland (48). It may be nomadic during the wet season,

TABLE 5.35. Ranked listing of zones that held more than 500 Oriental Pratincoles. (all counts are maxima)

| ZONE | NO. OF COUNTS | NO. OF INDIVIDUALS |
|---|---|---|
| Port Hedland Saltworks, WA | 12 | 10 000 |
| Pilbara coast, WA | 3 | 4 000 |
| Karumba plains, Qld | 1 | 700 |

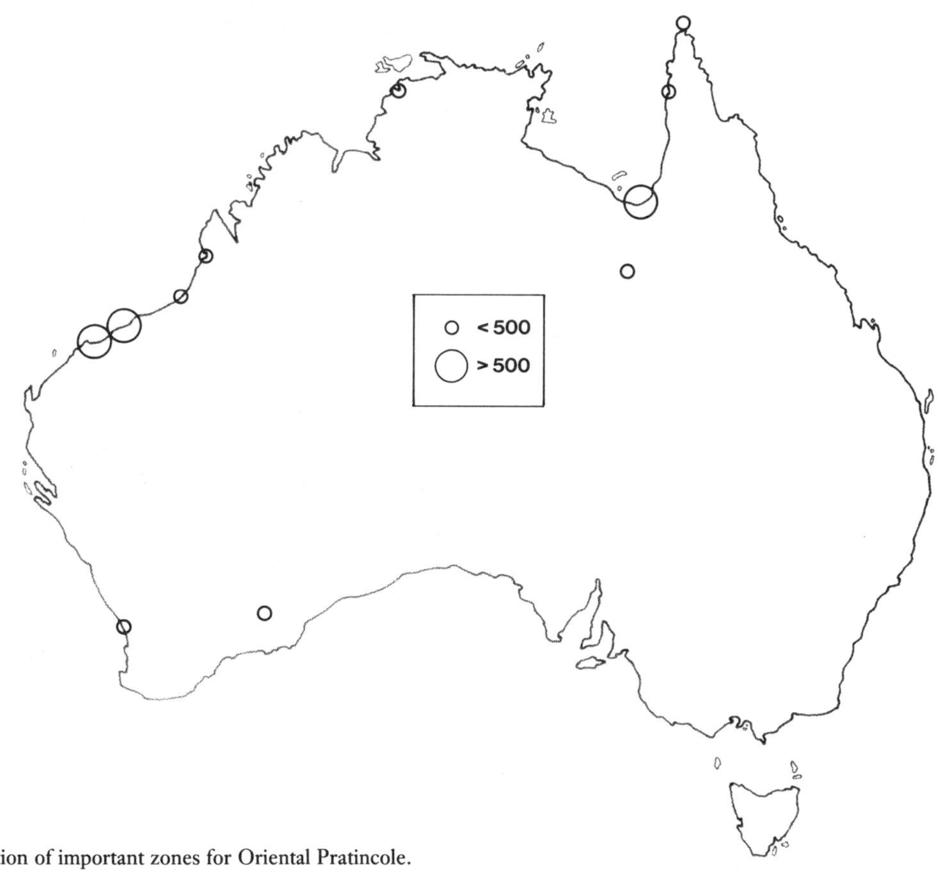


FIG. 5.70. Location of important zones for Oriental Pratincole.

moving to feed on plains after heavy rains and cyclones, which cause population explosions of grasshoppers, crickets and beetles.

It departs from Darwin (88) and north-western Australia by the first week in April.

## FEEDING

The dominant food of the Oriental Pratincole is insects, including grasshoppers, often caught in flight (300). Little detailed information is available.

## CONSERVATION

The available information suggests that a substantial proportion of the world's Oriental Pratincoles come to northern Australia each year. For its continued survival it needs extensive grassy plains in which to feed. Currently, there are no immediate threats to this habitat, but any intensive agricultural development over a wide area in the north should be assesed for its impact on this bird.

# AUSTRALIAN PRATINCOLE
## *Stiltia isabella*
(Plate 3)

**Other names**  Australian Courser; Swallow-plover; Roadrunner; Long-legged Pratincole; Prat.
No races recognised.

## DESCRIPTION

A medium-sized slender shorebird of open plains, often found away from water.

**Adult plumage**  Crown, nape, hindneck, sides of neck, mantle, scapulars and wing-coverts, rich orange-buff; lores, black; chin and throat, white; breast, paler orange-buff with *broken chestnut band on lower edge, extending onto dark brown flanks and belly*; vent, white. Iris, black; bill, red with black tip; legs, long and brown.

**In flight**  Rump, white; tail, square, black with white tip; legs trail behind tail; buoyant, swallow-like flight on long wings.

**Voice**  Tern-like 'quirrie-treet'.

**Juvenile plumage**  Similar to adult plumage, but feathers of upperparts edged buff; incomplete, mottled brown breast-bar; bill, completely dark.

**Field notes**  Found on open grassy plains in largest numbers in northern Australia; when standing, wings extend well beyond tail.

**Similar species**  Oriental Pratincole. The Australian Pratincole is more slender, longer-legged and paler than the Oriental Pratincole and has a different flight pattern and a square, not forked, tail.

## ASIAN RANGE

The Australian Pratincole occurs in Java, Sulawesi, southern Borneo, the islands north-west of Australia and New Guinea during the Australian winter (301, 440). (See Figure 5.71.) In Timor, Indonesia, about 50 000 have been observed in November (36). There are no confirmed breeding records outside Australia, but a young bird has been collected in south-eastern Irian Jaya, suggesting possible breeding there (301).

## STATUS AND DISTRIBUTION IN AUSTRALIA

The Australian Pratincole occurs on the coastal plains of northern Queensland and the Northern Territory,

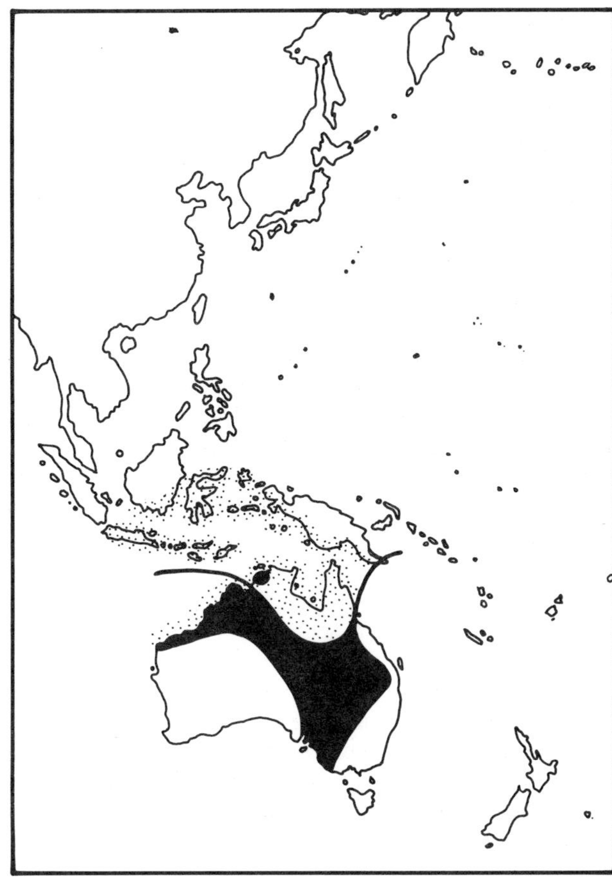

FIG. 5.71.   Breeding and non-breeding range of Australian Pratincole.

and in inland Australia from northern Victoria to the Gulf Country, eastern South Australia, Northern Territory and north-western Australia (38). The largest numbers occur in northern Australia. (See Figure 5.72, Table 5.36.)

*Atlas* project records (38) show that they occur most frequently in open plains, including tussock grassland, saltbush and drier agricultural land, generally avoiding large tracts of spinifex and *Acacia* scrub, except where the latter has been overgrazed in western New South Wales (38).

It occurs mostly away from wetlands, so its habitat has not been adequately surveyed during counts.

## MOVEMENTS IN AUSTRALIA

The Australian Pratincole is migratory, occurring in the southern half of the continent mainly in spring and

TABLE 5.36.   Ranked listing of zones that held more than 1000 Australian Pratincoles. (all counts are maxima)

| ZONE | NO. OF COUNTS | NO. OF INDIVIDUALS |
|---|---|---|
| Gulf country, NT | 2 | 12 500 |
| Kimberleys, WA | 1 | 1 240 |

NB.  This species occurs in large numbers away from most areas counted.

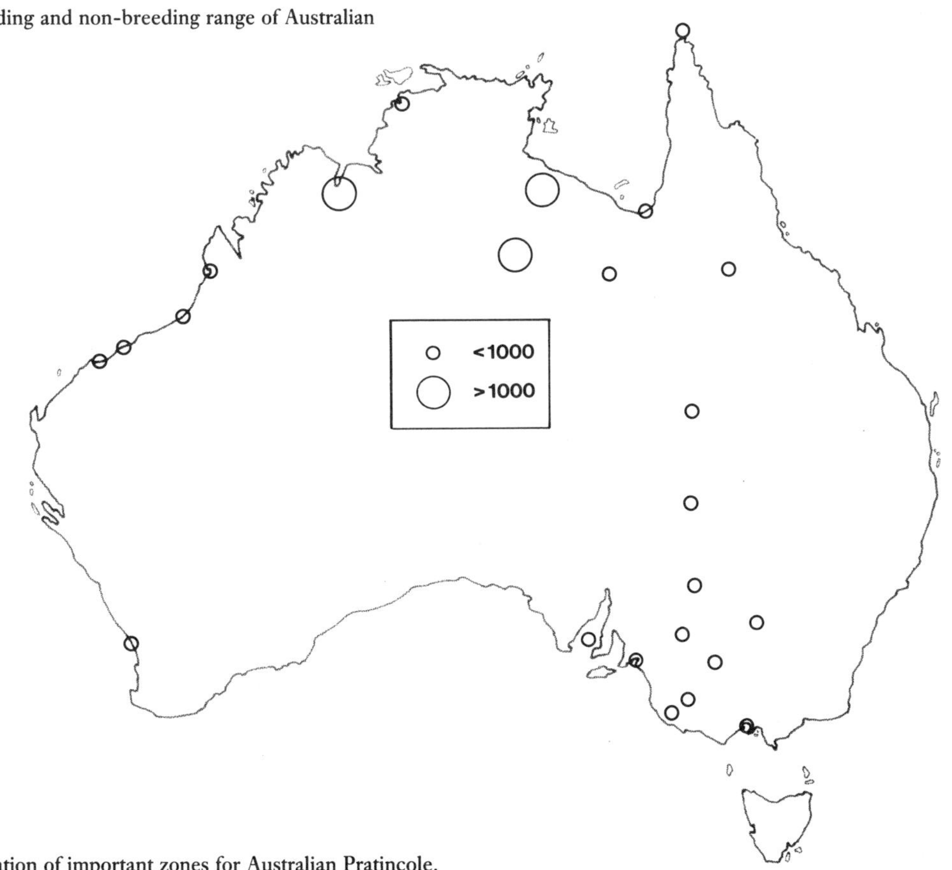


FIG. 5.72.   Location of important zones for Australian Pratincole.

summer and in the northern half and in the islands north of Australia mainly in winter (38). It arrives at Darwin between April and June and departs in November and December (88). In south-western New South Wales, it is present from October to February (195). Victorian *Atlas* project records are mainly between October and March (27). In western Queensland and the Northern Territory it occurs all year round, at times in hundreds (259). Over 12 000 were estimated to be in the country immediately south of the Gulf of Carpentaria in mid-May (358).

In many locations it appears only in wetter years (195, 289, 275), indicating that it moves around in response to rainfall.

## Breeding

Breeding of Australian Pratincoles has been recorded between August and December, mostly from September to November. It breeds in small, loose colonies, nests being spread thinly over as much as 2 square kilometres of open ground (290). Most breeding records in the *Atlas* were from south of 20°S (38).

The Australian Pratincole rarely excavates a scrape, preferring to lay its usual clutch of 2 eggs on a bare circular patch of soil among open gibber or gravel interspersed with low shrubs, herbs and grasses (290).

It usually nests within a kilometre or two of waterholes or dams. Incubation lasts about 21 days (274), and the young probably fly when between 4 and 5 weeks old. The main reason for nesting failure is probably predation by ravens, foxes, snakes and goannas; and nests are more vulnerable near water, where there are records of Black-tailed Native Hens taking eggs (290).

## Feeding

The Australian Pratincole feeds in open habitat, capturing prey from the surface of the ground or sometimes in the air while flying (289). It takes mostly insects, including grasshoppers, beetles, dragonflies, termites, bugs, caterpillars, earwigs, ants and cockroaches, as well as centipedes, spiders and plant seeds (289, 290, 472).

## Conservation

Largest numbers of Australian Pratincoles occur in northern Australia, where human settlement is sparse, so there are no immediate or serious threats to its future survival. The possibility that it is hunted when in Indonesia should be investigated.

# RARE SPECIES ACCOUNTS

Each year, individuals of some species of shorebirds stray from their usual migration routes and join up with the masses of shorebirds that regularly reach Australia. There are 26 of these rare species. Some of these are found each year, but others reach Australia only once or twice in a decade. Finding these rare species requires rigid application of the techniques of field identification given at the beginning of Chapter 5.

The following accounts of these rare species consist of a brief description of the adult non-breeding plumage of each species, comparing it with similar, more abundant species, and giving its usual breeding range, regular migration route and normal non-breeding range. The occurrence of each in Australia is briefly mentioned.*

## RINGED PLOVER

*Charadrius hiaticula*
(Plate 7)

**Other name**   Common Ringed Plover.

### DESCRIPTION

Between Red-capped and Double-banded Plover in size; *complete white 'ring' around neck;* dark brown cheek-patch and breast-band, latter continuing around hindneck; the adult has a *brown bill with a yellow base; conspicuous wing-bar,* markings show more contrast in breeding plumage and soft parts are brighter.

* The Golden Plover (*Pluvialis apricaria*) has been listed for Australia (65), but no confirmation is possible.

### RANGE

Breeds in arctic Eurasia, Greenland and western Canada, as well as on coasts and rivers of northern temperate Europe. Those of the race *tundrae* from Siberia probably visit Australia. The Semipalmated Plover (*C. semipalmatus*), which breeds in Alaska and western and central Canada, may yet be recorded in Australia; all birds should be checked carefully. The Ringed Plover has been recorded a number of times in Australia: New South Wales, Victoria, South Australia, Northern Territory and Queensland.

**References**   85, 242, 273, 281, 321, 377, 386.

## LITTLE RINGED PLOVER

*Charadrius dubius*
(Plate 7)

### DESCRIPTION

Similar to, though slightly smaller and less robust than the Ringed Plover; more *conspicuous yellow eye-ring;* stance more like Black-fronted Plover; in flight, *lacks the wing-bar of the Ringed Plover.*

### RANGE

A number of races occur: *curonicus* breeds throughout temperate and sub-arctic Eurasia east to China and Japan, migrating to central Africa and southern and south-eastern Asia in the non-breeding months; *dubius* is sedentary and occurs from the Philippines to New Guinea; *jerdoni* is sedentary and occurs in southern Asia (85). Australian records are probably those of the race *curonicus*, which have migrated from north-eastern Asia (211).

**References**   85, 211.

# CASPIAN PLOVER
## *Charadrius asiaticus*

### DESCRIPTION

Almost identical to the Oriental Plover, but slightly smaller; in flight, shows a *more distinct white wing-bar* and whitish underwings; *legs, slightly shorter* and dull greenish. Difficult to distinguish in the field.

### RANGE

Breeding range contiguous with that of Oriental Plover: breeds sporadically in south-western USSR, migrating to eastern and southern Africa (85). One confirmed Australian record: a specimen collected at Pine Creek, Northern Territory, in September 1896; two unconfirmed sight records near Darwin.

**References**   85, 269.

# EURASIAN CURLEW
## *Numenius arquata*

**Other names**   European or Common Curlew.

### DESCRIPTION

Resembles Eastern Curlew, but has a prominent white rump and back, and slightly shorter bill; larger than Whimbrel.

### RANGE

Breeds throughout temperate Europe east into central Siberia and north-western China, migrating to southern Europe, Africa and southern and eastern Asia. Two unconfirmed Australian sightings: Northern Territory (108) and south-western Australia (365).

**References**   85, 108, 365.

# UPLAND SANDPIPER
## *Bartramia longicauda*

**Other name**   Bartram's Sandpiper

### DESCRIPTION

Very distinctive; slightly smaller than Little Curlew, with a more *upright stance; shorter, straighter bill* and yellowish legs; *long tail extends well beyond tips of folded wing*; in flight, underwings are white, heavily barred dark brown.

### RANGE

Breeds from eastern Alaska, western Canada to northern central USA, migrating through central and eastern North America, Mexico, West Indies, Central America and northern South America, east of the Andes, to spend the non-breeding months in Brazil, Uruguay and Argentina (7). Accidental in Greenland, Europe and the Azores (85). Known in Australia from one specimen collected near Sydney in 1848.

**References**   85, 189.

# GREEN SANDPIPER
## *Tringa ochropus*

### DESCRIPTION

Very similar to Wood Sandpiper, but darker above, with *smaller white spots on upperparts*; broader barring on tail and *predominantly dark, not pale, underwings*.

### RANGE

Breeds across northern Eurasia from central northern Europe and Scandinavia, eastwards to far eastern Siberia; avoids high-arctic regions. It migrates to central Africa, the Arabian Peninsula, the Indian subcontinent and south-east Asia (85). There is one unconfirmed sight record at Darwin.

**Reference**   85.

# SPOTTED GREEN-SHANK
## *Tringa guttifer*

**Other names** Nordmann's Greenshank; Armstrong's Sandpiper.

## DESCRIPTION

Superficially resembles Greenshank, but has *shorter, pale green to yellow legs* and pure white underwing-coverts and axillaries; *stouter bill with more yellowish base.*

## RANGE

It is one of the rarest of the east-Asian shorebirds. Little is known of its distribution, and only small flocks have been sighted in south-east Asia. One unconfirmed sight record at Darwin.

**Reference** 286.

# REDSHANK
## *Tringa totanus*
(Plate 9)

**Other name** Common Redshank

## DESCRIPTION

Slightly smaller, stockier and darker than Greenshank, with *red legs and base to shorter bill.* In flight: shows *broad white trailing edge to secondaries and inner primaries.*

## RANGE

Breeds throughout northern temperate Eurasia, migrating to shores of western Europe, Africa and southern and south-eastern Asia. Probably occurs annually in small numbers in north-western Australia and the Northern Territory; occasionally single birds occur in south-western Australia and South Australia.

**References** 85, 106, 286, 397.

# LESSER YELLOWLEGS
## *Tringa flavipes*
(Plate 8)

## DESCRIPTION

Superficially resembles Marsh Sandpiper; slightly larger and more robust with *duskier upperparts; legs brighter and yellow to orange-yellow*; in flight, white *rump and uppertail-coverts cut-off sharply from grey back.*

## RANGE

Breeds in Alaska and central and western Canada, migrating through the USA to spend the non-breeding months on the Gulf of Mexico and Carribean coasts and south to Argentina (7). One record in Australia: Geelong, Victoria, 1983. Nine New Zealand records (134).

**Reference** 433.

# PIN-TAILED SNIPE
## *Gallinago stenura*

## DESCRIPTION

Almost identical to Latham's and Swinhoe's Snipe and difficult to distinguish from them in the field. Generally separated in the hand by tail characteristics (*stenura*: 24–26 tail feathers, central 10 broad; *megala*: 20–22; central 8 broad; *hardwickii*: 16–18).

## RANGE

Breeds in south-eastern Siberia and Mongolia, migrating to Burma, southern Thailand, Malaysia and Western Borneo (440). Occurs regularly in small numbers in the Pilbara region of northern Western Australia. Its occurrence farther south and east awaits confirmation because of the difficulties of field identification. Often found in dry habitats (cf. other 2 species).

**References** 38, 447, 448.

# SWINHOE'S SNIPE
*Gallinago megala*

**Other names**   Chinese Snipe; Pin-tailed Snipe.

## DESCRIPTION

Almost identical to Latham's Snipe in the field (see *G. stenura*).

## RANGE

Breeds in central Siberia and Mongolia and migrates through China to non-breeding areas in south-eastern China, Vietnam, the Philippines, Indonesia and New Guinea (146). Occurs annually from the Kimberley region of north-western Australia (446) to Mount Isa (445) and Cape York (133). Generally prefers habitat similar to that of Latham's Snipe: wetlands with rank vegetation.

**References**   38, 133, 146, 445, 446.

# HUDSONIAN GODWIT
*Limosa haemastica*

## DESCRIPTION

Very similar to Black-tailed Godwit from which it is distinguished in flight by its *black underwing-coverts and axillaries*, narrower white wing-bar and narrower white band across uppertail-coverts.

## RANGE

Breeds in coastal southern Alaska and parts of arctic Canada, migrating through central North America, eastern Mexico, Central America, the West Indies and north-eastern South America to non-breeding grounds in southern South America from Paraguay, southern Brazil and Uruguay to Argentina and coastal Chile (7). Single record for Australia: Hunter estuary, central New South Wales, 1982–83.

# PECTORAL SANDPIPER
*Calidris melanotos*
(Plate 12)

**Other name**   American Pectoral Sandpiper.

## DESCRIPTION

Very similar to Sharp-tailed Sandpiper: adult *lacks chestnut crown; throat and breast more heavily streaked, contrasting sharply with white lower breast and belly*; vent and undertail-coverts, pure white, compared with Sharp-tailed Sandpiper, which has dark streaks on its vent and undertail-coverts, yellower legs and base to *slightly longer bill*; slightly more upright stance. Has different call: 'trrrt-trrrt', harsher than call of Sharp-tailed Sandpiper. Rarely occurs on tidal mudflats, preferring freshwater marshes.

## RANGE

Breeds in high-arctic tundra from the Yamal Peninsula eastwards to the Bering Strait in Siberia and in arctic Alaska and Canada, migrating mostly through the USA and Mexico to spend the non-breeding months in South America (7, 144). Small numbers reach New Guinea and Australia each year; more frequent in south-eastern Australia.

**References**   73, 85, 122, 192, 193, 194, 209, 210, 399, 415, 418, 419, 441, 442, 493.

# COX'S SANDPIPER
*Calidris paramelanotos*
(Plate 13)

**Other name**   False Dunlin.

## DESCRIPTION

*Between Curlew Sandpiper and Sharp-tailed Sandpiper in general appearance*; more scalloped on uppperparts than former, less so than latter; browner than former, greyer than latter; bill, black, sometimes with dull yellow base, slightly down-curved, shorter than in former and longer than in latter; legs, olive rather than black as in Curlew Sandpiper and Dunlin; rump, white with grey line down centre; stance like that of Pectoral Sandpiper; very difficult to detect; inhabits coastal saltworks and mudflats, occasionally near-coastal freshwater wetlands, usually in the company of the common small sandpipers.

## RANGE

Unknown; described from only two specimens and about twenty sightings, all in Australia. All records are of single birds. A great mystery requiring much more research.

**References**   350, 379, 430, 431, 434, 435.

# BAIRD'S SANDPIPER
*Calidris bairdii*
(Plate 14)

## DESCRIPTION

Most likely to be confused with slightly larger Sharp-tailed Sandpiper or smaller Long-toed Stint. Slightly larger, browner and more scalloped on the upperparts than Red-necked Stint; *projection of wings beyond end of tail* gives impression of horizontal stance.

## RANGE

Breeds in arctic regions of far-eastern Siberia, Alaska and Canada, migrating through North America and Mexico to South America for the non-breeding months (7). There are no confirmed Asian records, but it has occurred in Tasmania, Victoria, Western Australia, Northern Territory and possibly New South Wales.

**References**   71, 283, 313, 436.

# WHITE-RUMPED SANDPIPER
*Calidris fuscicollis*
(Plate 14)

## DESCRIPTION

Between Red-necked Stint and Sharp-tailed Sandpiper in size; generally similar to Curlew Sandpiper except for shorter, straighter bill and shorter legs; like Baird's Sandpiper, it shows a marked *extension of the wing-tips beyond the end of the tail*; in flight it is the only small sandpiper in southern Australia other than Curlew Sandpiper that has *completely white uppertail-coverts*.

## RANGE

Breeds in far northern arctic Alaska and Canada, migrating through central and eastern North America,

eastern Mexico and the West Indies to South America east of the Andes, south to southern Argentina. Five Australian records.

**References**   428, 437.

# WESTERN SANDPIPER
*Calidris mauri*

## DESCRIPTION

Similar in appearance to, but slightly larger and heavier than, Red-necked Stint; longer, slightly down-curved black bill.

## RANGE

Breeds in far-eastern Siberia and western and northern Alaska (7, 144). Migrates down the western coast of North America to south-western USA, Mexico, Central America and northern South America (7). Three unconfirmed Australian records: two in New South Wales and one in Tasmania.

**References**   198, 359, 457.

# LITTLE STINT
*Calidris minuta*
(Plate 14)

## DESCRIPTION

In non-breeding plumage, Little Stint is almost impossible to distinguish from Red-necked Stint; its bill is slightly more slender. Australian records are of birds in breeding plumage, which is noticeably different from that of the Red-necked Stint: *lacks the extensive rufous wash on the neck* of Red-necked Stint. Many records are of birds in reverse moult cycles, showing breeding plumage during the non-breeding months.

## RANGE

Breeds in high-arctic regions of Scandinavia and western and central USSR as far east as the New Siberian Islands, migrating mostly to southern and eastern Africa, the Arabian Peninsula and the Indian subcontinent (85). There are Australian records from Western Australia, South Australia, Victoria and Tasmania.

**Reference**   450.

# LONG-TOED STINT
*Calidris subminuta*
(Plate 14)

**Other name**  Middendorf's Stint.

## DESCRIPTION

Slightly smaller and more slender than Red-necked Stint; plumage resembles that of Sharp-tailed Sandpiper; *legs, olive-green to olive-yellow*; in flight, toes trail beyond end of tail. Voice: 'kreee-kreee', singly or repeated.

## RANGE

Breeding range largely unknown: nests have been found in the middle of the Ob valley in western Siberia and in widely scattered places in eastern Siberia, generally in grassy, swampy areas and in mountain tundra (111, 144). Migrates down the east coast of Asia; nowhere abundant. Occurs annually, mostly in small numbers in Australia during the non-breeding months, occasionally at other times. More frequently reported, and in numbers, in Western Australia, e.g. 80 at Lake Forrestdale, near Perth, in February 1981.

**References**  47, 91, 94, 121, 286, 410, 420, 424.

# DUNLIN
*Calidris alpina*
(Plate 13)

**Other name**  Red-backed Sandpiper.

## DESCRIPTION

Like Curlew Sandpiper: stockier, shorter legs and neck; bill averages shorter and is *slightly down-curved at tip*; more heavily streaked on breast. In flight, it has *dark central line down white rump* (cf. Curlew Sandpiper). In breeding plumage, it becomes generally chestnut above and has a *black belly patch*; confusion with Cox's Sandpiper is possible.

## RANGE

Breeds in northern temperate and arctic Europe, arctic Siberia, Alaska, Canada, Greenland and Iceland (85). Those of the race *sakhalina* breed in north-eastern Siberia and migrate through far-eastern USSR to non-breeding areas north of the equator in coastal Korea, Japan, eastern China and Taiwan. Two unconfirmed Australian sightings thought to be this species rather than Cox's Sandpiper: north-eastern Queensland, South Australia (379).

**References**  262, 279, 379, 430.

# BUFF-BREASTED SANDPIPER
*Tryngites subruficollis*
(Plate 12)

## DESCRIPTION

Very distinctive small sandpiper about the same size as or a little smaller than Sharp-tailed Sandpiper; *conspicuous pale area around eye; very buff on head, throat and breast*; crown spotted brown; *bright yellow legs*; buff sides to rump and uppertail-coverts; more upright stance than Sharp-tailed Sandpiper.

## RANGE

Breeds in grassy arctic tundra of Wrangel Island in far-eastern Siberia (144) and in northern Alaska and Canada (7). Migrates through interior of North America and eastern Mexico to northern South America. Accidental in Europe, Japan and the Kurile Islands. There have been a handful of records in Australia: Victoria, eastern Western Australia, New South Wales. Prefers short saltmarsh and grassy areas.

**References**  233, 409, 429.

# RUFF
*Philomachus pugnax*
(Plate 12)

**Other names**  The male is called Ruff and the female Reeve.

## DESCRIPTION

Male is noticeably larger than female; superficially resembles Sharp-tailed Sandpiper with its generally brown scalloped upperparts, but is larger, has longer legs, neck and bill; *less marked around head and neck*; leg colour varies: lime-green to bright orange-red; in flight shows *conspicuous white sides to rump*.

## RANGE

Breeds throughout most of arctic and temperate regions of Eurasia, migrating predominantly south-westwards to non-breeding quarters in southern Europe and Africa (85). Ruffs and Reeves visit Australia annually in small numbers. Up to four have been seen together, but more usually they are encountered singly; found in most parts of Australia.

**References**  52, 64, 92, 154, 199, 271, 286, 452, 465.


# STILT SANDPIPER
*Micropalama himantopus*

## DESCRIPTION

Similar to Curlew Sandpiper, but slightly larger with longer *greenish or yellow legs*. Similar to Ruff, but has longer, *down-curved bill* and more conspicuous supercilium and white rump, which is barred in breeding plumage.

## RANGE

Breeds in eastern Alaska and northern Canada as far east as northern Ontario (7), migrating through central North America to non-breeding grounds in central South America as far south as northern Chile and Argentina. One record at Darwin, August 1980.

**Reference**  282.


# RED-NECKED PHALAROPE
*Phalaropus lobatus*
(Plate 8)

**Other name**  Northern Phalarope.

## DESCRIPTION

Small, stint-sized aquatic shorebird with a *short, black needle-like bill*; quite distinctive: clean grey-and-white appearance; *dark grey line through eye to ear-coverts*; lobed toes; often swims in shallow water, pecking at surface for food; sometimes very tame; in flight shows white wing-bar and dark grey central line to rump and tail; in breeding plumage shows bright chestnut sides to neck and breast, and a white throat.

## RANGE

Breeds throughout the arctic in shrubby and alpine tundra areas (85). Migrates to tropical seas where it spends most of the non-breeding months far from land; often seen in coastal and inland wetlands. Birds recorded in Australia probably breed in eastern Siberia and spend the non-breeding months at sea, north of New Guinea and Indonesia and in wetlands in Indonesia and the Philippines (440). It is an annual visitor, particularly at Port Hedland in northern Western Australia, where up to 12 have been recorded. It has also been recorded in south-western Australia, South Australia, Northern Territory, Victoria and New South Wales.

**References**  83, 123, 200, 294, 371, 412, 423, 432, 438.


# WILSON'S PHALAROPE
*Phalaropus tricolor*
(Plate 8)

## DESCRIPTION

Differs from Red-necked Phalarope by being larger, paler and longer-necked; *lacks dark line down rump and white wing-bar* of other phalaropes; grey mark through eye and ear-coverts not as conspicuous as in other phalaropes; resembles Marsh Sandpiper, but differs by having a grey, not white, back; long, needle-like black bill; swims readily; in breeding plumage shows dark red head and neck markings and chestnut scapulars.

## RANGE

Breeds in western and central Canada and northern USA, migrating through western USA, Mexico and Central America to western South America from Peru to Argentina (7). Unlike the other 2 species, they are not pelagic, preferring wetlands. They are accidental in Europe, Africa, the islands of the central Pacific Ocean and Australia. There are five records from Australia, all in Victoria.

**References**  421, 422.

# GREY PHALAROPE

*Phalaropus fulicarius*
(Plate 8)

**Other name**   Red Phalarope.

## DESCRIPTION

Very like Red-necked Phalarope, except for slightly larger size, *shorter bill, usually with yellowish base,* and more uniform grey upperparts. In breeding plumage, shows black crown, white eye-patch and rich chestnut underparts.

## RANGE

Breeds discontinuously through high-arctic areas of Eurasia, North America and Greenland as well as in Iceland. Its migration route and non-breeding areas are not accurately known, as it is almost exclusively oceanic at this time. It is thought to migrate across the Atlantic Ocean to waters off western Africa and across the Pacific to seas west of South America; both are plankton-rich zones of upwelling (85). There are four records from New Zealand (134) and one from Australia, near Swan Hill in north-western Victoria.

**Reference**   427.

# REGIONAL ACCOUNTS

In Chapters 5 and 6 the distribution of each species of shorebird in Australia was described. To complete the picture, however, this chapter outlines which shorebirds occur in each part of Australia. Important shorebird habitats and abundant species in them are identified, and the impact of human activities described for ten regions of Australia.

The regions are chosen somewhat arbitrarily for convenience, and correspond to State borders or, in the larger States, regions of similar habitat or with a similar collection of species. The descriptions of these regions serve to demonstrate the immense geographical variability in Australian shorebird habitats and in the mixture of species that use them. They are ordered anti-clockwise around the continent, starting with Victoria and ending with Tasmania.

## VICTORIA

Shorebirds use a diversity of wetlands in Victoria. Where the tidal range is highest — along the central coast — large numbers occur in marine embayments. Port Phillip Bay, Westernport Bay and Corner Inlet hold the largest numbers, but the less extensive mudflats of Anderson's Inlet and Shallow Inlet also hold smaller but important numbers of some species. Victoria's inland wetlands, particularly those of the west and north-west are also important, not only for Australian-breeding species, but also for some migrants, especially during southward migration. The most important areas are the Western District lakes, the numerous lakes of the Horsham area, Lake Buloke, Lake Hindmarsh, Lake Albacutya (when it contains water) and the wetlands of the Kerang—Swan Hill region in the north-west. Table 7.1 shows the relative importance of major coastal and inland wetlands in Victoria for shorebirds.

The most abundant of the migratory shorebirds in Victoria are the Red-necked Stints, Curlew Sandpipers, Sharp-tailed Sandpipers, Bar-tailed Godwits, Red Knots and Eastern Curlews. Victoria holds substantial proportions of the numbers in Australia. More Double-banded Plovers occur in winter in the coastal wetlands of Victoria than elsewhere in Australia. Banded Stilts, Red-necked Avocets, Black-winged Stilts and Red-capped Plovers are the most numerous of the residents, but comparatively large numbers of Red-kneed Dotterels and Black-fronted Plovers also occur there. These residents, which the exception of Banded Stilts, breed regularly on Victorian wetlands (38).

Human activities have considerably affected wetlands in Victoria to the advantage and disadvantage of shorebirds. In Port Phillip Bay, an average of 75 per cent of the shorebird population occurs on artificial wetlands in the Werribee Sewage Farm and the three evaporative saltworks at Geelong, Avalon and Altona. Less encouragingly, for instance, 79 per cent of the *shallow* freshwater marshes of western Victoria have been drained since European settlement 150 years ago, mostly for agricultural use (80). Such destruction has caused a substantial reduction in the area of wetlands available for the Red-kneed Dotterel and Latham's Snipe and for migratory shorebirds during southward migration. Victoria has the highest human population density of any State in Australia. Consequently, recreational and development pressure is increasing in coastal areas and could lead to the reduction of coastal populations of beach-nesting oystercatchers, Red-capped Plovers and Hooded Plovers.

**Key references**   21, 79, 80, 81, 96, 252, 264, 265, 297, 414, 417, 485.

TABLE 7.1. Summary of count results from Victoria.

| Species | 1 | 2 | 3 | 4 | 5 | 6 | 7 | 8 | 9 | 10 | 11 | 12 | 13 | 14 | 15 | 16 | 17 |
|---|---|---|---|---|---|---|---|---|---|---|---|---|---|---|---|---|---|
| Pied Oystercatcher | 30 | 20 | 870 | 14 | 160 | 110 |  | 40 |  |  |  |  |  |  |  |  |  |
| Sooty Oystercatcher | 4 | 1 | 270 | 16 | 8 | 11 |  | 11 |  |  |  |  |  |  |  |  |  |
| Grey Plover |  |  | 450 |  | 1 | 220 |  | 20 |  |  |  |  |  |  |  |  |  |
| Lesser Golden Plover |  |  | 90 | 70 | 50 | 220 |  | 20 | 20 |  |  |  |  |  |  |  |  |
| Red-kneed Dotterel |  | 4 |  |  |  | 160 |  | 2 | 4 | 120 | 30 | 90 | 30 | 60 | 20 | 12 | 2 |
| Hooded Plover | 30 | 75 | 80 | 60 | 20 | 30 | 7 | 210 |  |  |  |  |  |  |  |  |  |
| Mongolian Plover |  |  | 70 |  | 7 | 60 |  | 6 |  |  |  |  |  |  |  |  |  |
| Double-banded Plover | 15 | 250 | 730 | 550 | 500 | 1400 | 1 | 300 | 170 | 20 |  | 2 | 40 | 10 |  |  |  |
| Large Sand Plover |  |  | 20 |  | 4 | 3 |  |  |  |  |  |  |  |  |  |  |  |
| Oriental Plover |  |  | 53 |  |  |  |  |  |  |  |  |  |  |  |  |  |  |
| Red-capped Plover | 70 | 90 | 60 | 30 | 80 | 510 | 10 | 160 | 130 | 1200 | 140 | 230 | 450 | 30 | 12 | 350 |  |
| Black-fronted Plover | 2 | 12 |  |  | 1 | 70 |  | 6 | 3 | 120 | 40 | 40 | 3 | 60 | 9 | 30 | 25 |
| Black-winged Stilt |  | 120 |  | 12 |  | 510 | 50 | 30 | 230 | 330 | 10 | 200 | 300 | 10 | 30 | 7 |  |
| Banded Stilt |  |  | 1 |  |  | 600 |  | 20 | 2050 | 1560 | 15 | 160 | 980 |  | 15 |  |  |
| Red-necked Avocet |  |  |  |  |  | 980 |  | 16 | 40 | 600 | 560 | 480 | 310 | 50 | 20 | 40 |  |
| Ruddy Turnstone |  |  | 60 |  | 70 | 110 |  | 210 |  |  |  |  |  |  |  |  |  |
| Eastern Curlew | 20 | 4 | 1740 | 230 | 1020 | 160 |  |  |  |  |  |  |  |  |  |  |  |
| Whimbrel |  |  | 20 |  | 13 | 6 |  | 1 |  |  |  |  |  |  |  |  |  |
| Little Curlew |  |  |  |  |  |  |  |  |  |  |  |  | 3 |  |  |  |  |
| Grey-tailed Tattler |  |  | 20 |  | 20 | 30 |  | 4 |  |  |  |  |  |  |  |  |  |
| Greenshank |  | 50 | 220 | 240 | 260 | 460 |  | 20 | 1 | 14 |  | 40 | 16 | 10 |  | 1 |  |
| Marsh Sandpiper |  | 3 |  |  |  | 80 |  | 7 |  | 11 |  | 20 | 70 | 4 | 2 |  |  |
| Terek Sandpiper |  |  | 1 | 2 | 5 |  |  |  |  |  |  |  |  |  |  |  |  |
| Black-tailed Godwit |  | 20 | 2 |  |  | 30 |  |  |  | 1 |  | 2 | 6 |  |  |  |  |
| Bar-tailed Godwit | 50 | 2 | 7300 | 2 | 290 | 570 | 1 | 1 |  |  |  |  |  | 1 |  |  |  |
| Red Knot | 3 | 4 | 2740 |  | 23 | 730 |  | 20 |  |  |  |  |  |  |  |  |  |
| Great Knot | 1 | 12 | 210 |  | 6 | 480 |  |  |  |  |  |  |  |  |  |  |  |
| Sharp-tailed Sandpiper | 9 | 1090 | 220 | 380 | 310 | 8740 |  | 310 | 1510 | 3030 |  | 330 | 2260 | 100 | 20 | 290 |  |
| Red-necked Stint | 100 | 2120 | 15600 | 1940 | 3750 | 22600 |  | 970 | 1450 | 330 | 14 | 110 | 980 | 60 |  | 60 |  |
| Curlew Sandpiper |  | 60 | 6000 | 350 | 3100 | 17700 |  | 2 | 670 | 60 |  | 50 | 760 | 17 |  | 2 |  |
| Sanderling |  |  | 180 |  |  | 10 |  | 430 |  |  |  |  |  |  |  |  |  |
| Australian Pratincole |  |  |  |  |  | 2 |  |  |  | 10 |  | 9 | 3 |  |  |  |  |

1. East Gippsland coast   2. Gippsland lakes   3. Corner & Shallow inlets
4. Anderson's Inlet   5. Westernport Bay   6. Port Phillip Bay   7. Cape Otway coast
8. SW coast   9. Western District lakes   10. Horsham area lakes   11. NW lakes
12. Mildura area   13. Kerang–Swan Hill area   14. Bendigo area   15. Shepparton area
16. Central Victoria   17. NE Victoria

## New South Wales

The coast, the western plains, and the tablelands and mountains form three distinct regions in New South Wales, each with its characteristic wetlands for shorebirds. Largest numbers of shorebirds generally occur along the coast, principally in the larger river estuaries, such as those of the Richmond, Clarence, Hunter, George's (Botany Bay) and Shoalhaven rivers. These rivers have extensive flood plains and associated freshwater wetlands, whereas elsewhere the mountains approach to within a few kilometres of the coast. The tablelands and mountains have few wetlands suitable for shorebirds; in the south, near Canberra, lakes George and Bathurst hold large numbers at times, but, apart from these, shorebirds are confined to local sewage treatment works, such as those at Parkes, Mudgee and on the New England tableland. A number of large rivers flow across the western plains. Larger wetlands (e.g. Lake Cowal on the Lachlan River, the Macquarie Marshes, the Menindee Lakes on the Darling River and the overflow country of the far north-west) can hold large numbers of Australian-breeding shorebirds during floods. Migratory shorebirds use them while crossing the continent during southward migration. Table 7.2 shows the relative importance of wetlands for shorebirds in New South Wales.

The most abundant migratory shorebirds in coastal New South Wales are the Bar-tailed Godwits, Curlew Sandpipers, Red-necked Stints, Eastern Curlews, Leser Golden Plovers and Mongolian Plovers. Smaller numbers of Grey-tailed Tattlers, Ruddy Turnstones and Whimbrels also occur, and about 500 Black-tailed Godwits visit the Hunter estuary each year. The number of species is higher on the north coast. A substantial proportion of the Double-banded Plovers that visit Australia spend the winter on the New South Wales coast. Pied Oystercatchers and Red-capped Plovers are the most abundant resident shorebirds on the coast, but the former is not common; no more than about 200 were counted. The Sooty Oystercatcher breeds on islands along the coast. In inland areas, the Sharp-tailed Sandpipers, Greenshanks, and Marsh Sandpipers are the most common migratory shorebirds. Black-winged Stilts and Latham's Snipe occur on near-coastal swamps. With the exception of the Sharp-tailed Sandpipers, Australian-breeding shorebirds are the most abundant shorebirds on inland wetlands; large numbers of Red-necked Avocets, Black-winged Stilts, Red-capped Plovers and Red-kneed Dotterels occur at times. Black-fronted Plovers are widespread.

Wetlands in New South Wales have been greatly modified by human activities. On the central coast, urban and industrial development has reduced the area of coastal wetlands, but some of what remains is now reserved, e.g. southern Botany Bay and the Hunter estuary. On the north coast, unplanned canal-based, real-estate development and a lack of reserves will result in the destruction of a considerable area of wetland used by shorebirds. Intensive recreation on beaches has probably resulted in a decline in New South Wales of Pied Oystercatchers, Hooded Plovers and Red-capped Plovers (183, 254, 322). Inland, the flow of most rivers has been modified for irrigation and flood mitigation. The precise effects of this on shorebirds have not been documented. The creation of artificial wetlands (e.g. Tullakool Saltworks, local sewage treatment works) has benefited some species.

**Key references**   163, 190, 195, 298, 357, 470.


## Southern Queensland

Coastal southern Queensland from Bundaberg to the New South Wales border has several large wetlands. The most important for shorebirds are Hervey Bay, Great Sandy Strait and Moreton Bay. Smaller wetlands on the Sunshine Coast hold fewer shorebirds, and those of the Gold Coast have been so altered by human activities that few shorebirds remain. Little is known of the shorebirds of inland southern Queensland as few of the wetlands there were surveyed during this study. Table 7.3 shows the relative importance of wetlands for shorebirds in southern Queensland.

The most abundant shorebirds in coastal southern Queensland are Bar-tailed Godwits, Mongolian Plovers, Eastern Curlews, Red-necked Stints, Great Knots, Curlew Sandpipers and Whimbrels. There are also important concentrations of Lesser Golden Plovers, Grey-tailed Tattlers, Greenshanks and Terek Sandpipers. Hervey Bay, Great Sandy Strait and Moreton Bay hold nearly half the Eastern Curlews counted in Australia and nearly half the Bar-tailed Godwits, Mongolian Plovers and Grey-tailed Tattlers counted in eastern Australia. During the 1982–83 drought, Lake Numalla, near Cunnamulla in the south-west, held more than one thousand Red-necked Avocets and Sharp-tailed Sandpipers and hundreds of Red-kneed Dotterels, Black-winged Stilts, Banded Stilts, Marsh Sandpipers and Curlew Sandpipers.

The coastal wetlands of southern Queensland have either been destroyed or are rapidly being destroyed. No information is available on the distribution and abundance of shorebirds along the coast before this study, and it is not possible to assess what the impact of this destruction has been. The Nerang River estuary on the Gold Coast has been greatly altered by the construction of extensive canal-based housing estates. This development is spreading northwards and is now encroaching on the maze of tidal waterways and mangroves at the southern end of Moreton Bay. The Brisbane side of the bay is being altered in many places by reclamation of tidal wetlands for a new airport, wharf facilities, rubbish tips, housing estates and industrial development. Fortunately, Hervey Bay and Great Sandy Strait have not suffered a similar fate, but, at

TABLE 7.2. Summary of count results from New South Wales.

| Species | 1 | 2 | 3 | 4 | 5 | 6 | 7 | 8 | 9 | 10 | 11 | 12 | 13 | 14 | 15 |
|---|---|---|---|---|---|---|---|---|---|---|---|---|---|---|---|
| Pied Oystercatcher | 22 | 60 | 8 | 12 | 12 | 11 | 80 | | | | | | | | |
| Sooty Oystercatcher | 20 | 4 | | 17 | 13 | 30 | 50 | | | | | | | | |
| Grey Plover | 20 | | | | 1 | | 1 | | | | | | | | |
| Lesser Golden Plover | 390 | 70 | 410 | 60 | 210 | 48 | 120 | | | | | | | | |
| Red-kneed Dotterel | 4 | 15 | 20 | 9 | | 9 | | 20 | 20 | | 14 | 40 | 90 | 240 | 50 |
| Hooded Plover | | | | | | | 30 | | | | | | | | |
| Mongolian Plover | 690 | 40 | 130 | 130 | 30 | 2 | 20 | | | | | | | | |
| Double-banded Plover | 140 | 50 | 90 | 20 | 130 | 140 | 130 | | | | | | 3 | | |
| Large Sand Plover | 40 | | 2 | 1 | 2 | | | | | | | | | | |
| Red-capped Plover | 120 | 60 | 20 | 20 | 40 | 70 | 110 | 220 | 8 | | | 32 | 1520 | 450 | 6800 |
| Black-fronted Plover | 9 | 14 | 22 | 12 | 2 | | 2 | 7 | 40 | | 9 | 40 | 16 | 150 | 200 |
| Black-winged Stilt | 60 | 16 | 550 | 90 | 8 | 30 | 4 | 60 | 19 | 60 | 5 | 160 | 410 | 220 | 130 |
| Banded Stilt | | | | | | | | | | | | | 7 | 160 | 60 |
| Red-necked Avocet | | | 60 | | | | | | | | | 70 | 420 | 170 | 6850 |
| Ruddy Turnstone | 50 | 2 | 50 | 90 | 60 | 30 | 350 | | | | | | | | |
| Eastern Curlew | 390 | 530 | 490 | 100 | 120 | 30 | 350 | | | | | | | | |
| Whimbrel | 340 | 30 | 30 | 12 | 17 | | 25 | | | | | | | | |
| Grey-tailed Tattler | 380 | 70 | 100 | 23 | 60 | 15 | | | | | | | | | |
| Greenshank | 50 | 14 | 560 | 60 | 9 | 30 | 14 | | 1 | | | 50 | 150 | 4 | 26 |
| Marsh Sandpiper | 60 | 1 | 280 | 8 | | | | 40 | | | | 130 | 50 | 50 | 70 |
| Terek Sandpiper | 150 | | 30 | 40 | 20 | | | | | | | | | | |
| Black-tailed Godwit | 90 | | 470 | | | 2 | | | | | | 22 | | | |
| Bar-tailed Godwit | 1830 | 370 | 1300 | 450 | 470 | 10 | 620 | | | | | | | | |
| Red Knot | 310 | | 20 | 7 | 1 | 17 | 15 | | | | | | | | |
| Great Knot | 18 | | 25 | 30 | 7 | | 3 | 29 | 3 | | | 3 | | | |
| Sharp-tailed Sandpiper | 170 | 260 | 280 | 1830 | 30 | | 6 | 435 | | | | 1590 | 1180 | 370 | 7450 |
| Red-necked Stint | 390 | 110 | 69 | 240 | 360 | 40 | 40 | 40 | | | | | 60 | 250 | 200 |
| Curlew Sandpiper | 110 | 30 | 1570 | 240 | 200 | 7 | 1 | | | | | | 80 | | 70 |
| Sanderling | 30 | | | 40 | | 1 | | | | | | | | | |
| Broad-billed Sandpiper | | | 8 | | | | | | | | | | | | |

1. North coast    2. Port Stephens    3. Hunter estuary    4. Central coast    5. Botany Bay
6. Illawarra region    7. South Coast    8. Southern tablelands    9. Central tablelands
10. Northern tablelands    11. Namoi−Barwon area    12. Parkes area    13. Riverina
14. Lower Darling    15. Far North-west

TABLE 7.3 Summary of count results for southern Queensland.

| Species | 1 | 2 | 3 | 4 | 5 |
|---|---|---|---|---|---|
| Beach Thick-knee | 4 | | 1 | | |
| Pied Oystercatcher | 65 | | 140 | 5 | |
| Sooty Oystercatcher | | | 3 | | |
| Grey Plover | 7 | 8 | 2 | | |
| Lesser Golden Plover | 80 | 17 | 360 | | |
| Red-kneed Dotterel | | 1 | 30 | | 100 |
| Mongolian Plover | 1430 | 19 | 1770 | | |
| Double-banded Plover | 30 | 14 | 22 | | |
| Large Sand Plover | 20 | 7 | 120 | | |
| Red-capped Plover | 40 | 40 | 450 | | 100 |
| Black-fronted Plover | 7 | | 13 | | 120 |
| Black-winged Stilt | 17 | 1 | 370 | | 510 |
| Banded Stilt | | | 26 | | 500 |
| Red-necked Avocet | | | 40 | | 1000 |
| Ruddy Turnstone | 40 | | 20 | | |
| Eastern Curlew | 1970 | 50 | 1080 | 12 | |
| Whimbrel | 80 | 16 | 280 | 5 | |
| Little Curlew | | 1 | 2 | | |
| Grey-tailed Tattler | 180 | 8 | 3190 | | |
| Greenshank | 90 | 7 | 120 | 3 | 50 |
| Marsh Sandpiper | 6 | | 60 | 4 | 500 |
| Terek Sandpiper | 60 | | 90 | | |
| Black-tailed Godwit | 40 | | 260 | | |
| Bar-tailed Godwit | 7700 | 50 | 6300 | 130 | |
| Red Knot | 4 | | 100 | | |
| Great Knot | 20 | | 910 | | |
| Sharp-tailed Sandpiper | 60 | 6 | 570 | | 2010 |
| Red-necked Stint | 1460 | 40 | 1150 | 15 | |
| Curlew Sandpiper | 60 | | 620 | 2 | 200 |
| Broad-billed Sandpiper | | | 8 | | |

1. Hervey Bay—Great Sandy Strait
2. Sunshine Coast  3. Moreton Bay
4. Gold Coast  5. Far SW lakes

TABLE 7.4. Summary of count results for north-eastern Queensland.

| Species | 1 | 2 | 3 | 4 | 5 | 6 | 7 | 8 | 9 |
|---|---|---|---|---|---|---|---|---|---|
| Beach Thick-knee | 4 | 2 | | | 8 | 16 | | | |
| Pied Oystercatcher | 50 | 16 | 4 | | 30 | 150 | 60 | | |
| Sooty Oystercatcher | | | 7 | | | 40 | | 1 | |
| Grey Plover | 18 | | | | 13 | 80 | | | |
| Lesser Golden Plover | 120 | | 13 | 20 | 40 | 40 | | 8 | |
| Red-kneed Dotterel | | | | | 13 | 5 | | | 6 |
| Mongolian Plover | 700 | | 70 | 150 | 330 | 1090 | | 60 | |
| Double-banded Plover | | | | | 20 | 10 | | | |
| Large Sand Plover | 90 | | | 6 | 18 | 40 | | | |
| Oriental Plover | | | | | | 10 | | | |
| Red-capped Plover | 160 | | | 20 | 130 | 90 | | 220 | 2 |
| Black-fronted Plover | | | | 5 | 8 | 30 | | 2 | 30 |
| Black-winged Stilt | 5 | 40 | | 30 | 50 | 110 | | 9 | 940 |
| Red-necked Avocet | | | | | 5 | 2 | | | |
| Ruddy Turnstone | 2 | | 100 | 5 | 20 | 120 | | 100 | |
| Eastern Curlew | 70 | 2 | 1 | 40 | 100 | 710 | 430 | 30 | |
| Whimbrel | 18 | 3 | 14 | 30 | 70 | 140 | 50 | 7 | |
| Little Curlew | | | | | | 50 | | | 1 |
| Grey-tailed Tattler | 70 | | 90 | 17 | 80 | 80 | | 90 | |
| Greenshank | 15 | | | 60 | 50 | 100 | | 10 | 12 |
| Marsh Sandpiper | 1 | | | | 10 | 8 | | | 70 |
| Terek Sandpiper | 80 | | 3 | 20 | 9 | 10 | | | |
| Black-tailed Godwit | 19 | 50 | | 30 | 180 | 30 | | | 9 |
| Bar-tailed Godwit | 110 | | 170 | 200 | 300 | 890 | 1070 | 60 | |
| Red Knot | | | 80 | | 60 | 3 | | | |
| Great Knot | 6 | 80 | | 40 | 230 | 4050 | 4160 | 1 | |
| Sharp-tailed Sandpiper | 100 | | | 330 | 290 | 420 | | 8 | 210 |
| Red-necked Stint | 440 | | 14 | 410 | 150 | 300 | | 1 | |
| Curlew Sandpiper | 4 | | 8 | 20 | 30 | 10 | | 1 | |
| Sanderling | | | | | 12 | 2 | | | |
| Broad-billed Sandpiper | | | | | 10 | | | | |
| Oriental Pratincole | 1 | | | | | | | | |
| Australian Pratincole | 20 | | | | | | | | 15 |

1. Torres Strait—Cape York  2. Princess Charlotte Bay
3. Daintree—Melville  4. Cairns area  5. Townsville area
6. Mackay area  7. Broad Sound & Shoalwater Bay
8. Capricorn coast  9. Charters Towers area

the rate at which new land subdivisions are being made, there may in the near future be pressures for Gold Coast style development. Given the great importance to shorebirds of the wetlands of south coastal Queensland, there is an urgent need to plan development in a balanced way to protect key areas.

**Key reference**   9.

## NORTH-EASTERN QUEENSLAND

Shorebirds occur in two main regions of north-eastern Queensland: the mainland coast and the shores of islands and cays in the Great Barrier Reef. Only a few counts were made on the Great Barrier Reef, near the northern end between Cairns and Cooktown. Aerial surveys show that the mainland coast between Cairns and Torres Strait holds only a few thousand shorebirds, mostly at Princess Charlotte and Shelburne bays. Some areas, notably between Rockhampton and Gladstone and between Townsville and Bowen, were never comprehensively counted, yet there is suitable shorebird habitat along most of this coast. An area found to hold large numbers was the coast south from Mackay, including Broad Sound and Shoalwater Bay, where the tidal range is the highest along the east coast of Australia. The north-eastern Queensland coast is followed during migration by many shorebirds, but the location of major staging areas, if they exist, is not known. The relative importance of wetlands to shorebirds in north-eastern Queensland is shown in Table 7.4.

The most abundant migratory species along the mainland coast are Bar-tailed Godwits, Red-necked Stints, Eastern Curlews, Whimbrels, Mongolian Plovers, Turnstones and Great Knots. Smaller but important numbers of Grey-tailed Tattlers, Terek Sandpipers, Greenshanks and Lesser Golden Plovers also occur. Pied Oystercatchers and Red-capped Plovers are the most common Australian-breeding species. More Beach Thick-knees have been counted here than anywhere else in Australia, but totals have never more than 20. On near-coastal swamps, Sharp-tailed Sandpipers, Black-winged Stilts and Black-fronted Plovers occur regularly. The only wetlands west of the Great Dividing Range that have been counted are those near Charters Towers. In February 1985 they held Sharp-tailed Sandpipers, Marsh Sandpipers and Greenshanks, as well as over 1600 Black-winged Stilts. The cays and reefs of the Great Barrier Reef held mostly Sooty and Pied Oystercatchers, Lesser Golden Plovers (which are far less numerous on the nearby mainland coast), Turnstones, Whimbrels and Grey-tailed Tattlers. Wandering Tattlers may occur in small numbers throughout the reef, but problems in separating them from Grey-tailed Tattlers makes assessment of their status difficult. On the islands of Torres Strait, Mongolian Plovers, Red-necked Stints, Bar-tailed Godwits and Sharp-tailed Sandpipers occur in the hundreds.

The human population density along the coast of north-eastern Queensland is fairly low compared with other areas of eastern Australia. Consequently there have probably been no serious changes in the areas of wetland suitable for shorebirds. Future tourism, recreational and industrial development in the vicinity of large towns (e.g. Cairns, Townsville, Mackay and Gladstone) could cause local destruction of habitat. The shorebird populations of areas not covered during this study need to be determined.

**Key references**   9, 33, 114, 115, 119, 132, 152, 156, 170, 191, 208, 228, 483.

## THE GULF OF CARPENTARIA

The coast of the Gulf of Carpentaria runs for approximately 2000 kilometres between Cape York, Queensland, and Gove Peninsula, Northern Territory. It is one of three sectors of the Australian coast that together hold most of Australia's shorebirds. It contains two types of wetland used by shorebirds: the shore, comprising intertidal mudflats, tidal channels and mangrove woodlands; and the grassy plains, which extend for up to 100 kilometres inland. The intertidal mudflats are most extensive, often reaching 2 kilometres from the high-tide mark, along the coastline 150 kilometres either side of Karumba in the south-eastern corner of the Gulf. There, the tidal range is the highest in the Gulf, and up to 250 000 shorebirds occur there. Concentrations greater than 5000 were counted near the mouth of the McArthur River, in the south-western corner near the Roper River and in Blue Mud Bay. Table 7.5 gives the results of sample ground counts in the south-eastern corner.

The most abundant species on the coast are Great Knots, Black-tailed Godwits, Red-necked Stints and Mongolian and Large Sand Plovers. Red Knots were also present in tens of thousands in September 1983, but changes in species composition between September and April (see Table 7.5) suggest that such large numbers occur mostly during migration. More Black-tailed Godwits were counted there than anywhere else in Australia (see species account in Chapter 5). Smaller numbers of Eastern Curlews, Grey-tailed Tattlers and Terek Sandpipers were counted, and Red-capped Plovers and Pied Oystercatchers were the most abundant Australian-breeding shorebirds. Thousands of kilometres of tidal, mangrove-fringed creeks provide suitable habitat for the Whimbrel and Common Sandpiper and may hold large numbers. The plains behind the coast are an important habitat for Little Curlews, Oriental Pratincoles and Australian Pratincoles. Towards the end of the wet season these plains flood and are used as a stopover during northward migration by thousands of Sharp-tailed Sandpipers from southern Australia.

The Gulf of Carpentaria is one of the most remote, inaccessible parts of the northern Australian coast and is largely unchanged by human activities. There are no

TABLE 7.5. Sample counts from the Gulf of Carpentaria and summary of results from the Mount Isa area.

| Species | 1 | 2 | 3 | 4 | 5 |
|---|---|---|---|---|---|
| Beach Thick-knee | 20 | | 2 | | |
| Pied Oystercatcher | 120 | 230 | 20 | | |
| Grey Plover | 1550 | 14 | 9 | | |
| Lesser Golden Plover | 170 | 50 | | 1 | |
| Red-kneed Dotterel | | | | | 30 |
| Mongolian Plover | 4050 | 120 | 250 | | |
| Large Sand Plover | 4160 | 250 | 210 | | |
| Oriental Plover | | 30 | | 27 | 6 |
| Red-capped Plover | 3620 | 6 | 230 | | 20 |
| Black-fronted Plover | | | | 2 | 50 |
| Black-winged Stilt | 200 | 2 | 40 | 6 | 20 |
| Red-necked Avocet | | | | | 1 |
| Ruddy Turnstone | 230 | 1 | 1 | | 6 |
| Eastern Curlew | 690 | 24 | 40 | | |
| Whimbrel | 1100 | 40 | 7 | | |
| Little Curlew | | 6 | | 6390 | 180 |
| Grey-tailed Tattler | 980 | 13 | 6 | | |
| Common Sandpiper | 170 | 2 | | | |
| Greenshank | 140 | 60 | 1 | 16 | 1 |
| Marsh Sandpiper | 150 | 30 | | 7 | |
| Terek Sandpiper | 900 | 80 | 11 | | |
| Black-tailed Godwit | 49400 | 1150 | 150 | 1 | |
| Bar-tailed Godwit | 2210 | 60 | 6 | | |
| Red Knot | 79100 | 100 | 360 | | |
| Great Knot | 68000 | 1630 | 870 | | |
| Sharp-tailed Sandpiper | 7300 | 100 | | 3 | 30 |
| Red-necked Stint | 35200 | 1660 | 220 | 6 | |
| Curlew Sandpiper | 8500 | 150 | 350 | | |
| Sanderling | 180 | | | | |
| Broad-billed Sandpiper | 1740 | 100 | | | |
| Oriental Pratincole | | | | 706 | 7 |
| Australian Pratincole | | | | 38 | 80 |

1. Sept.—Oct. 1983   2. Dec. 1984 (coast)
3. June 1984   4. Dec. 1984 (plains behind coast)
5. Mount Isa area

immediate threats to shorebird habitat in the area, but future developments should be assessed for their impact on this outstanding area for shorebirds.

**Key references**   47, 148, 152, 170, 258.

## NORTHERN TERRITORY

The coast of the Northern Territory from Gove Peninsula to the Western Australian border has two main regions of intertidal mudflat that hold shorebirds: the north coast of Arnhem Land, where the tidal range is highest between Buckingham Bay and Maningrida, and the coast from Shoal Bay just north of Darwin south-westwards to Anson Bay. Apart from the Darwin area, the coastline of the Northern Territory was surveyed from the air only, so the species composition cannot be accurately assessed. There are about 65 000 shorebirds on the north coast of Arnhem Land and around 15 000 south-west from Shoal Bay. The wetlands behind the coast in western Arnhem Land were not surveyed, but they hold large numbers of shorebirds at times. In the far inland the only natural wetlands are on the Barkly Tableland, where migratory shorebirds were recorded during the *Atlas* project (38). Elsewhere, sewage treatment works attract small numbers of shorebirds, e.g. Tennant Creek, Alice Springs.

Some species were identified from the air, and Great Knots appeared to be most abundant in both coastal areas. On the Arnhem Land coast over 12 000 Black-tailed Godwits were counted in February 1984, as well as more than 100 Pied Oystercatchers. Counts of the Darwin area showed that Great Knots, Bar-tailed Godwits and Mongolian and Large Sand Plovers were most abundant. Smaller numbers of shorebirds were counted round Van Diemen Gulf, but there were few on Cobourg Peninsula or Bathurst and Melville islands. The coastal plains behind the coast are used by tens of thousands of Little Curlews just before the wet season, and ponds remaining from the previous wet season hold numerous migratory shorebirds, such as Greenshanks and Sharp-tailed Sandpipers (September—November). Once the wet season starts, these plains flood, and rank vegetation grows over them, making most areas unsuitable for shorebirds, though Black-winged Stilts and Greenshanks use the edges of flooded areas. The sewage treatment works inland attract small numbers of Wood Sandpipers, Common Sandpipers, Marsh Sandpipers, Greenshanks and Sharp-tailed Sandpipers during southward migration (234; this study).

Development on the Northern Territory coast is confined to Darwin and the Gove Peninsula. In Darwin, most swamps have been drained for mosquito control, causing a decline in the number of shorebirds away from beaches. This draining programme is continuing and is expected to have further deleterious effects (185). There are few sandy beaches for shorebirds to roost on round Darwin. Those that do exist are popular for beach recreation, which frequently disturbs roosting shorebirds. The coast of Gove Peninsula holds few shorebirds compared to the two main regions mentioned above. There are no immediate threats to shorebirds or their habitat in the Northern Territory away from Darwin.

**Key references**   88, 108, 148, 234, 425.

## North-western Australia

The coast of north-western Australia has the highest tidal range in Australia. It is the most important region for shorebirds in the continent, regularly holding over 500 000 birds. Allowing for the migration of individuals through the region over a year, the total number using it approaches 850 000, or nearly a half of Australia's migratory shorebirds. They use the extensive intertidal mudflats between Broome and Karratha throughout the non-breeding months (August–April), but the largest numbers during this study were counted between September and November in 1982 and 1983. Four areas hold most of the shorebirds: Roebuck Bay (Broome to Sandy Point), Eighty Mile Beach (Cape Missiessy to Cape Keraudren), the Leslie Saltworks, about 35 kilometres east of Port Hedland, and Shark Bay, south of Carnarvon. Lesser numbers (tens of thousands) occur on La Grange Bay, between the De Grey River and Dampier Peninsula (including Dampier Saltworks) and in Exmouth Gulf near North West Cape. Major concentrations of up to 30 000 Little Curlews occur on Roebuck Plains and Anna Plains (between Broome and Port Hedland), and the grassy plains between the De Grey River and Karratha hold thousands of Oriental Pratincoles (38). The swamps that form on Roebuck and Anna Plains in the wet season sometimes last all year, and in September 1982 held about 25 000 recently arrived Sharp-tailed Sandpipers. Also, hundreds of shorebirds occur on the sewage treatment works near most towns in the region and on wetlands associated with the rivers in the Kimberleys (e.g. Fitzroy, Ord) as well as the margins of Lake Argyle, near Kununurra.

Table 7.6 shows the maximum numbers of each species counted in parts of north-western Australia, identifying migratory species that stay throughout the non-breeding months and those that use the area only during migration. The region is of outstanding importance for 21 species of migratory shorebird or more than half of the species that regularly visit Australia. More shorebirds use it on the southward migration than on the northward migration. The most abundant species are Great Knots, Bar-tailed Godwits, Large Sand Plovers, Red Knots, Little Curlews, Red-necked Stints, Curlew Sandpipers, Oriental Plovers, Grey-tailed Tattlers, Oriental Pratincoles and Terek Sandpipers. Interesting records for the region included 6 Redshanks, 300 Wood Sandpipers, 1600 Broad-billed Sandpipers, 130 Asian Dowitchers (212) and 12 Red-

necked Phalaropes. It is the only region of Australia that regularly holds the Pin-tailed Snipe.

Roosting shorebirds are frequently disturbed by human activities on their high-tide roosts on the northern side of Roebuck Bay at Broome. At present this is not detrimental, but any tourist or residential development in the area must be assessed for its impact on shorebirds. Overgrazing has damaged saltflat vegetation behind the mangroves on the eastern shores of Roebuck Bay, and its effects need to be monitored. Shorebirds have probably benefited from the construction of saltworks at Port Hedland, Dampier, Lake Macleod near Carnarvon and at Useless Loop in Shark Bay. There are no serious threats to shorebird habitat in the region at present. Only one of the important sites on this exceptional stretch of coastline has yet been declared a reserve: the eastern and southern shores of Roebuck Bay from Crab Creek to Sandy Point.

**Key references**   19, 125, 247, 248, 315, 316.

## South-western Australia

In south-western Australia, shorebirds live on coastal wetlands and inland lakes and swamps. Generally, coastal wetlands are not extensive because the tidal range in the south-west is low, and densities of shorebirds are lower than elsewhere in coastal southern Australia. Peel Inlet, south of Perth, is the most important habitat in the region and it holds up to 19 000 shorebirds. The Swan estuary in Perth and nearby islands, as well as the Vasse estuary near Busselton, are other areas holding thousands of shorebirds. Lesser numbers occur on the beaches and inlets near Albany and Esperance. Away from the coast, the lakes of the Swan coastal plain round Perth and those near Esperance hold good numbers. Farther inland, not all lakes were surveyed, but those that were held substantial numbers of Australian-breeding shorebirds (213; this study) and the large lakes in the eastern goldfields near Kalgoorlie were, when they held water, important breeding sites for Banded Stilts (44, 236). Table 7.7 shows the relative importance of wetlands for shorebirds in the south-west.

The most abundant migratory shorebirds on the south-west coast are Red-necked Stints, Curlew Sandpipers, Sharp-tailed Sandpipers (mostly Peel Inlet and Vasse estuary), Red Knots, Great Knots and Greenshanks. The region also holds important concentrations of Grey Plovers, Turnstones and Sanderlings. Banded and Black-winged Stilts, Red-necked Avocets and Red-capped Plovers are the most abundant Australian-breeding species. This is to be expected, as most of the inland lakes of the region are saline. Lesser numbers of Red-kneed Dotterels and Black-fronted Plovers also occur inland. The south-west holds the largest numbers of Long-toed Stints in Australia: 80 have been counted on Lake Forrestdale

TABLE 7.6. Summary of count results from north-western Australia.

| Species | 1 | 2 | 3 | 4 | 5 | 6 | 7 |
|---|---|---|---|---|---|---|---|
| Pied Oystercatcher | 7 | 50 | 190 | | | | |
| Sooty Oystercatcher | 2 | | | 19 | | | 1 |
| Grey Plover | 40 | 20 | 1650 | 1300 | | | 6 |
| Lesser Golden Plover | 15 | 3 | 440 | 8 | | | 40 |
| Red-kneed Dotterel | | | | | | | 3000 |
| Mongolian Plover | | 60 | 5 | 100 | | | |
| Large Sand Plover | 230 | 20 | 30400 | 29900 | | | 20 |
| Oriental Plover | | 30000 | 18400 | 8750 | | | |
| Red-capped Plover | 50 | 1000 | 9600 | 3300 | | | 100 |
| Black-fronted Plover | 20 | | | | | | 60 |
| Black-winged Stilt | 440 | 1230 | 1 | 150 | | | 690 |
| Banded Stilt | 50 | 800 | 3 | | | | |
| Red-necked Avocet | 300 | 2700 | 5 | | | | |
| Ruddy Turnstone | 35 | 4 | 740 | 2060 | | | |
| Eastern Curlew | 50 | 1 | 480 | 960 | | | 9 |
| Whimbrel | 120 | 5 | 180 | 1020 | | | 6 |
| Little Curlew | | 20 | 12000 | 30000 | | | |
| Grey-tailed Tattler | 50 | 4 | 8500 | 3180 | | | |
| Common Sandpiper | | 5 | | 40 | | | |
| Greenshank | 230 | 150 | 2440 | 560 | | | 30 |
| Marsh Sandpiper | 40 | 500 | 140 | | | | 8 |
| Terek Sandpiper | 5 | 2 | 3000 | 820 | | | 30 |
| Black-tailed Godwit | 200 | 20 | 110 | 500 | | | 1 |
| Bar-tailed Godwit | 390 | 350 | 34300 | 65000 | | | |
| Red Knot | 250 | 80 | 80700 | 11200 | | | |
| Great Knot | 430 | 60 | 160000 | 22600 | | | |
| Sharp-tailed Sandpiper | 14 | 20000 | 25000 | 500 | | | 2 |
| Red-necked Stint | 1360 | 23000 | 60000 | 19800 | | | 1 |
| Curlew Sandpiper | 600 | 25000 | 60000 | 6000 | | | |
| Sanderling | | 30 | 100 | 1510 | | | |
| Broad-billed Sandpiper | 100 | 1800 | 55 | 110 | | | |
| Oriental Pratincole | 4000 | 10000 | | | | | |
| Australian Pratincole | | | 100 | | | | 1240 |
| Unidentified Shorebirds* | | | | | 30000 | 23800 | |

1. Pilbara coast    2. Port Hedland Saltworks    3. Eighty Mile Beach
4. Roebuck Bay    5. Port Hedland to Cape Keraudren (aerial count)
6. Lagrange Bay and nearby coast (aerial count)    7. Kimberley region

* aerial count, all species

TABLE 7.7. Summary of count results from south-western Australia.

| Species | 1 | 2 | 3 | 4 | 5 | 6 | 7 | 8 | 9 | 10 |
|---|---|---|---|---|---|---|---|---|---|---|
| Pied Oystercatcher | 7 | 30 | 30 | 9 | 40 | 14 | | | | |
| Sooty Oystercatcher | 140 | 60 | 11 | 16 | 7 | 15 | | | | |
| Grey Plover | | 50 | 130 | 70 | 100 | 30 | | | | |
| Lesser Golden Plover | 1 | | 40 | 6 | 2 | 30 | | | | |
| Red-kneed Dotterel | | | | 1 | 20 | | 17 | 20 | | |
| Hooded Plover | | 480 | | | | | | | | |
| Mongolian Plover | 1 | | 1 | | 2 | | | | | |
| Double-banded Plover | 8 | | | | | | | | | |
| Large Sand Plover | | | 20 | 10 | 2 | 20 | | | | |
| Oriental Plover | 1 | | | | | | | | | |
| Red-capped Plover | 190 | 230 | 91 | 1780 | 380 | 140 | 90 | 60 | 6 | 15 |
| Black-fronted Plover | | 11 | 5 | 20 | 40 | 11 | 13 | 8 | 3 | 8 |
| Black-winged Stilt | 3 | 50 | 7 | 850 | 1680 | 90 | 270 | 40 | 40 | 9 |
| Banded Stilt | | 2380 | 210 | 1660 | 240 | 20 | 400 | 100 | 210 | |
| Red-necked Avocet | | 100 | 6 | 300 | 370 | 30 | 90 | 30 | | |
| Ruddy Turnstone | | 120 | 1 | 1 | 280 | 50 | | | | |
| Eastern Curlew | | 1 | 1 | 14 | | 1 | | | | |
| Whimbrel | | | | 1 | 2 | 1 | | | | |
| Little Curlew | | | | | | 3 | | | | |
| Grey-tailed Tattler | 1 | 2 | 3 | 6 | 7 | 2 | | | | |
| Greenshank | | 60 | 130 | 160 | 70 | 7 | 5 | | | |
| Marsh Sandpiper | | | 1 | 20 | 5 | | | | 2 | |
| Terek Sandpiper | | 3 | | 1 | 2 | | | | | |
| Black-tailed Godwit | | | | 4 | 4 | | | | | |
| Bar-tailed Godwit | | 30 | 20 | 70 | 11 | 11 | | | | |
| Red Knot | | 50 | 50 | 8 | 80 | 4 | | | | |
| Great Knot | | 1 | 150 | 30 | 70 | | | | | |
| Sharp-tailed Sandpiper | 3 | 150 | 500 | 740 | 150 | 40 | 1 | 2 | | |
| Red-necked Stint | 30 | 870 | 1100 | 4100 | 2870 | 240 | 19 | 90 | | |
| Curlew Sandpiper | 2 | 90 | 350 | 420 | 710 | 60 | | | | |
| Sanderling | 1 | 30 | 100 | | 110 | 11 | | | | |
| Broad-billed Sandpiper | | | 1 | | | | | | | |
| Oriental Pratincole | | | | | 2 | | 1 | | | |
| Australian Pratincole | | | | | 1 | | | | | |

1.  Nullarbor coast    2.  Esperance coast    3.  Albany coast
4.  Peel Inlet & nearby coast    5.  Swan coastal plain    6.  Central W coast
7.  SE lakes    8.  SW lakes    9.  Mid-western lakes    10.  Nullarbor

(94) and over 60 on Lake McLarty. Wood Sandpipers occur in larger numbers in the south-west than elsewhere in southern Australia, and up to 60 have been seen on the Vasse estuary. Over 400 Hooded Plovers occur on the lakes near Esperance, more than at any other place in Australia.

Most inland wetlands in the south-west have been affected by rising saline groundwater tables caused by vegetation clearance. Some coastal lakes and wetlands, notably Peel Inlet, have been severely affected by fertiliser-laden runoff from nearby farmland. The effects of these changes on shorebird populations are unknown. Disturbance from human activities affects shorebirds in the Swan estuary. In the Perth metropolitan area, many lakes are sprayed with the insecticide Abate to reduce midges. This spraying is known to have caused the deaths of at least 100 shorebirds on one occasion at Lake Forrestdale (94). On beaches at Esperance, disturbance from off-road vehicles has caused a decline in the number of Hooded Plovers in the area (214).

**Key reference**   213.


## South Australia

Shorebirds in South Australia are confined mostly to the coast. There are few inland wetlands except in the southern part of the State and these generally dry up in summer. St Vincent and Spencer gulfs, the Coorong and Murray mouth lakes and the lakes and coast of the south-east, as well as sheltered bays on the western coast of the Eyre Peninsula and on Kangaroo Island hold thousands of shorebirds. Numbers on the Coorong and Murray mouth lakes (max. 236 000 in February 1982) are exceeded only at two other locations in Australia. In St Vincent and Spencer gulfs, most shorebirds occur at the northern ends, where the tidal range is highest. The wide sandy beaches west of Ceduna were not comprehensively counted and may hold large numbers of Hooded Plovers and Sanderlings. On the islands in Spencer Gulf and off the Eyre Peninsula large numbers of oystercatchers and Ruddy Turnstones may occur. The salt lakes in inland South Australia (Torrens, Eyre, Callabonna, Grace and Frome) and the lakes of the eastern Lake Eyre drainage basin hold thousands of stilts and avocets when they fill temporarily with water (17; this study). South Australia's coastline holds more shorebirds than any other part of southern Australia, and the State's shorebird habitats are of outstanding value in both national and international contexts. Table 7.8 shows the relative importance of wetlands for shorebirds in South Australia.

The most abundant migratory species in South Australia are Red-necked Stints, Curlew Sandpipers and Sharp-tailed Sandpipers. Smaller but important numbers of Grey and Lesser Golden Plovers, Ruddy Turnstones, Greenshanks, Bar-tailed Godwits, Red

Knots and Sanderlings also occur. Of the Australian-breeding species, Banded Stilts were most abundant (e.g. over 100 000 counted in February 1982), followed by Red-capped Plovers, Red-necked Avocets and Pied and Sooty Oystercatchers. Only small numbers of Eastern Curlews (less than 300) and Double-banded Plovers reach South Australia each year.

There have been many changes to South Australia's wetlands since European settlement. In the south-east, many freshwater wetlands have been drained for agricultural use, causing a considerable loss of habitat suitable for Latham's Snipe, Painted Snipe and other non-coastal shorebirds. In 1940 the Murray mouth lakes were separated from the sea by a series of barrages at their lower end in order to reduce the salinity of water used for irrigation along the nearby reaches of the Murray River (35). The effect of this on shorebirds is not known, but it has led to the colonisation of much of the shoreline by reeds (*Phragmites communis*), and the water-level no longer fluctuates as it did. These changes have probably reduced the area of open mudflat for shorebirds. The commercial saltworks at Price and St Kilda on St Vincent Gulf and at Whyalla on Spencer Gulf are used by large numbers of roosting shorebirds and are good feeding areas for stilts and avocets. Wetlands on the western Eyre Peninsula and Kangaroo Island are still relatively intact. On the beaches of the south-east there is disturbance from off-road vehicles, but the impact of this on shorebirds remains to be investigated.

**Key references**   17, 32, 39, 40, 42, 158, 275.


## Tasmania

Most shorebird habitat in Tasmania is coastal. The most important areas for shorebirds are the Derwent estuary and Pittwater near Hobart, the coast near Cape Portland in the north-east, and the lagoons and coast of Flinders and King islands. Lesser numbers occur at inlets along the east coast, in the Tamar estuary near Launceston, on the north-west coast round Stanley, and on Ocean Beach near Strahan. A few shorebirds occur on the central lakes. The extensive intertidal mudflats in Robbins Passage in the north-west of the State were not comprehensively counted, but limited observations suggest that the area could hold up to 3000 shorebirds (354). Table 7.9 shows the relative importance of areas for shorebirds in Tasmania.

The most abundant migratory species are Red-necked Stints, Curlew Sandpipers, Ruddy Turnstones, Eastern Curlews and Bar-tailed Godwits. Lower but important numbers of Lesser Golden Plovers, Greenshanks and Sanderlings visit annually. Sharp-tailed Sandpipers occur in unexpectedly low numbers compared to nearby Victoria. Of the Australian-breeding species, Pied Oystercatchers are most abundant; more were counted in Tasmania than elsewhere in Australia. Red-capped Plovers and Sooty Oystercatchers are also

TABLE 7.8. Summary of count results from South Australia.

| Species | 1 | 2 | 3 | 4 | 5 | 6 | 7 | 8 | 9 | 10 | 11 | 12 |
|---|---|---|---|---|---|---|---|---|---|---|---|---|
| Pied Oystercatcher | | | 450 | 200 | 20 | 240 | | 630 | 14 | | | |
| Sooty Oystercatcher | | 13 | 440 | 100 | 120 | 110 | 9 | 18 | 5 | | | |
| Grey Plover | | 2 | 1280 | 740 | 390 | 120 | | | 1 | | | |
| Lesser Golden Plover | | | 18 | 21 | 60 | 30 | 2 | 290 | 200 | | | |
| Red-kneed Dotterel | | | 5 | 2 | 11 | | 16 | 7 | 15 | 30 | 50 | 1 |
| Hooded Plover | | | 37 | 17 | | 75 | | 10 | 140 | | | |
| Mongolian Plover | | | 40 | 7 | 20 | | 4 | | | | | |
| Double-banded Plover | | | 30 | | 100 | 50 | | 30 | 90 | 6 | | |
| Large Sand Plover | | | 1 | 8 | 30 | 2 | | | 1 | | | |
| Oriental Plover | | | 178 | 10 | 5 | | | 18 | | | | |
| Red-capped Plover | 50 | 22 | 1040 | 1910 | 3590 | 810 | 420 | 5700 | 360 | 92 | 72 | |
| Black-fronted Plover | | | 140 | 11 | 4 | 40 | 40 | 14 | 7 | 10 | 40 | 40 |
| Black-winged Stilt | 36 | | 50 | 23 | 210 | 350 | 400 | 990 | 50 | 100 | 3 | |
| Banded Stilt | 30000 | | 1340 | 1900 | 6100 | 1510 | 210 | 77000 | 5900 | 2190 | 24 | |
| Red-necked Avocet | 95000 | | 310 | 28 | 740 | 9 | 580 | 5400 | 1 | 19 | 23 | |
| Ruddy Turnstone | | | 780 | 470 | 180 | 440 | | 1 | 900 | | | |
| Eastern Curlew | | | 20 | 46 | 130 | 40 | | 24 | 3 | | | |
| Whimbrel | | | 1 | 1 | 90 | | | | 1 | | | |
| Grey-tailed Tattler | | | 30 | 16 | 60 | 10 | 24 | | 18 | | | |
| Greenshank | | 1 | 580 | 320 | 1130 | 260 | 130 | 720 | 220 | 28 | 3 | |
| Marsh Sandpiper | | 1 | 60 | 25 | 40 | | 60 | 2 | 1 | 1 | | |
| Terek Sandpiper | | | | 2 | 2 | | | | | | | |
| Black-tailed Godwit | | | | 1 | 90 | | 120 | | | | | |
| Bar-tailed Godwit | | | 154 | 600 | 1240 | 31 | | 15 | | | | |
| Red Knot | | | 710 | 4750 | 2000 | 300 | | 170 | 25 | | | |
| Great Knot | | | 230 | 1300 | 520 | 800 | | 4 | 4 | | | |
| Sharp-tailed Sandpiper | 4000 | | 1740 | 2730 | 17000 | 3150 | 2630 | 55700 | 850 | 494 | 4 | |
| Red-necked Stint | 3 | 26 | 8500 | 7600 | 24000 | 5600 | 2010 | 63800 | 6800 | 450 | 80 | |
| Curlew Sandpiper | 600 | | 920 | 2500 | 2500 | 840 | 320 | 40000 | 1880 | 20 | 6 | |
| Sanderling | | | 510 | 100 | 100 | 4 | | 930 | 70 | | | |

1. Lake Eyre North, Sept. 1984   2. Nullarbor coast   3. West coast, Eyre Peninsula
4. Spencer Gulf   5. St Vincent Gulf   6. Kangaroo Island   7. Murray mouth lakes
8. The Coorong   9. SE coast   10. SE inland lakes   11. Inland Eyre Peninsula
12. Murray lands

TABLE 7.9. Summary of count results from Tasmania.

| Species | 1 | 2 | 3 | 4 | 5 | 6 | 7 | 8 | 9 | 10 | 11 | 12 | 13 | 14 | 15 | 16 | 17 | 18 |
|---|---|---|---|---|---|---|---|---|---|---|---|---|---|---|---|---|---|---|
| Pied Oystercatcher | 8 | 260 | 480 | 150 | 17 | | 100 | 250 | 120 | 130 | 150 | 180 | 10 | 50 | 24 | 14 | | 10 |
| Sooty Oystercatcher | | 20 | 40 | 2 | | | 2 | 30 | 50 | 50 | 2 | 70 | 30 | 480 | 30 | | | 18 |
| Lesser Golden Plover | | | 180 | | | | 30 | 6 | 30 | 26 | 30 | 30 | 40 | 18 | 17 | | | |
| Hooded Plover | ← | | 194 | | | → | | | 17 | | | 18 | | 60 | 30 | 60 | | |
| Mongolian Plover | | | 2 | | | | | | 9 | 1 | | 7 | | | | | | |
| Double-banded Plover | 5 | 40 | 360 | 40 | | | 30 | 50 | 680 | 50 | 30 | 19 | 370 | 120 | 30 | 170 | 20 | |
| Red-capped Plover | 3 | 120 | 170 | 50 | 5 | 30 | 50 | 100 | 200 | 20 | 7 | 20 | 150 | 80 | 40 | 100 | 40 | |
| Black-fronted Plover | 2 | | 30 | 8 | | 9 | 4 | | | | | | 20 | | | | 3 | |
| Ruddy Turnstone | | | | | | | | 4 | 130 | 140 | | 13 | 830 | 130 | 100 | | | |
| Eastern Curlew | 2 | | 150 | 12 | | | 13 | 11 | 3 | 100 | 70 | 9 | 1 | 60 | 100 | | | |
| Whimbrel | | | | 1 | | | 1 | | | 6 | 1 | | | | 1 | | | |
| Grey-tailed Tattler | | | 1 | 1 | | | | | 3 | 6 | | | | 2 | 6 | | | |
| Greenshank | | 1 | 70 | | | | 70 | 5 | 10 | 20 | | | 12 | 70 | 20 | | | |
| Marsh Sandpiper | | | | | | | | | | | | | 5 | | | | | |
| Bar-tailed Godwit | | | 70 | 14 | | | | 20 | 16 | 50 | | | | 80 | 50 | | | |
| Red Knot | | | 4 | 6 | | | | | 11 | 4 | | | | 9 | 1 | | | |
| Sharp-tailed Sandpiper | | | 20 | | | | 8 | | 5 | | | | 50 | | | | 7 | |
| Red-necked Stint | | 120 | 3050 | 460 | | | 80 | 500 | 1580 | 480 | 120 | 170 | 610 | 1550 | 310 | 650 | 40 | |
| Curlew Sandpiper | | | 1280 | 1 | | | 140 | 1 | 700 | 160 | 1 | 500 | 100 | 2640 | 100 | 1 | | |
| Sanderling | | | | | | | | | | | | | 6 | 160 | 20 | 330 | | |

1. SE coast    2. Bruny Island    3. Derwent estuary & Pittwater    4. Prosser    5. Tasman Peninsula
6. Midlands    7. Freycinet area    8. St Helen's area    9. Cape Portland coast    10. Tamar estuary
11. Central N coast    12. NW coast    13. King Island    14. Furneaux Islands    15. Central NW coast
16. Ocean Beach, Strahan    17. Central lakes    18. SW coast

numerous. Tasmania appears to be the stronghold for Hooded Plovers in eastern Australia (331). Many of the Double-banded Plovers that visit Australia in winter come to the north-east and south-east coasts of Tasmania. Masked Lapwings occur in flocks not uncommonly numbering 200 to 300 birds, and they may be more abundant in Tasmania than on the mainland.

Most areas of shorebird habitat in Tasmania are not immediately threatened, but urban and industrial development round Hobart may lead to detrimental changes to shorebird habitat in the Derwent estuary and Pittwater, which are the most important areas for shorebirds in Tasmania. Disturbance by human recreational activities is affecting large parts of the Tasmanian coast. This may be having deleterious effects on locally breeding shorebirds.

**Key references**  34, 72, 314, 335, 353, 454, 478, 479, 480, 481.

# SHOREBIRDS AND THE FUTURE

The majority of the world's shorebirds rely on wetlands for their survival, especially those on the coast. Australian economic history has followed the development of its overseas trade, and a number of key trading ports have become today's large metropolitan centres. This has resulted in 80 per cent of the human population of Australia living within 100 kilometres of the coast. This concentration of population places increasing pressures on Australia's coastal wetlands.

Economically, the Asian region is the fastest growing in the world, and urbanisation and industrialisation could lead to increased alteration and destruction of coastal and other wetland ecosystems with consequent deleterious effects on the populations of shorebirds that use them. Similar changes are occurring where shorebirds breed. The consequences of continued economic growth throughout the Asia—Australasia region have the potential for wiping out a substantial proportion of the habitat for shorebirds that migrate annually between the northern breeding grounds and Australia. If their habitat is destroyed, shorebirds must alter their habits. They may be unable to do this and might perish.

This chapter illustrates how human activities — including hunting and disturbance of shorebirds, destruction and alteration of their habitat, and pollution — have affected shorebirds on their non-breeding grounds in Australia, where they migrate through Asia, and on their breeding grounds. An exhaustive account of this is not possible in Australia, because in most cases the impact of human activities on wetlands and shorebirds has never been documented.

## HUNTING

In the nineteenth and early twentieth centuries, shorebirds were game and were hunted regularly in large numbers in Western countries. For example, Bent (28) said of the Marbled Godwit (*Limosa fedoa*) of North America: '. . . it is rapidly disappearing and before many years it may join the ranks of those that are gone but not forgotten. Although shy at times, it is often foolishly tame and is then easily slaughtered. It is large enough to appeal to the sportsman as legitimate game, and it makes a plump and toothsome morsel for the table.' Indeed, for some North American shorebirds, hunting was so destructive that it is thought that population levels have not yet fully recovered (324). In the case of the Eskimo Curlew (*Numenius borealis*), hunting in North America led to its probable extinction (last record: Manitoba, Canada, May 1980: 7). Hunting of most species of shorebirds in North America has been prohibited for many years now.

In Australia, species such as the Masked and Banded Lapwings, Lesser Golden Plover and Painted Snipe were once game birds (62). Latham's Snipe is still legal game in southern Queensland. It was only in 1983 that a moratorium on shooting it was enacted in other States pending the results of research on its biology. Today, most shorebirds are fully protected by law throughout Australia.

Such is not the case in Asia, where one of the major threats to shorebirds is from hunting. It has been estimated that between 250 000 and 1 500 000 shorebirds are trapped or shot for food each year during migration (346). Since only about 2 million shorebirds visit Australia during the non-breeding months, this figure is alarming.

More research is needed to document the size of kills. Additionally, the recruitment and mortality rates of shorebird populations must be studied to determine how much hunting they can sustain without declining. It has been pointed out (346) that unless such population parameters are known and harvest rates set accordingly and enforced, the facts now to hand, although limited, suggest that a reduction in some populations is imminent.

## DISTURBANCE

Recreational activities ranging from power-boating to strolling along beaches can adversely affect shorebirds. Furthermore, with the increasing sophistication of recreational activities, demands are increasing for development of facilities, such as boat marinas, requiring direct modification or destruction of shorebird habitat.

Shorebirds in coastal areas require undisturbed feeding grounds. Feeding time is more critical to smaller shorebirds (see Chapter 3), and frequent disturbance can make an area unsuitable for these species. Coastal shorebirds also require undisturbed roosting sites. In many instances they may use a variety of roosting sites depending on tide height and weather conditions. For example, at Shallow Inlet in Victoria, small shorebirds roost on the large sandy spit at the mouth of the inlet over the highest tides of the cycle, but on lower tides they roost in a number of saltmarsh sites within the inlet (224). They choose saltmarsh sites that give the best shelter from strong winds.

If a site is disturbed at a critical time, then the behaviour of shorebirds must change. In some parts of Britain, where human disturbance of shorebirds can be heavy, shorebirds have adopted the habit of 'aerial roosting' (175). Over the high tide they fly up to about 1000 metres above the ground, then glide down and repeat this procedure until the disturbance stops or feeding grounds are exposed by the tide again. This behaviour is energy-consuming, and it is not known if it has an adverse effect during bad weather or just before migration, when energy reserves are essential.

Disturbance from people on beaches that are popular for recreation can prevent successful breeding in resident species such as Hooded Plover, Red-capped Plover and Pied Oystercatcher, which nest on beaches. Unfortunately, in southern Australia, breeding in resident coastal shorebirds coincides with the warmer months of the year, when beach recreation is most popular. On Phillip Island in Victoria, 12 pairs of Hooded Plovers that attempted nesting successfully raised a single chick over two years (102). This poor breeding success was due to the interruption of breeding activities by bathers and people walking along the beach or to inadvertent destruction of nests and eggs by them.

FIG. 8.1.   Aerial view of fish ponds near Surabaya on the north coast of Java, Indonesia (cf. Figure 2.4). Reclamation of intertidal areas is seen as a solution to shortages of productive land in many Asian countries. Photo: M. Barter

## HABITAT DESTRUCTION AND ALTERATION

The alteration or complete destruction of wetlands inhabited by shorebirds for industrial, urban and port development, and dredging for shipping and marinas clearly reduces the area of habitat available to shorebirds (Figure 8.1). Too little is known of the precise effects of these habitat changes on shorebirds.

There has been considerable reclamation of saltmarsh in the Swan estuary in Perth and in Botany Bay in Sydney. This has probably caused the decline in numbers of Sharp-tailed Sandpipers using these areas, as this species often prefers to roost in saltmarsh (20, 322). This may not be serious in terms of the total population of the species, as it regularly moves between areas of suitable habitat. However, for other species, such changes may have more serious effects. Adult Red-necked Stints and Curlew Sandpipers return to the same area of habitat each year and rarely move out of it, except on migration (98), and may not be able to adapt to habitat destruction as easily.

The growing population of Asia and its rapid economic development in recent years have led to considerable destruction of coastal wetlands used by shorebirds. This is continuing at an increasing rate (346). Many of these countries, especially the densely populated ones such as China and Indonesia, need land for food production and see the reclamation of intertidal areas as a partial solution to land shortage (Figure 8.2). Table 8.1 shows the extent of habitat thought to have been destroyed already in Asia and that which is threatened in the coming decade (346). In those Asian countries for which information is available, an average of about 50 per cent of shorebird habitat has been destroyed and a further 20 per cent is threatened. These findings are startling in view of the almost complete lack of knowledge of shorebird distribution and migration routes in the region.

Little research is being done to assess the effects of habitat changes wrought by human activities on shorebird populations. Questions need to be answered before the effect of such changes can be predicted. Are wetlands holding their 'carrying capacity' of shorebirds? Do shorebirds behave in such a way (e.g. territorially) that 'carrying capacity' is not exceeded? If it is exceeded, will remaining food supplies be depleted? Do shorebirds move out of an altered area to others where shorebirds occur, thereby affecting populations elsewhere? Without research into these questions it will not be possible to ameliorate the effects on shorebirds of future economic growth.

FIG. 8.2.   Coastal wetland reclaimed for industrial development at Westernport Bay, Victoria. The effects on shorebirds of such a reduction in feeding area are not known. Photo: J. Hooper

Fig. 8.3. The shallow saline lagoons of saltworks provide excellent feeding and roosting habitat for thousands of shorebirds: Avalon Saltworks, Port Phillip Bay, Victoria, Photo: J. Hooper

TABLE 8.1. The percentage of shorebird habitat in Asian countries thought to have been destroyed and that threatened in the coming decade (346).

| Country | Habitat Type | Estimated % Habitat Degraded at Present | Additional % Threatened in Next Decade |
|---|---|---|---|
| USSR | Tundra, Taiga, Steppe | NI | NI |
| China | Steppe, Marshes, Coastal Mudflats | NI | NI |
| Japan | Marshes, Coastal Mudflats | NI | NI |
| Taiwan | Coastal Mudflats | 50 | NI |
| Hong Kong | Coastal Mudflats | 60 | 20 |
| Philippines | Mangrove/Mudflats | 70 | 20 |
| Thailand | Mangrove/Mudflats | 50 | 30 |
| W Malaysia | Mangrove/Mudflats | 40 | 40 |
| E Malaysia | Mangrove/Mudflats | 10 | 20 |
| Singapore | Marshes, Mangrove/Mudflats | 90 | 8 |
| Indonesia | Marshes, Mangroves/Mudflats | 10 | 30 |
| Australia | Marshes, Mangroves/Mudflats | 10 | NI |

NI = No information.

In some cases natural habitat has been destroyed, but replaced by a land-use that enhances the suitability of an area for shorebirds. In Port Phillip Bay, for instance, the construction of three evaporative saltworks and a large sewage treatment works is probably responsible for this area being so important for shorebirds. The lagoons and irrigated fields provide excellent feeding conditions and are used by about 75 per cent of the shorebirds that occur in the bay. (See Figures 8.3 and 8.4.) Saltworks have also been constructed near Adelaide in South Australia, at Tullakool in inland New South Wales (where no permanent wetlands existed before), and near Dampier and Port Hedland in north-western Australia, with similar results. Three of the ten most important areas for shorebirds in Australia owe their significance in part to such artificial changes.

Changes that are advantageous to shorebirds are not necessarily advantageous for other groups of birds. For instance, the destruction of saltmarsh for the sewage treatment works and saltworks around Port Phillip Bay has probably been responsible for the decline of what is now one of Australia's most endangered birds: the Orange-bellied Parrot (*Neophema chrysogaster*) (496).

Shorebird habitat can be altered by changes in the balance between the inflow and outflow of water in a wetland, often leading to changes in its salinity and

167

FIG. 8.4.    Irrigated pasture and ponds in the Werribee Sewage Farm near Melbourne are a haven for many shorebirds. Photo: J. Hooper

water-level. This can be deliberate (e.g. flood mitigation and irrigation schemes) or unanticipated (e.g. the rising of regional ground-water tables due to clearance of natural vegetation). These changes have an impact on aquatic plants and animals and hence on the suitability of an area for shorebirds. In parts of south-western Australia, numbers of Black-winged Stilts on wetlands have increased as a result of salinisation due to rising groundwater tables (299), but freshwater species, such as the Long-toed Stint and Wood Sandpiper, have declined and no longer occur there (214). This is an example of a local change. Such changes are likely to be affecting shorebirds in many of the more settled parts of Australia. Little research has been done, however, to document the extent or effects of these changes on wetlands.

## POLLUTION

Human activities can cause changes to the intertidal invertebrate communities upon which shorebirds depend for food. These can come about from harvesting and pollution. In Australia, little harvesting of mud-dwelling plants and animals occurs, but pollution, especially in the vicinity of large coastal urban centres,

can alter the species composition of intertidal invertebrate communities.

On the coast of the Werribee Sewage Farm in Port Phillip Bay near Melbourne, the input to nearby coastal waters of huge quantities of organic waste has considerably modified the intertidal mudflat invertebrate community. Detailed study (16) has shown that at Werribee microalgal growth on the surface of the mud is five times greater than that of an area in cleaner water away from the sewage treatment works. This, together with organic particles from the effluent, has resulted in the dominance of a few species of deposit-feeding invertebrates, notably polychaete worms and small amphipod crustacea. These occur in very high densities on the intertidal mudflats adjacent to the farm, and the polychaete worms are prey for the large numbers of small migratory shorebirds that occur there in exceptionally high densities (97). This is an example of shorebirds being favoured by organic pollution. The other side of the coin may be a drop in shorebird species diversity. At another, unpolluted site in Port Phillip Bay (Swan Bay), the diversity of shorebirds is higher. Consistent historic data on shorebird numbers and species composition are not available from Port Phillip Bay, so it is impossible to say what the long-term effect of such pollution has been.

The seriousness of the effects of other types of pollutants on shorebirds and their food supplies (e.g. petrochemicals, synthetic chemicals, heavy metals and mineral salts) is not known.

## PROBLEMS ON THE BREEDING GROUNDS

When breeding, migratory shorebirds occur in rather different habitats to those they use in their non-breeding range or during migration. They are spread over large tracts of country in comparatively low densities, and this makes them less susceptible to habitat changes in limited areas than they would be in Asia or Australia. However, longer-term changes are occurring in the breeding populations of shorebirds due to contraction or expansion of their breeding range. It is difficult to determine the causes of such changes, because many factors affect the survival of shorebirds, not just those on the breeding grounds. For example, the observed changes in breeding range and population of many shorebirds in Finland were originally explained in terms of climatic variation. More recently, however, this view has changed and human activities in breeding habitats are now considered to be partly responsible (186).

Drainage of wetlands, regulation of watercourses (particularly the construction of reservoirs that flood marshes), forest clearance, and dramatic changes in ecological conditions brought about by modern agricultural practices are listed by Hilden (186) as being responsible for changes in the breeding populations of many species of shorebirds in Finland. These changes are probably occurring in many areas of Siberia as well. Interestingly enough, some changes can be advantageous for certain species. Forest clearance in Finland has extended the area of open tundra-like habitat, which has been colonised by Golden Plover (*Pluvialis apricaria*) and Whimbrel. In Japan, forest clearance in mountain areas has created additional breeding habitat for Latham's Snipe, although urban-industrial development has greatly depleted the area of lowland breeding habitat (146). In south-eastern USSR, modern agricultural methods, drainage of swamps and hunting have resulted in a decline in the population of Eastern Curlew ((338). Over a similar period a decline in this species has been noted in south-eastern Australia at the extremities of its non-breeding range; in Tasmania and South Australia (63). The decline in Australia may be due more to what is happening in the USSR than to what is happening in Australia. Generally, changes in the breeding habitat of shorebirds lead to a decline in the diversity and numbers of breeding shorebirds (186).

## CONCLUSION

The five-year project reported in this book has shown that Australia holds almost the entire world populations of many species and races of shorebirds. Little was known of them before this study, showing that discoveries about the natural world are still possible in these times of high technology.

Human activities have already led to changes in the population and distribution of shorebirds in Australia. Fortunately, not all these activities have been detrimental, and no species has yet become extinct. That some changes have been beneficial for certain species shows that shorebirds can adapt in some situations and provides some cause for optimism, but not for complacency. A number of important shorebird habitats identified in this study are threatened, and others still remain to be formally protected from future degradation.

To the north, in Asia, the picture is less clear. Alteration and destruction of coastal and other wetland ecosystems are proceeding rapidly with consequent deleterious effects on the populations of migratory shorebirds that use them. This could continue in the absence of any detailed knowledge of the location of important shorebird habitats. New and exciting discoveries are yet to be made in Asia. Only by knowing what shorebirds require will governments in Asia and the Pacific have the basic information needed to develop effective conservation plans for shorebirds. As migrants pass through many countries, the development of those plans requires international co-operation.

It is hoped that this book will stimulate an interest in shorebird biology and encourage further research and conservation in Australia and Asia. Only by understanding the life-story of shorebirds will they continue to fly unhindered in impressive numbers along one of the world's most important bird migration routes.

# BIBLIOGRAPHY

1. AAGARD, C.J. 1930. The Common Birds of Bangkok. Copenhagen: Backhausen.

2. ALCASID, G.L. 1969. Northern Philippines. In MAPS Annual Report, 1968. H. E. McClure (ed.). San Francisco: US Army R&D Group, Far East.

3. ALCORN, R. 1985. 50 000 Banded Stilts. RAOU Newsletter 66:4.

4. ALCORN, R. pers. comm.

5. ALI, S., and S. D. RIPLEY. 1980. Handbook of the Birds of India and Pakistan. Vol. 2. 2nd ed. Delhi: OUP.

6. ALLAN, P. 1967. VORG Spur-winged Plover Survey. Aust. Bird Bander 5:30−2.

7. AMERICAN ORNITHOLOGISTS' UNION. 1983. Check-list of North American Birds. AOU. 6th ed.

8. AMERSON, A.B. 1969. Ornithology of the Marshall and Gilbert Islands. Atoll Res. Bull. 127:1−348.

9. AMIET, L. 1957. A wader survey of some Queensland coastal localities. Emu 57:236−54.

10. ANDERSON, D.A. 1981. Observations of Birds at Ujelang and other northern Marshall Islands atolls. Micronesica 17:198−212.

11. ANON. 1978. Report of the Bird Migration Research Centre. April 1973−March 1974. Tokyo: Yamashina Institute for Ornithology.

12. ANON. 1979. Report of the Bird Migration Research Centre. 1 February 1978 − 31 January 1979. Tokyo: Yamashina Institute for Ornithology.

13. ARMSTRONG, J.S. 1932. Handlist to the Birds of Samoa. London: John Bale, Sons & Danielsson Ltd.

14. ASH, J.S. 1984. Bird Observations on Bali. Bull. Brit. Orn. Club 104:24−35.

15. AUSTRALIAN BIRD BANDING SCHEME. pers. comm.

16. AXELRAD, D.M., G. C. B. POORE, G. H. ARNOTT, J. BAULD, R. R. C. EDWARDS and N. J. HICKMAN. 1981. The effects of treated sewage discharge on the biota of Port Phillip Bay, Victoria, Australia. In B. J. NEILSON and L. E. CRONIN (eds), Estuaries and Nutrients. Humana Press.

17. BADMAN, F.J., and I. A. MAY. 1983. Waders in Northern South Australia, S. Aust. Orn. 29: 29−39.

18. BAKER, R.H. 1951. The avifauna of Micronesia, its origin, evolution and distribution. University of Kansas Publ. Mus. Nat. Hist. 3: 1−359.

19. BAMFORD, M.T. 1983. Waders at Useless Loop in September 1981. Stilt 4: 13−14.

20. BAMFORD, M.T. pers. comm.

21. BARKLA, J., R. LOYN and KINHILL PLANNERS, 1980 Study of Waders at Point Wilson. Melbourne: ICI Australia Ltd.

22. BARTER, M.A. 1984. Weight variations and migration strategy of Curlew Sandpiper, *Calidris ferruginea*, wintering in Tasmania. Occ. Stint 3: 7−18.

23. BARTER, M.A. pers. comm.

24. BATHAM, J., and A. BATHAM. 1973. Field notes on birds observed on the 'motus' of two atolls (Penryhn and Suwarrow) in the northern Cook Islands, 1968. Notornis 20: 97−101.

25. BAYLISS-SMITH, T.P. 1972. The Birds of Ontong Java and Sikaiana, Solomon Islands. Bull. Brit. Orn. Club 92: 1−10.

26. BAYLY, I.A.E., and W. D. WILLIAMS. 1973. Inland Waters and Their Ecology. Melbourne: Longman-Cheshire.

27. BENNETT, S., *et al.* (in prep.) Atlas of Victorian Birds.

28. BENT, A.C. 1962. Life Histories of North American Shore-Birds. Parts 1 & 2. New York: Dover Publications Inc.

29. BERNEY, F.L. 1904. North Queensland notes on some migratory birds. Emu 4: 43−7.

30. BERTRAM, B.C.R. 1978 Living in groups: predators and prey. In J. R. Krebs and N. B. Davies (eds), Behavioural Ecology: an evolutionary approach. Oxford: Blackwell.

31. BERULDSEN, G.R. 1972. Unusual feeding method by Stilts and Avocets. Aust. Bird Watcher 4: 129−30.

32. BERULDSEN, G.R. 1972. Notes on waders in South Australia. Aust. Bird Watcher 4: 147−50.

33. BINGHAM, P. 1977. Birds of Heron Island, Queensland, October 4−11, 1975. Aust. Bird Watcher 7: 99−100.

34. BIRD OBSERVERS' ASSOCIATION OF TASMANIA. 1982. Birds and their habitats in the South Arm area. Occ. Stint 1:38−47.

35. BIRD, E.C.F. 1962. The utilisation of some Australian coastal lakes. Aust. Geogr. 8:199−206.

36. BISHOP, D. pers. comm.

37. BLACKSHAW, K.T. 1978. Notes and observations of birds in northern Taiwan, August 1973−May 1975. Quart. J. Taiwan Mus. 31: 297−316.

38. BLAKERS, M., S. J. J. F. DAVIES and P. N. REILLY. 1984. The Atlas of Australian Birds. Melbourne: Melbourne University Press.

39. BOEHM, E. F. 1960. Notes on some South Australian waders. Part 1. Emu 60: 211−18.

40. BOEHM, E. F. 1964. Notes on some South Australian waders. Emu 64: 276−82.

41. BRANSBURY, J. 1983. A Study of the Hooded Plover in the South-east of South Australia. Final Report to the Department of the Environment. June 1982−May 1983. Adelaide.

42. BRANSBURY, J. 1985. Waders of littoral habitats in south-eastern South Australia. S.Aust.Orn. 29:180−7.

43. BULL, P.C. 1948. Field notes on waders in the south-west Pacific with special reference to the Russell Islands. Emu 47:165−76.

44. BURBIDGE, A.A., and P. J. FULLER. 1982. Banded Stilt breeding at Lake Barlee, Western Australia. Emu 82: 212−16.

45. BURTON, P.J.K. 1974. Feeding and the Feeding Apparatus in Waders: a study of anatomy and adaptations in

the Charadrii. London: Trustees of the British Museum (Natural History).

46. CALDER, W.A. 1974. Consequences of body size for avian energetics. In R. A. Paynter (ed.), Avian Energetics. Cambridge, Massachusetts: Nuttall Ornithological Club.

47. CARRUTHERS, R.K. 1966. Waders in the Gulf Country. Aust. Bird Watcher 2:211−14.

48. CARRUTHERS, R.K. 1968. Notes on an influx of Oriental Prantincoles at Mount Isa. Emu 68: 216−17.

49. CARTER, M.J. 1983. Visible wader migration into northern Australia. Stilt 4:18−19.

50. CARTER, M.J. pers. comm.

51. CARTER, M.J., G. BOWKER and A. C. ISLES. 1976. Lage flock of Sanderling at Port Fairy, Victoria. Aust. Bird Watcher 6: 173−7.

52. CARTER, M.J., and F. T. H. SMITH. 1968. A Victorian sight record of the Ruff (*Philomachus pugnax*). Aust. Bird Watcher 3: 100−2.

53. CHAFER, C. 1984. Northern distribution of the Hooded Plover in NSW. Stilt 5: 27.

54. CHALMERS, M.L. 1986. An Annotated Checklist of the Birds of Hong Kong. Hong Kong: Hong Kong Bird Watching Society.

55. CHANDLER, L.G. 1959. The first banding of birds in Victoria. Aust. Bird Watcher 1: 11−14.

56. CHANG, James Wan-Fu. 1980. A Field Guide to the Birds of Taiwan.

57. CHAPLIN, G. pers. comm.

58. CHENG Tso Hsia. 1976. A Distribution List of Chinese Birds. 2nd ed. Peking: Science Press.

59. CLAPP, R.B. 1968. Additional new birds from the Phoenix and Line islands. Ibis 110: 573−5.

60. CLAPP, R.B., and F. C. SIBLEY. 1967. New records of birds from the Phoenix and Line islands. Ibis 109: 1222−5.

61. CLELAND, J.B. 1911. Examination of contents of stomachs and crops of Australian Birds. Emu 11: 79−95.

62. CLELAND, J.B., J. H. MAIDEN, W. W. FROGGATT, E. W. FERGUSON and C. T. MUSSON. 1918. The Food of Australian Birds: an investigation into the character of the stomach and crop contents. Science Bulletin. No. 15. Department of Agriculture, NSW.

63. CLOSE, D.H., and O. M. G. NEWMAN. 1984. The decline of the Eastern Curlew in south-eastern Australia. Emu 84: 38−40.

64. COLEMAN, A. 1973. A Ruff at McGrath's Hill, Windsor. Birds 8: 43.

65. CONDON, H.T. 1975. Checklist of the Birds of Australia. Part 1. Melbourne: RAOU.

66. CONDON, H.T., and A. R. McGILL. 1960. A Field Guide to the Waders. 2nd ed. Melbourne: Bird Observers' Club.

67. CONNOR, M.A. 1981. Common Sandpipers perched four metres up a tree. Aust. Bird Watcher 9: 32.

68. CONNORS, P.G. 1983. Taxonomy, distribution and evolution of Golden Plovers (*Pluvialis dominica* and *Pluvialis fulva*). Auk 100: 607−20.

69. CONSIDINE, M.T. 1979. The Feeding Behaviour of the Sooty Oystercatcher (*Haematopus fuliginosus*) on Rocky Shores in Victoria. Unpublished Honours Thesis, Dept of Zoology, Monash University. 105 pp.

70. CONSIDINE, M.T. 1982. Notes on the seasonal movements of the Sooty Oystercatcher (*Haematopus fuliginosus*) in southern Victoria. VWSG Bulletin 5: 11−15.

71. COOK, J.E., and J. W. WAUGH. 1977. A Baird's Sandpiper at Botany Bay. Aust. Birds 12: 32−5.

72. COOPER, R. 1982. Little Musselroe Bay. Occ. Stint 1: 59−64.

73. COOPER, R.M. 1973. Pectoral Sandpiper at Dangar's Lagoon, Uralla. Birds 8: 44.

74. COOPER, R.P. 1964. The Oriental Pratincole near Melbourne. Aust. Bird Watcher 2: 111−12.

75. COOPER, R.P. 1970. The Sanderling on Wilson's Promontory. Aust. Bird Watcher 3: 247−51.

76. CORBEN, C. 1972. Oriental Pratincoles in south-east Queensland. Sunbird 3: 6−8.

77. CORNELIUS, J. 1985. Ruddy Turnstone predation on Sooty Tern eggs. Stilt 6: 333−4.

78. CORNELIUS, J. pers. comm.

79. CORRICK, A.H. 1981. Wetlands of Victoria II. Wetlands and waterbirds of South Gippsland. Proc. Roy. Soc. Vict. 92: 187−200.

80. CORRICK, A.H. 1982. Wetlands of Victoria III. Wetlands and waterbirds between Port Phillip Bay and Mount Emu Creek. Proc. Roy. Soc. Vic. 94: 69−87.

81. CORRICK, A.H., and S. J. COWLING. 1975. A survey of the wetlands of Kerang, Victoria. Fisheries and Wildlife Paper (Victoria). No. 5.

82. CORRICK, A.H., and F. I. NORMAN. 1980. Wetlands and waterbirds of the Snowy River and Gippsland Lakes catchment. Proc. Roy. Soc. Vict. 91: 1−15.

83. COX, J.B. 1973. A further occurrence of the Red-necked Phalarope in South Australia. S.Aust.Orn. 26: 116−17.

84. COX, J.B. 1974. Some birds of north-eastern Eyre Peninsula, South Australia. S.Aust.Orn. 26: 142−4.

85. CRAMP, S. (ed.). 1983. Handbook of the Birds of Europe, the Middle East and North Africa. The Birds of the Western Palearctic. Vol. 3. Waders to Gulls. London, Oxford, OUP.

86. CRAWFORD, D.N. 1971. First records of Redshank in Australia. Emu 72: 112.

87. CRAWFORD, D.N. 1972a. First Australian record of Asiatic Dowitcher. Emu 72: 112−13.

88. CRAWFORD, D.N. 1972b. Birds of the Darwin area, with some records from other parts of the Northern Territory. Emu 72: 131−48.

89. CRAWFORD, D.N. 1975. Numbers of waders and waterbirds in relation to salinity in the saltfields of Adelaide, South Australia. S.Aust.Orn. 26: 193−5.

90. CRAWFORD, D.N. 1978. Notes on Little Curlew on the subcoastal plains, Northern Territory. Aust. Bird Watcher 7: 270−2.

91. CURRY, P.J. 1979a. Long-toed Stints, Sanderling and other waders at Lake Violet, central Western Australia, in mid-summer, 1977−78. W. Aust. Nat. 14: 109−13.

92. CURRY, P.J. 1979b. A Ruff, *Philomachus pugnax*, at Bibra Lake, June 1978. W. Aust. Nat. 14: 113−15.

93. CURRY, P.J. 1981. The Forrestdale phenomenon. Stilt 1:5.

94. CURRY, P.J. 1984. Long-toed stints on the way out? RAOU Newsletter 60:11.

95. DANN, P. 1979. Food robbing of Bar-tailed Godwits by Silver Gulls in Westernport Bay, Victoria. Corella 3:84−5.

96. DANN, P. 1980. The Status of the Waders of Westernport Bay in Relation to Other Areas in Coastal Victoria. Melbourne: Ministry for Conservation.

97. DANN, P. 1981a. Resource allocation among three congeneric species of sandpiper. Stilt 1:3.

98. DANN, P. 1981b. The evidence for feeding ground fidelity of waders in southern Victoria. VWSG Bulletin 3: 13−14.

99. DANN, P. 1981c. Breeding of the Banded and Masked Lapwings in southern Victoria. Emu 81: 121−7.

100. DANN, P. 1983. Feeding behaviour of four species of calidridine sandpiper at Lake Reeve, East Gippsland. VWSG Bulletin 7:1−6.

101. DANN, P. 1986. The Ecology of Birds Feeding in Intertidal Areas in Southern Victoria. Unpublished Ph.D. thesis, Dept of Zoology, University of Melbourne.

102. DANN, P. pers. comm.

103. DANN, P., and J. DORSEY (in prep.). The feeding ecology of three congeneric species of sandpiper in southern Victoria.

104. DAVIES, J.L. 1972. Geographic Variation in Coastal Development. Edinburgh: Oliver & Boyd.

105. DAVIES, N.B. 1976. Food, flocking and territorial behaviour of the Pied Wagtail (Motacilla alba yarrellii) in winter. J. Animal Ecology 45:235−54.

106. DAY, F.A.G. 1984. A Redshank in South Australia. S.Aust.Orn. 29: 115−17.

107. DEIGNAN, H.G. 1946. Correction and additions to the published records of Siamese birds. Auk 63:243−5.

108. DEIGNAN, H.G. 1966. Birds of the Arnhem Land expedition. In R. L. Specht (ed.). Records of the American-Australian Scientific Expedition to Arnhem land. Melbourne: Melbourne University Press.

109. DELACOUR, J., and P. JABOUILLE. 1925. Researches ornithologiques dans la province de Quangtri (centre Annam) et quelques autres régions de l'Indochine Français. Archives d'histoire naturelle. 1:1−197.

110. DELACOUR, J., and E. MAYR. 1946. Birds of the Philippines. New York: Macmillan.

111. DEMENT'EV, G.P., and N. A. GLADKOV (eds). 1969. Birds of the Soviet Union. Vol. 3. Jerusalem: Israel Programme for Scientific Translations.

112. DICK, W.J.A. 1979. Results of the WSG project on the spring migration of Siberian Knot, Calidris canutus, 1979. WSG Bulletin 27: 8−13.

113. DICK, W.J.A., M. W. PIENKOWSKI, M. WALTNER & C. D. T. MINTON. 1976. Distribution and geographical origins of Knot, Calidris canutus, wintering in Europe and Africa. Ardea 64:22−47.

114. DOMM, S. 1977. Sea-birds and waders of the Lizard Island Area. Sunbird 8:1−8.

115. DOMM, S., and H. F. RECHER. 1973. Birds of One Tree Island with notes on their yearly cycle and feeding ecology. Sunbird 4:63−86.

116. DOORNBOS, G. 1984. Nakdong Estuary Barrage and Reclamation project: preliminary results on the ornithological part of the environmental programme. NEDECO: Busan, Korea.

117. DOROGOI, I.V. 1982 [Materials on the Biology of the Knot on Wrangel Island.] Vestnik of Zoology 5:65−9. (In Russian.)

118. DOUGHTY, C.J., and M. J. CARTER. 1977. American race (dominica) of the Eastern Golden Plover (Pluvialis dominica) in Westernport Bay, Victoria. Aust. Bird Watcher 7:23−4.

119. DRAFFAN R.D.W., S. T. GARNETT, and G. MALONE. 1983. Birds of Torres Strait. Emu 83:207−34.

120. DU PONT, J.E. 1976. South Pacific Birds. Delaware Museum of Natural History Monograph Series. No. 3.

121. ECKERT, J. 1965. The Long-toed Stint, Calidris subminuta: a new record for South Australia. Emu 65: 125−8.

122. ECKERT, J. 1967. The Pectoral Sandpiper in South Australia. S.Aust.Orn. 24:135−6.

123. ECKERT, J. 1969. Red-necked Phalarope in South Australia. Emu 69:184−6.

124. EDGAR, A.T. 1974. Farewell Spit, March 1974. Notornis 21:250−9.

125. EDWARDS, E. 1982. October wader count at Roebuck Bay, Broome, 1981. Stilt 2:27−8.

126. ELKS, R., and B. MORGAN. 1969. The Little Whimbrel in south-eastern Queensland. Qld Bird Notes 5:1−2.

127. ENGBRING, J., and R. P. OWEN. 1981. New bird records for Micronesia. Micronesica 17:186−92.

128. ESCOTT, C.J., and D. A. HOLMES. 1980. The avifauna of Sulawesi, Indonesia: faunistic notes and additions. Bull. Brit. Orn. Club 100:189−94.

129. ETCHACOPAR, R.D., and F. HUE. 1978. Les oiseaux de Chine, de Mongolie et de Coree, non-passereaux. Papeete: Les éditions du Pacific.

130. EVANS, P.R. 1964. Wader measurements and wader migration. Bird Study 11:23−38.

131. EVANS, P.R. 1974. Exploratory investigation of feeding ecology and behaviour of shorebirds in Westernport Bay. A report on work carried out between September and mid-November, 1974. Westernport Environmental Study, Project W38/74. Melbourne.

132. EWART, A. 1973. Bird observations at the Plantation Creek Estuary, Ayr, North Queensland. Sunbird 4: 58−61.

133. EY, A. 1984. Queensland records of Chinese Snipe (Gallinago megala). Sunbird 14:14−15.

134. FALLA, R.A., R. B. SIBSON and E. G. TURBOTT. 1981. The New Guide to the Birds of New Zealand. Auckland, London: Collins.

135. FAVOLORO, N.J. 1944. Notes on the two resident Victorian Plovers. Emu 43: 145−53.

136. FAVOLORO, N.J. 1949. Notes on the Red-capped Dotterel in inland localities. Emu 49:13−18.

137. FENNELL, C.M., and B. F. KING. 1964. New occurrences and recent distributional records of Korean birds. Condor 66:239−46.

138. FINCH, B.W. 1979. The Palearctic wader presence in the Port Moresby Area, 1979. PNG Bird Soc. Newsletter 162:4−17.

139. FINCH, B.W. 1980. Bensbach. PNG Bird Soc. Newsletter 171−2: 10−33.

140. FINCH, B.W. 1982. Annotated list of the birds recorded in the zones: Vanapa-Veimouri and Kanosa−Cape Suckling. PNG Bird Soc. Newsletter 195−6:29−39; 197−98:11−14.

141. FINCH, B.W., and J. B. COX. 1974. Another Oriental Pratincole in South Australia. S.Aust.Orn. 26:166.

142. FISCHER, W. von. 1983. Ein Beitrag zum Vorkommen, Durchzug und zur Ubersommerung von Limikolen (Charadriiformes) in Vietnam. Bietr. Vogelkd, Leipzig 29:297−305.

143. FLETCHER, A.J.F. pers. comm.

144. FLINT, V.E., R. L. BOEHME, Y. V. KOSTIN and A. A. KUZNETSOV. 1984. A Field Guide to Birds of the USSR. Princeton: Princeton University Press.

145. FOREST, B.S. 1982. More waders that perch. Aust. Bird Watcher 9:159.

146. FRITH, J.J., F. H. J. CROME and B. K. BROWN. 1977. Aspects of the biology of the Japanese Snipe, *Gallinago hardwickii*. Austr. J. Ecology 2:341−68.

147. GALLOWAY, R.W. 1982. Distribution and physiographic patterns of Australian mangroves. In B. F. Clough (ed.), Mangrove Ecosystems in Australia. Canberra: ANU Press.

148. GARNETT, S.T. 1983. Report on the fifth aerial survey of migrating wading birds between Weipa and Millingimbi, 9−13 February 1983. Stilt 4: 15−17.

149. GARNETT, S.T. 1985. Nesting behaviour of the Bush Thick-knee. Stilt 7:24−5.

150. GARNETT, S.T. pers. comm.

151. GARNETT, S.T., and I. CARRUTHERS. 1982*a*. Error in aerial surveys of wading birds. Stilt 3:5−7.

152. GARNETT, S.T., and I. CARRUTHERS. 1982*b*. Report on the second aerial survey of north-eastern Australia, February 1982. Stilt 3:18−20.

153. GARNETT, S.T., and C. D. T. MINTON. 1985. Notes on the movements and distribution of Little Curlew, *Numenius minutus*, in northern Australia. Aust. Bird. Watcher 11:69−73.

154. GARRETT, C. 1980. Ruff sighted on Eyre Peninsula. S.Aust.Orn. 28:112.

155. GIBSON, D. 1981. Migrant birds at Shemya Island, Aleutian Islands, Alaska. Condor 83:65−77.

156. GILL, H.B. 1970. Birds of Innisfail and hinterland. Emu 70: 105−16.

157. GLENISTER, A.G. 1971. The Birds of the Malay Peninsula, Singapore and Penang. Kuala Lumpur: OUP.

158. GLOVER, B. 1954. Recent observations on South Australian migratory waders. Emu 54:51−7.

159. GLOVER, B. 1958. An inland record of the Common Sandpiper. Emu 58:132.

160. GLUSTCHENKO, Y.N., and Y. B. SHIBNEV. 1979. [New materials on the Asiatic Dowitchers on the Khanka Lake] pp. 67−74. In Bird Biology of Southern Far East USSR. Vladivostok. (In Russian.)

161. GORE, M.E.J. 1968. A checklist of the birds of Sabah, Borneo. Ibis 110:165−96.

162. GORE, M.E.J., and P. WON. 1971. The Birds of Korea. Seoul: Royal Asiatic Society.

163. GOSPER, D.G. 1981. Survey of birds on floodplain estuarine wetlands on the Hunter and Richmond rivers in northern NSW. Corella 5:1−18.

164. GOSS-CUSTARD, J.D. 1976. Variation in the dispersion of Redshank, *Tringa totanus*, on their winter feeding grounds. Ibis 119: 257−63.

165. GOSS-CUSTARD, J.D. 1977. Predator responses and prey mortality in Redshank *Tringa totanus* (L.), and a preferred prey, *Corophium volutator* (Pallas). J. Animal Ecology 46: 21−35.

166. GOSS-CUSTARD, J.D., R. E. JONES and P. E. NEWBERY. 1977. The ecology of the Wash. I. Distribution and diet of wading birds (Charadrii). J. App. Ecology 14:681−700.

167. GREEN, G.H., J. J. D. GREENWOOD and C. S. LLOYD. 1977. The influence of snow conditions on the date of breeding of wading birds in north-east Greenland. J. Zool. Lond. 183:311−28.

168. GREEN, R. 1956. Notes on the inland breeding of the Red-capped Dotterel. Emu 56:140−3.

169. GRIFFIN, D.R. 1964. Bird Migration. London: Heinemann.

170. GUARD, R., and S. T. GARNETT. 1982. Report on the first aerial survey of waders in north-eastern Australia, December 1981. Stilt 3:16−17.

171. HACHISUKA, M. 1931. The Birds of the Philippine Islands. London: Witherby.

172. HACHISUKA, M., and T. UDAGAWA. 1951. Contributions to the ornithology of Formosa. Part 2. Quart. J. Taiwan Mus. 4:1−180.

173. HADDEN, D. 1981. Birds of the North Solomons. Wau Ecology Institute Handbook. No. 8.

174. HAILS, C.J. (unpubl.). The Serangoon Estuary Nature Reserve. Singapore.

175. HALE, W.G. 1980. Waders. Collins New Naturalist Series. London: Collins.

176. HARDING, E. 1982. Birds of Nissan and Pinipel Island, North Solomons Province, PNG Bird Soc. Newsletter 195−6: 4−12.

177. HARDY, A.R., and C. D. T. MINTON, 1980. Dunlin migration in Britain and Ireland. Bird Study 27:81−92.

178. HARRINGTON, B.A., and D. C. TWICHELL. 1982. Untying the enigma of the Red Knot. Living Bird Quarterly 1:4−7.

179. HEMMINGSEN, A.M., and J. A. GUILDAL. 1968. Observations on birds in north-eastern China, especially the migration at Pei-Tai-Ho beach. 2. Special Part. Spolia Zool. Mus. Huan 28:1−326.

180. HENDERSON, D. 1982. The Banded Stilt in Tasmania. Tas. Bird Rep. 11:14−15.

181. HENDERSON, D. pers comm.

182. HENSHAW, H.W. 1910. Migration of the Pacific Plover to and from the Hawaiian Islands. Auk 27: 245−62.

183. HERMES, N. 1980. Endangered animals. Parks & Wildlife. Sydney: National Parks and Wildlife Service.

184. HERON, S.J. 1978. Waders of the New Guinea Region. PNG Bird Soc. Newsletter 149−50:8−15.

185. HERTOG, A. pers. comm.

186. HILDEN, O. 1983. Recent population changes of waders in Finland and their causes. In P. R. Evans, H.

Hafner and P. L'Hermite. (eds), Shorebirds and Large Water-birds Conservation. Brussels: Commission of the European Communities.

187. HINDWOOD, K.A. 1948. The Broad-billed Sandpiper in Australia. Emu 48:155–7.

188. HINDWOOD, K.A. 1949. Waders at Boat Harbour, near Sydney. Emu 48:323–4.

189. HINDWOOD, K.A. 1950. The Upland Plover, or Bartram's Sandpiper in Australia. Emu 50:91–6.

190. HINDWOOD, K.A., and E. S. HOSKIN. 1954. The waders of Sydney (County of Cumberland). Emu 54:217–55.

191. HINDWOOD, K.A., K. KEITH and D. L. SERVENTY. 1963. Birds of the South West Coral Sea. Div. of Wildl. Res. Tech. Pap. No. 3. Melbourne, CSIRO.

192. HOBBS, J.N. 1958a. A Pectoral Sandpiper and other northern waders in south-western New South Wales. Emu 58:56–9.

193. HOBBS, J.N. 1958b. A further record of the Pectoral Sandpiper in south-western New South Wales and some wader notes. Emu 58:412–14.

194. HOBBS, J.N. 1959. Pectoral Sandpiper in NSW. Emu 59:210.

195. HOBBS, J.N. 1961. The birds of south-west New South Wales. Emu 61:21–55.

196. HOBBS, J.N. 1972. Breeding of the Red-capped Dotterel at Fletcher's Lake, Dareton, NSW. Emu 72:121–5.

197. HOBBS, J.N. 1973. The Oriental Pratincole in NSW. Aust. Bird Watcher 5:1–4.

198. HOBBS, J.N. 1974. Western Sandpiper in New South Wales. Aust. Birds 9:21–4.

199. HOLMES, G. 1972. The Ruff in New South Wales. Aust. Bird Watcher 4:233–4.

200. HOLMES, G. 1979. First record of Red-necked Phalarope in New South Wales. Aust. Birds 13:75–6.

201. HOLMES, R.T. 1966. Feeding ecology of the Red-backed Sandpiper (Calidris alpina) in arctic Alaska. Ecology 47:32–45.

202. HOLMES, R.T., and F. A. PITELKA. 1964. Breeding behaviour and taxonomic relationships of the Curlew Sandpiper. Auk 81:362–79.

203. HOLMES, R.T., and F. A. PITELKA. 1968. Food overlap among co-existing sandpipers on northern Alaskan tundra. Systematic Zoology 17:305–18.

204. HOLYOAK, D.T. 1976. Records of waders in the Cook Islands. Notornis 23:1–3.

205. HOOGERWERF, A. 1967. Notes on the island of Bawean (Java Sea) with special reference to the birds. Nat. Hist. Bull. Siam Soc. 22:16–103.

206. HOOGERWERF, A. 1969. On the ornithology of the Rhino Sanctuary Udjung Kulon in West Java (Indonesia). Nat. Hist. Bull. Siam Soc. 23:9–65.

207. HOOGERWERF, A. 1971. On a collection of birds from the Vogelkop, near Manokwari, north-western New Guinea. Emu 71:1–12.

208. HOPKINS, N. 1947. Birds of Townsville. Emu 47:331–47.

209. HOSKIN, E.S. 1972. New records of the Pectoral Sandpiper. Birds 7:17–18.

210. HUMPHRIES, C. 1984. An observation of a Pectoral Sandpiper near Ulladulla, NSW. Aust. Birds 19:9–10.

211. JAENSCH, R.P. 1982. Little Ringed Plover at Bool Lagoon. S. Aust. Orn. 28:201–4.

212. JAENSCH, R.P. 1983. The Asian Dowitcher in north-western Australia. Stilt 4:2–5.

213. JAENSCH, R.P. 1986. Waterbird Usage of Wetland Nature Reserves in South-western Australia. Perth: WA Dept of Conservation and Land Management.

214. JAENSCH, R.P. pers. comm.

215. JENKINS, J.M. 1978. Two new bird records for Guam. Micronesica 14:361.

216. JENKINS, J.M. 1981. Seasonality and relative abundance of Guam shorebirds. Micronesica 17:181–4.

217. JESSOP, A.E. pers. comm.

218. JOHNSON, O.W. 1973. Reproductive condition and other features of shorebirds resident at Enewetak Atoll during the boreal summer. Condor 75:336–43.

219. JOHNSON, O.W. 1977. Plumage and molt in shorebirds summering at Enewetak Atoll. Auk 94:222–30.

220. JOHNSON, O.W. 1979. Biology of shorebirds summering on Enewetak Atoll. Stud. Avian. Biol. 2:193–205.

221. JOHNSTON, D.W., and R. W. McFARLANE. 1967. Migration and bioenergetics of flight in the Pacific Golden Plover. Condor 69:156–68.

222. JONES, E.L. 1983. A preliminary note on wader counts at Stockyard Point, Westernport Bay, Victoria, 1973–1983. VWSG Bulletin. 7:21–33.

223. JONES, E.L. 1984. Shallow Inlet, Victoria, as a wader resort. VWSG Bulletin. 8:26–39.

224. JONES, E.L. 1985. The influence of tide-height on waders' choice of roost sites. VWSG Bulletin. 9:2–6.

225. JONES, J. 1945. The Banded Stilt. Emu 45:1–36, 110–18.

226. KEAST, J.A. 1949. Field notes on the Grey-tailed Tattler. Rec. Aust. Mus. 22:207–11.

227. KIESER, J.A., and F. T. H. SMITH. 1982. Field identification of the Pectoral Sandpiper, Calidris melanotos. Aust. Bird Watcher 9:137–9.

228. KIKKAWA, J. 1976. The birds of the Great Barrier Reef. In Biology and Geology of Coral Reefs. Vol. 3. New York: Academic Press.

229. KING, B., M. WOODCOCK and E. C. DICKINSON. 1975. A Field Guide to the Birds of South-east Asia. London: Collins.

230. KINSKY, F.C. 1970. Annotated Checklist of the Birds of New Zealand. Wellington: Reed.

231. KINSKY, F.C., and J. C. YALDWIN. 1981 The Bird Fauna of Niue Island, South West Pacific, with Special Notes on the White-tailed Tropic Bird and Golden Plover. Nat. Mus. NZ Misc. Series. No. 2.

232. KITSON, A. 1979. Notes on the waders of Mongolia. WSG Bulletin. 27:34–5.

233. KLAPSTE, J. 1977. An occurrence of the Buff-breasted Sandpiper (Tryngites subruficollis) near Swan Hill, Victoria. Aust. Bird Watcher 7:56–9.

234. KLAPSTE, J. 1978. Bird observations at Alice Springs sewage farm, NT. Aust. Bird Watcher 7:148–51.

235. KLAPSTE, J. 1984. Occurrence of the Long-billed Dowitcher on Bali, Indonesia, and other observations. Aust. Bird Watcher 10:186−95.

236. KOLICHIS, N. 1976. New breeding records of the Banded Stilt in Western Australia. W. Aust. Nat. 13:114−19.

237. KOSLOWA, E. V. 1932. Birds of south-west Transbaikalia, northern Mongolia and central Gobi. Part 3. Ibis 13(2): 576−96.

238. KOSLOWA, E.V. 1966. Discovery of the nesting of the Sharp-tailed Sandpiper. Emu 66:252.

239. LABUTIN, Y.V., V. V. LEONOVITCH and B. N. VEPRINTSEV. 1982. The Little Curlew, *Numenius minutus*, in Siberia. Ibis 124: 302−19.

240. LACK, D. 1963. Migration across the southern North Sea studied by radar. Part 4. Ibis 105:1−54.

241. LAND CONSERVATION COUNCIL, VICTORIA. 1977. Final Recommendations − Melbourne Study area. Melbourne: Land Conservation Council.

242. LANE, B.A. 1981. A Victorian record of the Ringed Plover. Geel. Nat. 18:8−11.

243. LANE, B.A. 1982. A preliminary review of the status of Pied and Sooty Oystercatchers in Victoria. VWSG Bulletin. 5:6−10.

244. LANE, B.A. 1986. The sub-species of Mongolian Plover (*Charadius mongolus*) in Australia. Stilt 8:14−17.

245. LANE, B.A., and B. S. FOREST. 1984. Preliminary results from banding Latham's Snipe (*Gallinago hardwickii*) in southern Victoria. VWSG Bulletin 8:2−11.

246. LANE, B.A., and A. E. JESSOP. 1983. Distribution and habitat of the Red-capped Plover in Victoria. VWSG Bulletin 7:35−40.

247. LANE, B.A., and A. E. JESSOP. 1985*a*. Report on the 1985 north-west Australian wader studies expedition. Stilt 6:2−16.

248. LANE, B.A., and A. E. JESSOP. 1985*b*. Tracking of migrating waders in north-western Australia using meteorological radar. Stilt 6:17−28.

249. LANE, B.A., R. H. LOYN and KINHILL PLANNERS. 1980. Avifauna study. ICI Point Wilson development, Progress Report, June to November 1979. Melbourne: ICI (Australia) Ltd.

250. LANE, B.A., and C. D. T. MINTON. (in prep.) Primary moult, weight and biometrics of Sharp-tailed Sandpipers (*Calidris acuminata*) in southern Victoria.

251. LANE, B.A., and C. D. T. MINTON. (in prep.) Birds of Lake Eyre in 1984.

252. LANE, B.A., M. SCHULZ and K. L. WOOD. 1984. Birds of Port Phillip Bay. Coastal Unit Technical Report. No. 1. 159 pp. Melbourne: Victorian Government Printer.

253. LANE, S.G. 1978. The Asian Dowitcher on the east coast of Australia. Sunbird 9:13.

254. LANE, S.G. 1985. The Hooded Plover − northern distribution in eastern Australia. Stilt 6: 35.

255. LA TOUCHE, J. 1931−34. A Handbook of the Birds of Eastern China. Vols I-II. London: Taylor & Francis.

256. LEA, A.M., and J. T. GRAY. 1935. The food of Australian birds. Part 1. Emu 34:275−92.

257. LEKAGUL, B., and E. W. CRONIN. 1974. Bird Guide of Thailand. Bangkok: Assoc. for the Conservation of Wildlife.

258. LIDDY, J. 1955. Waders at Mount Isa, Queensland. Emu 55:297−302.

259. LIDDY, J. 1959. The Australian Pratincole in north-west Queensland. Emu 59: 136−40.

260. LIEDEL, K. 1982. Verbreitung und okologie des step penschlammlaufers. Mitt. zool. Mus. Berlin. Bd 58, Suppl.: Ann Orn. 6:147−62.

261. LINCOLNE, G. 1970. Little Whimbrels. Bird Observer 459:3.

262. LINDSEY, A., and T. R. LINDSEY. 1978. A sight record of the Dunlin on the Cairns Waterfront. Sunbird 9:1−2.

263. LOWE, V.T. 1963. Observations on the Painted Snipe. Emu 62: 221−37.

264. LOYN, R.H. 1975. Report on the Avifauna of Westernport Bay. Project Report 4.4.8. Westernport Bay Environmental Study. Melbourne: Ministry for Conservation.

265. LOYN, R.H. 1978. A survey of birds in Westernport Bay, Victoria. 1973−74. Emu 78:11−19.

266. MA Yi-Ching. 1984. On the Waders (Charadrii) of Harbin Area, North-East China. Paper presented to 10th Asian Section Conference, International Council for Bird Preservation, Kandy, Sri Lanka.

267. McCLURE, H.E. 1974. Migration and Survival of the Birds of Asia. Bangkok: Applied Scientific Research Corporation of Thailand.

268. McCLURE, H. E., and P. LEELAVIT. 1972. Birds Banded in Asia during the MAPS Programme, by Locality, from 1963 through 1971. San Francisco: US Army R&D Group, Far East.

269. McCRIE, N. 1984. Further records of the Oriental Plover and a reassessment of some problems in field identification. S. Aust. Orn. 29:106−7.

270. MacDONALD, S.D., and D. F. PARMELEE. 1962. Feeding behaviour of the Turnstone in arctic Canada. Br. Birds 55:241−4.

271. McEVEY, A.R. 1963. The Ruff: an addition to the Australian list. Emu 63:35−9.

272. McGILL, A.R. 1960. The Little Whimbrel. Emu 60:89−94.

273. McGILL, A.R. 1969. The Ringed Plover in Australia. Aust. Bird Watcher 3:198−201.

274. MACGILLIVRAY, W. 1924. A contribution to the life-story of the Australian Pratincole. Emu 24: 81−5.

275. McGILP, J.N. 1923. Birds of Lake Frome district, South Australia. Emu 22:237−43, 274−87.

276. McGILP, J.N. 1934. The Nesting of the Painted Snipe (*Rostratula australis*). S. Aust. Orn. 12:167−9.

277. McGILP, J.N., and A. M. MORGAN. 1931. The nesting of the Banded Stilt (*Cladorhynchus leucocephalus*). S. Aust. Orn. 11:37−53.

278. MACKAY, R.D. 1970. The Birds of Port Moresby and District. Melbourne: Nelson.

279. McKEAN, J.L. 1976. A Dunlin sighting in South Australia. S. Aust. Orn. 27:101.

280. McKEAN, J.L. 1978. Some remarks on the taxonomy of Australasian oystercatchers, *Haematopus* spp. Sunbird 9:3−6.

281. McKEAN, J.L. 1980. A sight record of the Ringed Plover (*Charadrius hiaticula*) in the Northern Territory. Aust. Bird Watcher 8:236−7.

282. McKEAN, J.L. 1981. First record of the Stilt Sandpiper from Australasia. Notornis 28:48−9.

283. McKEAN, J.L. 1984. A Northern Territory sighting of the Baird's Sandpiper. Aust. Bird Watcher 10:169.

284. McKEAN, J., M. C. BARTLETT and C. H. PERRINS. 1975. New records from the Northern Territory. Aust. Bird Watcher 6:45−6.

285. McKEAN, J.L., I. J. MASON & L. W. O'CONNOR. 1975. Birds not previously recorded from Timor. Emu 75:62−4.

286. McKEAN, J.L., H. A. F. THOMPSON and J. A. ESTBERGS. 1976. Records of uncommon migrant waders near Darwin, NT. Aust. Bird Watcher 6:143−8.

287. MACKWORTH-PRAED, C.W., and C. H. B. GRANT. 1960. Birds of Eastern and North-Eastern Africa. 2nd ed. London: Longmans.

288. McLACHLAN, G.R., and R. LIVERSIDGE. 1978. Roberts' Birds of South Africa. Cape Town: John Voelcker Bird Book Fund.

289. MACLEAN, G.L. 1973. A review of the biology of the Australian desert waders, *Stiltia* and *Peltohyas*. Emu 73:61−70.

290. MACLEAN, G.L. 1976a. A field study of the Australian Pratincole. Emu 76:171−82.

291. MACLEAN, G.L. 1976b. A field study of the Australian Dotterel. Emu 76: 207−15.

292. MACLEAN, G.L. 1977. Comparative notes on Black-fronted and Red-kneed Dotterels. Emu 77: 199−207.

293. MACLEAN, S.F., Jnr. 1980. The detritus-based trophic system. In Brown, J., P. C, Miller, L. L. Tieszen and F. L. Bunnell (eds), An Arctic Ecosystem − the coastal tundra at Barrow, Alaska. Stroudsberg: Dowden, Hutchinson & Ross, Inc.

294. McNAMARA, J.A. 1976. Two further records of the Red-necked Phalarope. S. Aust. Orn. 27:142.

295. McNAMARA, J.A. 1980. Nocturnal feeding of the Inland Dotterel. Emu 80:39−40.

296. McNEIL, R., and F. CADIEUX. 1972. Numerical formulae to estimate flight range of some North American shorebirds. Bird Banding 43:107−13.

297. MARTINDALE, J.D. 1982. A Study of Wading Birds in Corner Inlet. Arthur Rylah Institute for Environmental Research. Tech. Rep. No. 4.

298. MARTINDALE, J.D. 1984. A Study of Estuarine Birds in the Lower Clarence River. National Parks and Wildlife Service, New South Wales.

299. MASTERS, J.R., and A. L. MILHINCH. 1974. Birds of the shire of Northam, about 100 km east of Perth, WA. Emu 74:228−44.

300. MATHEWS, G.M. 1909. On the birds of north-west Australia. Emu 9:1−16, 53−65.

301. MAYR, E. 1941. List of New Guinea Birds. New York: American Museum of Natural History.

302. MAYR, E. 1945. Birds of the South-west Pacific. New York: Macmillan.

303. MEAD, C.J. 1974. Bird Ringing. BTO Guide. No. 16. Tring: BTO.

304. MEAD, C.J., and B. R. WATMOUGH, 1976. Suspended moult of trans-Saharan migrants in Iberia. Bird Study 23:187−96.

305. MEDWAY, LORD, and D. R. WELLS. 1976. Birds of the Malay Peninsula. Vol. 5. London: H. F. & G. Witherby Ltd.

306. MELVILLE, D.S. 1980. Birds at Kai Tak Airport, Hong Kong. Hong Kong: Agriculture and Fisheries Dept.

307. MELVILLE, D.S. 1981. Spring measurements, weights and plumage status of *Calidris ruficollis* and *C. ferruginea* in Hong Kong. WSG Bulletin. 33:18−21.

308. MELVILLE, D.S. 1982. Sight record of the Eastern Curlew (*Numenius madagascariensis*) in central Thailand. Nat. Hist. Bull. Siam. Soc. 30:47−8.

309. MELVILLE, D.S. (unpubl.) A study of waders in the inner Gulf of Siam.

310. MELVILLE, D.S. pers. comm.

311. MELVILLE, D.S., and P. D. ROUND. 1982. Further records of the Asian Dowitcher, *Limnodromus semipalmatus*, from Thailand, with notes on its distribution and identification. Nat. Hist. Bull. Siam. Soc. 30:199−204.

312. MERTON, D.V. 1970. Kermadec Islands expedition reports: a general account of birdlife. Notornis 17:147−88.

313. MILLEDGE, D.R. 1968. The first recorded occurrence of Baird's Sandpiper, *Calidris bairdii*, in Australia. Emu 68:1−5.

314. MILLEDGE, D.R. 1983. Migratory waders of the Furneaux Group Islands. Occ. Stint 2:42−8.

315. MINTON, C.D.T. 1982. AWSG wader expedition in north-west Australia, August/September 1982. Stilt 3:2−4.

316. MINTON, C.D.T., and J. D. MARTINDALE. 1982. North-west Australia wader studies expedition, 1981. Stilt 2:14−26.

317. MINTON, C.D.T., and R. J. PIERCE. 1986. Migration studies of Double-banded Plovers. Stilt 8:27.

318. MOREAU, R.E. 1961. Problems of Mediterranean-Saharan migration. Ibis 103:373−427, 580−623.

319. MOREAU, R.E. 1967. Water birds over the Sahara. Ibis 109:232−59.

320. MOREAU, R.E. 1972. The Palearctic-African Bird Migration Systems. London & New York: Academic Press.

321. MORRIS, A.K. 1975. The Birds of Gosford, Wyong and Newcastle (County of Northumberland). Aust. Birds 9:37−76.

322. MORRIS, A.K. pers. comm.

323. MULLER, P. 1969. Wind and Waders − Southland, January 1969. Notornis 16:126−38.

324. MYERS, J.P. 1983. Conservation of migrating shorebirds: staging areas, geographic bottlenecks and regional movements. Amer. Birds 37:23−5.

325. MYERS, J.P., O. HILDEN and P. TOMKOVICH. 1982. Exotic *Calidris* species of the Siberian tundra. Ornis Fennica 59:175−82.

326. NAARDING, J.A. 1983. Latham's Snipe (*Gallinago*

*hardwickii*) in Southern Australia. Tasmanian National Parks and Wildlife Service. Wildlife Div. Tech. Rep. 1983/1.

327. NAARDING, J.A. 1984. Latham's Snipe (*Gallinago hardwickii*) in Australia and Japan. Tasmanian National Parks and Wildlife Service. Wildlife Division Tech. Rep. 1984/1.

328. NETTLESHIP, D.N. 1973. Breeding ecology of Turnstones (*Arenaria interpres*) at Hazen Camp, Ellesmere Island, NWT. Ibis 115:202−17.

329. NEWMAN, O.M.G. 1982*a*. Dispersal of Pied Oystercatchers in the Hobart area. Occ. Stint 1:51−8.

330. NEWMAN, O.M.G. 1982*b*. Sooty Oystercatcher, evaluation of Hobart area count data, 1964−72. Occ. Stint 1:65−72.

331. NEWMAN, O.M.G. 1982*c*. Hooded plover: is Tasmania the real stronghold? Stilt 3:8−9.

332. NEWMAN, O.M.G. 1983. Pied Oystercatcher−Replacement Clutches. Occ. Stint 2:49−52.

333. NEWMAN, O.M.G. (in press.) Hooded Plover Incubation Period. Occ. Stint.

334. NEWMAN, O.M.G. pers. comm.

335. NEWMAN, O.M.G., and A. W. J. FLETCHER. 1982. Fluctuations in Hobart area wader populations, 1964−1981. Tas. Bird Rep. 10:4−11.

336. NEWMAN, O.M.G., and R. M. PATTERSON. 1984. A population survey of the Hooded Plover (*Charadrius rubricollis*) in Tasmania, October 1982. Occ. Stint 3:1−6.

337. NEWMAN, O.M.G., R. M. PATTERSON and M. BARTER. 1985. A study of the northward migration from southern Tasmania of Red-necked Stint, *Calidris ruficollis*, and Curlew Sandpiper, *C. ferruginea*, using colour-dyed birds. Stilt 7:18−20.

338. NOKIKOV, V. 1976. [Rare, vanishing and little-known birds of the USSR.] Ryzai: Gkskyi State Enterprises. (In Russian.)

339. NORTON-GRIFFITHS, M. 1967. Some ecological aspects of the feeding behaviour of the Oystercatcher, *Haematopus ostralegus*, on the edible mussel, *Mytilus edulis*. Ibis 109:412−24.

340. NOSKE, R. 1975. The Common Sandpiper in New South Wales. Aust. Birds 10:1−9.

341. OLIVER, J. 1982. The geographic and environmental aspects of mangrove communities: climate. In B. F. Clough, (ed.), Mangrove Ecosystems in Australia. Canberra: ANU Press.

342. OLIVER, W.R.B. 1974. New Zealand Birds. Wellington, Sydney, London: Reed.

343. OPIE, M.J. 1974. A further sighting of the Oriental Pratincole. S. Aust. Orn. 26:166−7.

344. ORNITHOLOGICAL SOCIETY OF JAPAN. 1974. Checklist of Japanese Birds. Tokyo: OSJ.

345. OWEN, R.P. 1977. New bird records for Micronesia and major island groups in Micronesia. Micronesica 13:57−63.

346. PARISH, D. 1985. Threats to wader populations in east Asia. pp. 211−14 in Proceedings of Third East Asian Bird Protection Conference. Tokyo: WBSJ.

347. PARISH, D. pers. comm.

348. PARISH, D., and D. R. WELLS. 1984. Interwader '83 Report. Kuala Lumpur: Interwader Publication No. 1.

349. PARK, P. 1983. Orielton Lagoon and Sorell Wader areas. Occ. Stint 2:15−33.

350. PARKER, S.A. 1982. A new sandpiper of the genus *Calidris*. S. Aust. Nat. 56:63.

351. PATON, P.A. 1982. Biota of the Coorong. Adelaide: S.Aust. Dept of Env. & Plan.

352. PATON, P.A. 1984. Diet of waders in the Coorong, S.A. Stilt 5:27.

353. PATTERSON, R.M. 1982. A survey of the wader population of Barilla Bay. Occ. Stint 1:21−8.

354. PATTERSON, R.M. pers. comm.

355. PEARSON, D.J. 1981. The wintering and moult of Ruffs, *Philomachus pugnax*, in the Kenyan rift valley. Ibis 123:158−82.

356. PEARSON, T.H. 1968. The feeding biology of seabird species breeding on the Farne Islands, Northumberland. J. Animal Ecology 37:521−52.

357. PEGLER, J. 1980. A wader survey of the northern shores of Port Stephens and the lower Myall River. Aust. Birds. 14:68−72.

358. PEGLER, J. pers. comm.

359. PEGLER, J., and A. R. McGILL. 1979. Identification of a Western Sandpiper, *Calidris mauri*, and a comparison of this species with other small calidrids. Aust. Birds 14:30−2.

360. PIENKOWSKI, M.W. 1983. A poor breeding season in 1983 for some waders on the Taimyr Peninsula. WSG Bulletin. 39:34.

361. PIENKOWSKI, M.W., and W. J. A. DICK. 1975. The migration and wintering of Dunlin, *Calidris alpina*, in north-west Africa. Ornis Scand. 6:151−67.

362. PIENKOWSKI, M.W., C. S. LLOYD and C. D. T. MINTON. 1979. Seasonal and migrational weight changes in Dunlins. Bird Study 26:134−48.

363. PIERCE, R.J. 1983. The Charadriiformes of a high-country river valley. Notornis 30:169−85.

364. PIERSMA, T. 1985. Wader studies, and water-birds, in the Nakdong Estuary, South Korea, in September 1984. Haren: Report to NEDECO for the Nakdong Estuary Barrage and Reclamation Project, Environmental Programme.

365. PIZZEY, G. 1980. A Field Guide to the Birds of Australia. London: Collins.

366. POORE, G.C.B., A. H. CORRICK and F. I. NORMAN. 1979. Food of three waders at Lake Reeve, Victoria. Emu 79:228−9.

367. PORTENKO, L.A. 1981. Birds of the Chukchi Peninsula and Wrangel Island. Vol. 1. New Delhi: Amerind.

368. PRATER, A.J. 1979. Trends in accuracy of counting birds. Bird Study 26:198−200.

369. PRATER, A.J. 1981. Estuary Birds of Britain & Ireland. Calton: T. & A. D. Poyser.

370. PRATER, A.J., J. H. MARCHANT and J. VOURINEN. 1977. Guide to the Identification and Ageing of Holarctic Waders. BTO Guide 17. Tring: BTO.

371. PRINCE, J. 1983. A Red-necked Phalarope at Pelican Point. W. Aust. Nat. 15:124.

372. PURCHASE, D. 1981. Wader banding in Australia, East Asia and the Pacific. Stilt 1:4.

373. PUTTICK, G.M. 1978. The diet of the Curlew Sandpiper at Langebean Lagoon, South Africa. Ostrich 49:158−67.

374. PYKE, G.H., H. R. PULLIAM, and E. L. CHARNOV. 1977. Optimal foraging: a selective review of theory and tests. Quart. Rev. Biol. 52:137−54.

375. RECHER, H.F. 1966. Some aspects of the ecology of migrant shorebirds. Ecology 47:393−407.

376. RIDE, W.D.L., and A. NEWMAN. 1967. List of Northern Territory Birds. W. Aust. Museum Spec. Publ. 4.

377. RIX, C.E. 1977. New records of Ringed Plover in Australia. Aust. Bird Watcher 7:40−4.

378. ROBERTS, G. 1977. Central Australian records of the Grey-tailed Tattler. S. Aust. Orn. 27:192.

379. ROBERTS, G. 1983. A sighting of the Dunlin, *Calidris alpina*, in north Queensland and a review of Australian Dunlin records. Sunbird 13:15−19.

380. ROBERTS, P.E. 1957. Notes on birds of the Cumberland Islands. Emu 57:303−10.

381. ROBERTSON, H.A., and M. D. DENNISON. 1979. Feeding and roosting behaviour of some waders at Farewell Spit. Notornis 26:73−88.

382. ROBINSON, H.C., and C. B. KLOSS. 1921. The birds of south west and peninsula Siam. J. Nat. Hist. Soc. Siam. 5.

383. ROGERS, K.G. 1982. Moult, biometrics and sexing of the Eastern Curlew. VWSG Bulletin. 5:23−6.

384. ROGERS, K.G. 1984. Bar-tailed Godwit morphometrics. VWSG Bull. 8:23−5.

385. ROYAL AUSTRALASIAN ORNITHOLOGISTS UNION, Nest Record Scheme.

386. RUSSILL, N. and J. 1980. Ringed Plover at Botany Bay, Aust. Birds 14:54.

387. SAGAR, P. 1984. National Wader Count − November 1983. OSNZ Newsletter 30:6−8.

388. SAGAR, P. 1985. National Wader Count − November 1984. OSNZ Newsletter 35:4−5.

389. SCHMIDT-KOENIG, K., and W. T. KEETON. 1978. Animal Migration, Navigation, and Homing. Berlin, New York: Springer-Verlag.

390. SCHODDE, R., and I. MASON. 1980. Nocturnal Birds of Australia. Melbourne: Lansdowne Editions.

391. SCHODDE, R., G. F. VAN TETS, C. R. CAMPION and G. S. HOPE. 1975. Observations on birds at glacial altitudes on the Carstenz Massif, western New Guinea. Emu 75: 65−72.

392. SCHULZ, M. 1985. Notes on the diet of the Ruddy Turnstone, *Arenaria interpres*. VWSG Bulletin 9:27−9.

393. SCHULZ, M. pers. comm.

394. SCHULZ, M., A. GRANT and L. LUMSDEN. 1984. Some aspects of the feeding behaviour and diet in the Hooded Plover, *Charadrius rubricollis*, during the non-breeding season. Stilt 5:2−8.

395. SCHULZ, M., and L. LUMSDEN. 1983. Fluctuations in Hooded Plover numbers at Venus Bay in 1981 and 1982. VWSG Bulletin 7:11−12.

396. SCHULZ, M., and K. A. MENKHORST. 1984. A survey of the waders of south-west Tasmania. Stilt 5:21−4.

397. SEDGWICK, L.E. 1976. Redshank on north-west coast. W. Aust. Nat. 13:123.

398. SELANDER, R.F. 1966. Sexual dimorphism and differential niche utilisation in birds. Condor 68:113−51.

399. SERVENTY, D.M. 1960. A record of the Pectoral Sandpiper in Western Australia. Emu 60:68−9.

400. SERVENTY, D.M., and H. M. WHITTELL. 1976. Birds of Western Australia. 5th ed. Perth: UWA Press.

401. SEVERINGHAUS, S.R., and S. R. BLACKSHAW. 1976. A New Guide to the Birds of Taiwan. Taipei: Mei Ya Publications Inc.

402. SHAPIRO, M.A. 1975. Westernport Bay Environmental Study, 1973−74. Melbourne: Ministry for Conservation.

403. SHARLAND, M.S.R. 1942. Black-fronted Dotterel in Tasmania. Emu 42:52−3.

404. SHAW, T.H. 1938. The avifauna of Tsingtao and neighbouring districts. Bull. Fan Mem. Inst. Biol., Zool. Ser. vii: 133−222.

405. SKINNER, N.J. 1983. The occurrence of waders at Suva Point, Fiji, Notornis 30:227−32.

406. SMART, J.B. 1973. Notes on the occurrence of waders in Fiji. Notornis 18:267−79.

407. SMART, M. (ed). 1976. Proc. International Conference on the Conservation of Wetlands and Waterfowl. Slimbridge: International Waterfowl Research Bureau.

408. SMITH, A.P. 1977. Observations of birds in Brunei. Sarawak Mus. J. 25:235−69.

409. SMITH, F.T.H. 1962*a*. An Australian sight record of the Buff-breasted Sandpiper. Aust. Bird Watcher 1:185−92.

410. SMITH, F.T.H. 1962*b*. A probable record of the Long-toed Stint. Aust. Bird Watcher 1:214−15.

411. SMITH, F.T.H. 1962*c*. A Victorian record of the Oriental Pratincole. Aust. Bird Watcher 2:4−5.

412. SMITH, F.T.H. 1963*a*. An Australian sight record of the Red-necked Phalarope (*Phalaropus lobatus*). Aust. Bird Watcher 2:1−4.

413. SMITH, F.T.H. 1963*b*. The Pectoral Sandpiper (*Erolia melanotos*) near Melbourne, Victoria. Aust. Bird Watcher 2:9−17.

414. SMITH, F.T.H. 1964*a*. Wader observations in southern Victoria, 1962−1963. Aust. Bird Watcher 2:70−84.

415. SMITH, F.T.H. 1964*b*. The Pectoral Sandpiper in mid-southern Victoria. Aust. Bird Watcher 2:104−8.

416. SMITH, F.T.H. 1966*a*. Little Whimbrels and other waders at a north Queensland Airport. Aust. Bird Watcher 2:221−2.

417. SMITH, F.T.H. 1966*b*. Wader records and observations in mid-southern Victoria, 1963−1965. Aust. Bird Watcher 2:246−66.

418. SMITH, F.T.H. 1967. Wader records and observations in mid-southern Victoria, 1963−1965. Part 2. Aust. Bird Watcher 3:19−29.

419. SMITH, F.T.H. 1968*a*. The Pectoral Sandpiper in mid-southern Victoria. Aust. Bird Watcher. 3:122−8.

420. SMITH, F.T.H. 1968*b*. The Long-toed Stint, *Erolia subminuta*, in southern Victoria. Aust. Bird Watcher 3:132−40.

421. SMITH, F.T.H. 1969*a*. Wilson's Phalarope. Aust. Bird Watcher 3:157−61.

422. SMITH, F.T.H. 1969*b*. Wilson's Phalarope, A third sight record for Australia. Aust. Bird Watcher 3:161−2.

423. SMITH, F.T.H. 1969*c*. The Red-necked Phalarope near Melbourne. Aust. Bird Watcher 3:166−7.

424. SMITH, F.T.H. 1969*d*. Additional records of the Long-toed Stint. Aust. Bird Watcher 3:167−8.

425. SMITH, F.T.H. 1971. Little Whimbrels in NT. Bird Observer 471:5−6.

426. SMITH, F.T.H. 1974. A Victorian record of the Asiatic Dowitcher. Aust. Bird Watcher 5:111−19.

427. SMITH, F.T.H. 1976*a*. An Australian record of the Grey Phalarope. Aust. Bird Watcher 6:292−9.

428. SMITH, F.T.H. 1976*b*. An Australian sight record of the White-rumped Sandpiper. Aust. Bird Watcher 6:317−20.

429. SMITH, F.T.H. 1977. Another Victorian record of the Buff-breasted Sandpiper. Aust. Bird Watcher 7:59−60.

430. SMITH, F.T.H. 1981. A retraction of Victorian Dunlin records. Aust. Bird Watcher 9:43.

431. SMITH, F.T.H. 1982. Dunlins undone. Stilt 2:10−11.

432. SMITH, F.T.H. 1983*a*. A Red-necked Phalarope, *Phalaropus lobatus*, at Point Lonsdale, Victoria. Aust. Bird Watcher 10:99−101.

433. SMITH, F.T.H. 1983*b*. An Australian record of the Lesser Yellowlegs. Aust. Bird Watcher 10:122−6, 143.

434. SMITH, F.T.H. 1984*a*. Further to 'A retraction of Victorian Dunlin records'. Aust. Bird Watcher 10:240−1.

435. SMITH, F.T.H. 1984*b*. Victorian records of a sandpiper new to science. Aust. Bird Watcher 10:264−5.

436. SMITH, F.T.H., and R. J. SWINDLEY. 1975. A Victorian record of Baird's Sandpiper. Aust. Bird Watcher 6:35−40.

437. SMITH, F.T.H., R. J. SWINDLEY and J. G. BARKLA. 1978. A second Australian record of the White-rumped Sandpiper. Aust. Bird Watcher 7:194−7.

438. SMITH, G.T. and D. A. SAUNDERS. 1980. The Red-necked Phalarope − a new migrant for the south of Western Australia. W. Aust. Nat. 14:237−8.

439. SMYTHIES, B.E. 1953. The Birds of Burma. London: Oliver & Boyd.

440. SMYTHIES, B.E. 1981. The Birds of Borneo. 3rd ed. Sabah Society & Malayan Nature Society.

441. SPEECHLEY, B.A. 1966. Sydney records of the Pectoral Sandpiper, *Erolia melanotos*. Emu 66:1−5.

442. SPENCE, T. 1973. A third specimen of the Pectoral Sandpiper. W. Aust. Nat. 12:168.

443. STANLEY, P.I., and C. D. T. MINTON. 1972. The unprecedented westward migration of Curlew Sandpipers in autumn, 1969. Brit. Birds 65:365−80.

444. STARKS, J.R., and B. A. LANE. (in prep.) Northward migration of waders in Australia, February-April 1985. Stilt.

445. STORR, G.M. 1973. List of Queensland Birds. Spec. Publs West Aust. Mus. No. 5.

446. STORR, G.M. 1980. Birds of the Kimberley Division, Western Australia. Spec. Publs West. Aust. Mus. No. 11.

447. STORR, G.M., and R. E. JOHNSTONE. 1977. First Australian record of the Pintail Snipe (*Gallinago stenura*). W. Aust. Nat. 13:207−8.

448. STORR, G.M., and R. E. JOHNSTONE. 1979. A second Australian specimen of the Pintail Snipe (*Gallinago stenura*). W. Aust. Nat. 14:157.

449. STRESEMANN, E. 1936. A nominal list of the birds of the Celebes. Ibis 78(2):356−69.

450. STRUDWICK, J. 1980. Little Stint, *Calidris minuta*, at Melbourne and Metropolitan Board of Works Farm, Werribee. Geel. Nat. 17:16−17.

451. SUMMERS, R.W., and M. WALTNER. 1979. Seasonal variations in the mass of waders in southern Africa, with special reference to migration. Ostrich: 50:21−37.

452. SYMPSON, R.T. 1969. A sighting of an unusually large Ruff, Aust. Bird Watcher 3:174−5.

453. THOM, B.G., and J. CHAPPELL. 1975. Holocene sea levels relative to Australia. Search 6:90−3.

454. THOMAS, D.G. 1968. Waders of Hobart. Emu 68:95−125.

455. THOMAS, D.G. 1969. Breeding Biology of the Australian Spur-winged Plover. Emu 69:81−102.

456. THOMAS, D.G. 1970*a*. Fluctuation of numbers of waders in south-eastern Tasmania. Emu 70: 79−85.

457. THOMAS, D.G. 1970*b*, Western Sandpiper in Tasmania. Emu 70: 88−9.

458. THOMAS, D.G. 1970*c*. Wader migration across Australia. Emu 70: 145−54.

459. THOMAS, D.G., and A. J. DARTNALL. 1970*a*. Pre-migratory deposition of fat in the Red-necked Stint. Emu 70:87.

460. THOMAS, D.G., and A. J. DARTNALL. 1970*b* Difference in size between the sexes of the Curlew Sandpiper. Emu 70:89.

461. THOMAS, D.G., and A. J. DARTNALL. 1971*a*. Ecological aspects of the feeding behaviour of two calidritine sandpipers wintering in south-eastern Tasmania. Emu 71:20−6.

462. THOMAS, D.G., and A. J. DARTNALL. 1971*b*. Moult of the Red-necked Stint. Emu 71:49−53.

463. THOMAS, D.G., and A. J. DARTNALL. 1971*c*. Moult of the Curlew Sandpiper in relation to its annual cycle. Emu 71:153−8.

464. THOMPSON, H.A.F. 1977. Notes on birds in the Darwin and northern areas of the Northern Territory. Sunbird 8:83−91.

465. THOMPSON, H.A.F. 1978. Further records of Palearctic species in Darwin. Sunbird 9:54−9.

466. THOMPSON, M.C. 1973. Migratory patterns of Ruddy Turnstones in the central Pacific. Living Bird 12:5−23.

467. THOMPSON, M.C., and C.D. HACKMAN. 1968. Birds of the Tokelau Islands. Notornis 15:109−17.

468. TOMKOVICH, P.S. 1982. [Peculiarities of the Autumn migration of the Sharp-tailed Sandpiper.] Bull. Moscow Soc. Naturalists 87:56−61. (In Russian.)

469. TRELOAR, K., and L. UNDERWOOD. 1982. A South Australian record of the Little Ringed Plover. S. Aust. Orn. 28:204.

470. VAN GESSELL, F.W.C. 1976. An introduction to some aspects of bird banding and the study of migrant shorebirds on Kooragang Island. Hunter Nat. Hist. Aug. 1976: 159−68.

471. VAN TETS, G.F., A. H. D'ANDRIA and E. SLATER. 1967. Nesting distribution and nomenclature of Australasian Vanelline Plovers. Emu 67:85−93.

472. VAN TETS, G.F., W. J. M. VESTJENS, A. H. D'ANDRIA and R. BARKER. 1977. Guide to the Recognition and Reduction of Aerodrome Bird Hazards. Canberra: Australian Government Publishing Service.

473. VAURIE, C. 1964. A survey of the birds of Mongolia. Bull. Am. Mus. Nat. Hist. 127 (3).

474. VEITCH, C.R. 1978. Waders of the Manukau Harbour and Firth of Thames. Notornis 25:1−24.

475. VEITCH, C.R. 1979. Kaipara Harbour − Easter 1978. Notornis 26:289−96.

476. VICTORIAN WADER STUDY GROUP. pers. comm.

477. VOWLES, R.S., and G. A. VOWLES. (unpubl.) A study of the waders in South-West Brunei.

478. WAKEFIELD, W.C. 1984. The wader habitats of Moulting Lagoon, Tasmania. Occ. Stint 3:26−34.

479. WALL, L.E. 1953. Some notes on migrant waders in Southern Tasmania. Emu 53: 80−6.

480. WALL, L.E. 1982. Marion Bay and Blackman Bay. Occ. Stint 1:48−50.

481. WALL, L.E. 1983. Shorebirds at George's Bay, St Helen's. Occ. Stint 2:53−5.

482. WARD, P., and A. ZAHAVI. 1973. The importance of certain assemblages of birds as 'information centres' for food finding. Ibis 115:517−34.

483. WARHAM, J. 1962. Bird islands within the Barrier Reef and Torres Strait. Emu 62:99−111.

484. WEBSTER, M., and K. PHILLIPPS. 1976. A New Guide to the Birds of Hong Kong. Hong Kong: Sino-American Publishing Co.

485. WHEELER, W.R. 1955. Charadriiformes at the Laverton Saltworks, Victoria, 1950−1953. Emu 55:279−95.

486. WHITBOURNE, E. 1963. Waders observed in the Northern Territory. Aust. Bird Watcher 2: 44−7.

487. WHITE, C.H.N. 1975. Migration of palearctic waders in Wallacea. Emu 75:337−9.

488. WILD BIRD SOCIETY OF JAPAN. Wader counts − 1978 (1978), Wader counts − 1979 (1979), Wader counts − 1980 (1980) Tokyo. WBSJ (1982). Results of the Nationwide Counts of waders in 1981. Strix 1:56−60. WBSJ (1983) Results of the Nationwide counts of Waders and Waterfowl Conducted by the WBSJ. Strix 2:131−9. WBSJ (1984). Strix 3:101−12.

489. WILDASH, P. 1968. Birds of South Vietnam. Rutland, Vt: Charles Tuttle & Co.

490. WILDER, G.D., and H. W. HUBBARD. 1924. List of the birds of Chihli province. J. North China Branch, Roy. Asiatic Soc. 55:156−239.

491. WILLIAMS, T.C., and J. M. WILLIAMS. 1977. Orientation of transatlantic migrants. In K. Schmidt-koenig and W. T. Keeton (eds), Animal Migration, Navigation and Homing. Berlin, New York: Springer-Verlag.

492. WILLIAMS, T.C., J. M. WILLIAMS: L. C. IRELAND and J. M. TEAL. 1977. Autumnal bird migration over the western North Atlantic Ocean. Amer. Birds 31:251−67.

493. WREN, J. 1981. Sighting of Pectoral Sandpiper at Queerah, North Queensland. Aust. Bird Watcher 9:13.

494. YOSHII, M. 1975. Eleventh Annual Report of the bird-banding scheme, 1 April 1971 to 31 March 1972. Misc. Reps. Yamashina Inst. Orn. 7:452−75.

495. YOSHII, M. 1977. Thirteenth Annual Report of the Japanese Bird-banding scheme, 1 April 1973 to 31 March 1974. Misc. Reps. Yamashina Inst. Orn. 9:286−324.

496. YUGOVIC, J.Z. 1984. The Grey Glasswort (*Halosarcia halocnemoides*) in coastal Victoria and some implications for the Orange-bellied Parrot. Vic. Nat. 101:234−9.

497. ZHOU Shi-E. 1984. The survey of Charadriiformes on Chongming Island, Shanghai, China. Paper presented to 10th Asian Section Conference, International Council for Bird Preservation, Sri Lanka. April 1984.

498. SAGAR. P., pers. comm.

# ACKNOWLEDGEMENTS

This study of shorebirds and consequently this book would not have been possible without the assistance of many people and organisations. They are listed below in chronological order, not in order of priority.

Dr Clive Minton was instrumental in getting the study under way. In his capacity as member of the RAOU Research Committee, leader of expeditions to the north-west and dedicated counter and bander of shorebirds, his enthusiasm generated a greater interest in the study of shorebirds among many bird-watchers and brought many people into the ranks of those who supported the study.

The shorebird counters produced much of the information on which this book is based. It was a pleasure to work with so many keen people all over Australia. They are listed in the following section.

John Martindale was the first co-ordinator of the study, and he made my task easier. When I took over, in 1981, the first network of regional organisers and counters had been already established by him.

State and regional organisers devoted much time and energy to organising teams of counters, collecting and mailing count-sheets to RAOU headquarters and answering queries from this end. A special thank you to: Alan McBride and Alan Morris (NSW), Richard Alcorn, Margaret Cameron, Val Curtis, Peter Dann and Chris Sonter (Vic), Marion Crouther, Stephen Garnett, Roger Guard, Bill Horton, Bob and Julie Lake, and Peter Woodall (Qld), David Close, Niven McCrie and Peter Penney (SA), Mike Bamford, Ken Mills and Doug Watkins (WA), David Henderson, Mike Newman and Bob Patterson (Tas.), Mike Fleming, Tony Hertog and Anne Kerle (NT).

The RAOU Council, Research Committee and Publications Committee supported the study.

Simon Bennett did all the computer programming, and Angela Jessop and Jon Starks were scientific assistants on the study. I am most grateful for their dedication and hard work over the years, making the task of storing and analysing the information so much easier. The staff of the RAOU assisted in many ways: Margaret Blakers, Kay Busija, Margaret Considine, Stephen Davies, Burna Dunn, Roger Jaensch, Kerrie Milburn-Clarke, Julie Strudwick, and typists Bernadette Paxman, Mary Siketa, Hazel Dann and Carolyn Muir.

Volunteer Staff at RAOU headquarters and in the Perth office were a great help: Cecily Allen, Lisa Barter, Bob Cross, Frank Davidson, Alexandra Djurovich, Ros Jessop, Shapelle McNee, Brenda Murlis, Mick Murlis, Annie Rogers, Dan Rogers, Tom Smith, Elizabeth Turnbull, Doug Watkins and Tony Zidarich. Librarian Trish White worked hard to ensure that all publications that I requested were available.

Major financial support for the study came from the Australian National Parks and Wildlife Service and the Utah Foundation. Financial assistance for special projects was provided by Australia-Japan Foundation, BP (Aust.) Ltd, Cheetham Salt Ltd, CRA Ltd, Dampier Salt Co. Ltd, East-West Airlines, Hamersley Iron Ltd, Hooker Corporation, Humes Ltd, ICI (Aust.) Ltd, The M.A. Ingram Trust, MIM Holdings Ltd, Santos Ltd, Shell Co. of Australia, Swan Television Ltd, Vamgas Ltd, Victorian Fisheries and Wildlife Division WA Fisheries and Wildlife Department, and World Wildlife Fund, Australia. Donations to support work were made by I. Sinclair and H. N. B. Wettenhall.

Logistical support for many field activities was provided by: WA Fisheries and Wildlife Department, NSW National Parks and Wildlife Service, NT Conservation Commission, Victorian Fisheries and Wildlife Division and Tasmanian National Parks and Wildlife Service.

Bird groups and naturalists' clubs organised field trips and publicised the study: Bendigo Field Naturalists' Club, Bird Observers' Association of Tasmania, Bird Observers' Club of Australia, Canberra Ornithologists' Group, Geelong Field Naturalists' Club, Horsham Field Naturalists' Club, Hunter Bird Observers' Club, Millicent and Mount Gambier Field Naturalists' clubs, NSW Field Ornithologist Club, Queensland Ornithological Society, South Australian Ornithologists' Association, Sunraysia Bird Observers' Club, Tasmanian Shorebird Study Group, Victorian Ornithological Research Group, Victorian Wader Study Group, WA Group of the RAOU and the WA Wader Study Group.

Pilot Don Jeans flew aerial surveys in many parts of Australia. The Australian Coastal Surveillance Organisation provided places on their aircraft for observers to conduct aerial surveys.

Property owners permitted access to their land.

The Bureau of Meteorology permitted the use of their radar in northern Australia to track migrating shorebirds.

David Purchase and Kim Lowe provided information from the Australian Bird Banding Scheme.

I am grateful to Peter Dann for writing Chapter 3.

Stephen Davies, Ken Rogers, Annie Rogers and Dan Rogers read draft manuscripts of the book and made many helpful suggestions for improvements. David Melville commented on all the sections on the Asian range of species. The following people read parts of the typescript and made helpful comments: Mark Barter, Margaret Cameron, David Close, Margaret Considine, Marion Crouther, Peter Dann, Doug Dow, David Eades, Alan Fletcher, Stephen Garnett, Tony Hertog, Roger Jaensch, Ros Jessop, Sylvia Jones, Leo Joseph, Kim Lowe, Richard Loyn, Stephen Marchant, Clive Minton, Alan Morris, Mike Newman, Priscilla Park, Bob Patterson, Grant Pearson, Sue Robinson, Martin Schulz, Clare Speedie, Doug Watkins and Dennis Watson.

Sally Pribble drew all the maps and made helpful suggestions on the presentation of information. Bruce Quinn drew the figures. Jim Hooper took most of the black-and-white photographs and Mark Barter and Clive Minton provided additional material. Brian Chudleigh and Don Jeans took the photographs on the dust jacket.

The following people provided reference material and technical advice during the preparation of the colour plates: Kevin Bartram, Simon Bennett, Bird Observers' Club, Ron Brown, Mike Carter, Andrew Corrick, Rex Davies, David Eades, Bruce Male, Peter Menkhorst, Clive Minton, Charles Silviera and Bob Swindley. Belinda Gillies and Rory O'Brien allowed access to the ornithological collections of the Museum of Victoria and arranged loans of material.

The Victorian Fisheries and Wildlife Division allowed Jeff Davies to complete the colour plates in good time.

Finally, I am most grateful to the Rogers Family of Ninks Road for their hospitality in providing a quiet writing retreat and for their support in many ways during the preparation of this book.

I apologise to anyone whose name may inadvertently have been left out.

## Study participants

### NEW SOUTH WALES AND AUSTRLIAN CAPITAL TERRITORY

T. Alley, R. Annels, W. Barden, A. Barnett, J. Barnett, R. Bigg, M. Blanch, K. Brandwood, M. Brewer, J. Brickhill, J. Burfoot, A. Burton, W. Cambridge, M. Carey, C. Chafer, G. Clancy, M. Clayton, J. Cook, D. Creer, F. Creer, P. Croft, J. Dalby, A. Dampney, P. Davie, S. Debus, P. Disher, S. Dorey, R. Draffan, M. Dwyer, R. Edwards, W. Emery, R. Estreich, E. Felton, P. Fielding, A. Gibson, J. Gibson, A. Gillies, R. Gordon, D. Gosper, S. Green, C. Hamilton, S. Hamonet, R. Harmer, J. Higgins, J. Hobbs, K. Holmes, B. Howie, P. Humphries, Hunter Bird Observers' Club, K. Hutton, J. Imrie, R. Imisides, J. Izzard, D. Johnson, R. Jordan, T. Korn, N. Kurtz, S. Lane, R. Langdown, D. Larkins, C. Lentfer, M. Lenz, A. Lindsey, T. Lindsey, E. Lisser, M. Maher, P. Maher, P. Mannel, B. Mannes, H. Mannes, S. Marchant, K. Marriot, N. Maxwell, A. McBride, P. Metcalfe, J. Miller, R. Moffatt, A. Morris, D. Moulton, F. Moulton, J. O'Shea, M. O'Shea, J. Pegler, J. Perry, A. Pidgeon, R. Pidgeon, T. Poynton, T. Quested, K. Rochford, A. Rose, A. Ross, M. Rowe, N. Russill, J. Russill, A. Salter, B. Salter, N. Schrader, J. Scrivens, D. Secomb, N. Sheppard, D. Smedley, J. Smith, L. Smith, J. Tarr, I. Taylor, E. Thomas, D. Turner, M. Tyler, V. Tyler, B. Versey, G. Vincent, N. Wakeling, P. Walker, J. Waugh, J. Willows, D. Winterbottom, J. Young.

### VICTORIA

J. Abbott, E. Abel, R. Alcorn, C. Allen, C. Appleby, G. Appleby, J. Barkla, L. Barrow, L. Barter, M. Barter, K. Bartram, G. Baverstock, R. Baverstock, J. Beatty, M. Beck, I. Bell, S. Bennett, J. Berry, L. Billing, N. Billing, P. Bingham, Bird Observers' Club, K. Bode, R. Borgett, J. Bowden-Perry, G. Bowker, M. Brady, O. Brewster, J. Brouwer, R. Brown, A. Burbidge, T. Burbridge, M. Cameron, J. Campbell, M. Carter, G. Cerini, C. Chandler, A. Chapman, M. Chapman, R. Chennell, W. Coles, A. Collins, J. Collins, L. Conole, M. Considine, J. Cook, G. Cornwall, A Corrick, I. Crawford, A. Curry, V. Curtis, P. Dann, J. Davies, J. N. Davies, R. Davies, J. Dawson, J. Dearneley, G. Deason, V. Dedman, D. Deerson, X. Dennett, R. Dennis, A. Djurovich, C. Doughty, J. Drummond, R. Drummond, J. Duclos, D. Eades, C. Eastwood, M. Elliget, S. Evans, F. Farr, B. Forest, D. Franklin, F. Garrett, G. Gayner, D. Gerard, G. Gibbs, B. Gillies, J. Gooch, G. Goods, M. Griffith, H. Guyatt, A. Hargraves, P. Hargraves, H. Harris, B. Harvey, P. Hermans, M. Hewish, A. Hickey, R. Hill, R. Horwood, G. Hosken, A. Howard, J. Ipsen, A. Isles, M. Jennings, W. Jennings, A. Jessop, R. Jessop, J. Johnstone, E. Jones, S. Jones, P. Kadwell, D. Keller, R. Keller, J. Kelly, P. Kennedy, B. Kentish, D. King, V. King, J. Kirton, J. Klapste, P. Klapste, T. Kloot, H. Kroger, M. Kroger, F. Kurdle, B. Lane, P. Lansley,

D. Legg, J. Leslie, J. Lindner, H. Livingston, K. Lowe, T. Lowe, R. Loyn, L. Lumsden, B. Male, I. Manseagh, J. Martindale, R. McAuley, G. McCarthy, E. McCulloch, N. McFarlane, N. McIntyre, P. Menkhorst, D. Middleton, R. Middleton, W. Middleton, R. Millson, C. Minton, N. Minton, P. Minton, R. Minton, B. Mitchell, W. Mitchell, J. Moor, P. Moore, I. Morgan, C. Morley, C. Morris, L. Mumford, B. Murlis, M. Murlis, L. Naismith, J. Neaves, F. Noelker, K. Norris, G. Nott, J. Ozols, G. Parkes, N. Pamment, A. Pascoe, B. Pascoe, P. Peake, J. Pearse, M. Pegler, J. Perry, R. Perry, J. Phillips, D. Pialt, J. Pickford, G. Pizzey, G. Possingham, J. Pratt, D. Premier, T. Pescott, D. Quick, J. Quick, K. Quick, D. Quinn, J. Raines, A. Raymond, P. Reilly, H. Rich, A. Roberts, D. Roberts, D. Robertson, D. Robinson, P. Robinson, A. Rogers, D. Rogers, K. Rogers, M. Rogers, P. Russell, I. Savage, P. Schultz, M. Schulz, R. Semmens, R. Shaw, R. Sherwood, C. Silviera, S. Simons, M. Simpson, T. Simpson, F. Smith, T. Smith, C. Sonter, J. Starks, P. Starks, D. Stewart, A. Stirling, J. Strudwick, Sunraysia Bird Observers' Club, A. Sutton, A. Swan, R. Swindley, M. Tarrant, P. Tarrant, D. Tease, D. Thomas, M. Thomas, P. Thomas, D. Tonkinson, B. Traill, G. Tribe, W. Tuck, S. Turnbull, K. Turner, D. Twaites, J. Vincent, S. Wadsworth, A. Walker, R. Walters, J. Ward, G. Webb, G. Werren, J. Wettenhall, N. Wettenhall, G. Wilson, J. Wilson, E. Witham, W. Witham, J. Withell.

### QUEENSLAND

M. Andrews, P. Berry, Y. Bertreaux, L. Bone, T. Brickhill, P. Britten, I. Carruthers, R. Chandler, D. Cheb, V. Cooper, C. Corben, J. Cornelius, G. Corrin, M. Crawford, M. Crouther, G. Czechura, R. Elvish, E. Eves, A. Finch, S. Garnett, A. Griffin, R. Guard, A. Haffenden, P. Harris, M. Hawken, R. Henderson, S. Holohan, J. Hopkinson, H. Horton, W. Horton, R. Johnson, J. Lake, R. Lake, P. Lander, G. Leach, A. Lloyd, C. Lloyd, D. Magarry, N. Marr, G. McBurney, J. McCabe, A. Morton, J. Moverley, T. Murphy, R. Naiger, T. Norris, J. Noyce, G. Oxbrough, J. Pacey, G. Palmer, D. Pearse, E. Pearse, R. Pearson, P. Peeders, D. Reid, E. Reid, D. Schultz, D. Seabright, N. Shorten, G. Smith, A. Smythe, E. Sticklen, D. Sullivan, J. Sullivan, P. Sutton, A. Taplin, M. Templeton, P. Veerman, I. Venables, L. Walter, R. Walter, D. Watson, G. Watson, J. Wieneke, I. Weston, T. Weston, J. Wheeler, P. Woodall, J. Wren, R. Wyatt.

### SOUTH AUSTRALIA

R. Allen, H. Anderson, C. Ashton, C. Baxter, I. Beale, L. Bigg, M. Bonnin, J. Bourne, T. Bradley, J. Bransbury, G. Carpenter, N. Cheshire, R. Cleggett, D. Close, T. Cox, W. Cunningham, J. Cusak, P. Doherty, L. Dunkerley, J. Eckert, R. Eckert, L. Edington, J. Fairly, P. Fargher, C. Garrett, M. Gersch, P. Gibbons, C. Gill, B. Glover, D. Harper, J. Hatch, M. Heininger, M. Hinscliffe, S. Hocking, I. Hopton, P. Horton, G. Jackson, R. Jaensch, L. Joseph, R. Kernot, R. Kreisl, A. Lashmar, A. Lees, B. Levings, A. Mackey, G. Markey, B. Mart, J. Matthew, I. May, A. McArthur, N. McCrie, A. McIntyre, M. McKelrey, J. McNamara, W. Moore, T. Morgan, J. Mullins, V. Natt, J. Needle, D. Palmer, P. Paton, C. Pawsey, L. Pedler,

P. Penney, L. Potts, J. Ramsey, P. Roach, D. Rowley, K. Shurcliffe, F. Smeaton, B. Snell, A. Spiers, W. Treloar, L. Underwood, F. Vickery, D. Vincent, R. Watmore, D. Wiesner.

## Western Australia

B. Allan, J. Allan, G. Baker, M. Bamford, H. Bekle, B. Bellairs, G. Boland, A. Bornman, R. Breeden, L. Broadhurst, W. Brooke, B. Buchanan, R. Burden, R. Burking, P. Clay, D. Congreve, P. Congreve, P. Curry, Alister Cuthbert, Ann Cuthbert, S. Davies, A. Daw, A. De Rebeira, P. De Rebeira, D. Doust, N. Duff, D. Dureau, J. Dymond, E. Edwards, S. Farrer, J. Fleay, P. Fuller, R. Goodale, V. Hatt, R. Hewitt, D. Hobcroft, T. Hordacre, P. Howden, R. Hughes, R. Hunt, R. Jaensch, D. James, S. Keeling, J. Kelly, R. King, B. Kneebone, K. Lance, J. Lane, P. Mack, P. Maisack, J. Malone, K. Marshall, L. Masters, W. McCarthy, S. McNee, M. Millard, Y. Millard, K. Mills, E. Milne, D. Montague, K. Morris, A. Motherwell, O. Mueller, D. Nash, B. Newby, J. Nilson, W. Okell, G. Pearson, A. Reigan, E. Richmond, G. Roberts, F. Robinson, R. Rogers, G. Rogerson, A. Rose, S. Rose, W. Russell, J. Scott, E. Sedgewick, V. Smith, T. Stone, J. Talbot, B. Telford, F. Thomas, R. van Delft, D. Watkins, J. White, P. Wilmot, B. Wykes, W. Zadow.

## Tasmania

R. Ashby, J. Bayly-Stark, J. Biggs, S. Breen, G. Bromfield, P. Brown, R. Cooper, P. Duckworth, P. Fielding, A. Fletcher, L. Hannavy, K. Harris, D. Henderson, E. Hiller, T. Johns, W. Jones, R. Jordan, M. Lord, N. Mooney, A. Moscal, M. Newman, C. O'Brian, P. Park, R. Patterson, B. Penney, D. Rounsevell, A. Towney, W. Wakefield, L. Wall, R. Walters, D. Whitchurch, C. White, R. Wilson, E. Woehler.

## Northern Territory

K. Fisher, L. Fisher, M. Fleming, S. Garnett, A. Hertog, J. Kerle, J. McKean, D. Peacock, D. Percival, H. Thompson, F. van Gessell, G. Wills, S. Wills.

For a complete list of plates see page IX